AF866268

Water Quality Examination: T
Investigation of Heavy Metal
Contamination, Nutrients, an
Organic Pollutants in Aquati
Environments for an Enhance
Prevention and Purification P

Water Quality Examination: The Investigation of Heavy Metal Contamination, Nutrients, and Organic Pollutants in Aquatic Environments for an Enhanced Prevention and Purification Process

Guest Editor

Thomas Dippong

MDPI

Basel • Beijing • Wuhan • Barcelona • Belgrade • Novi Sad • Cluj • Manchester

Guest Editor
Thomas Dippong
Chemistry and Biology
Technical University of Cluj
Napoca
Baia Mare
Romania

Editorial Office
MDPI AG
Grosspeteranlage 5
4052 Basel, Switzerland

This is a reprint of the Special Issue, published open access by the journal *Water* (ISSN 2073-4441), freely accessible at: https://www.mdpi.com/journal/water/special_issues/SY370X012T.

For citation purposes, cite each article independently as indicated on the article page online and as indicated below:

Lastname, A.A.; Lastname, B.B. Article Title. *Journal Name* **Year**, *Volume Number*, Page Range.

ISBN 978-3-7258-8423-0 (Hbk)
ISBN 978-3-7258-8424-7 (PDF)
https://doi.org/10.3390/books978-3-7258-8424-7

© 2026 by the authors. Articles in this reprint are Open Access and distributed under the Creative Commons Attribution (CC BY) license. The reprint as a whole is distributed by MDPI under the terms and conditions of the Creative Commons Attribution-NonCommercial-NoDerivs (CC BY-NC-ND) license (https://creativecommons.org/licenses/by-nc-nd/4.0/).

Contents

About the Editor . . . . **vii**

Preface . . . . **ix**

Thomas Dippong, Cristina Mihali and Alexandra Avram
Evaluating Groundwater Metal and Arsenic Content in Piatra, North-West of Romania
Reprinted from: *Water* **2024**, *16*, 539, https://doi.org/10.3390/w16040539 . . . . **1**

Ovidiu Nasca, Thomas Dippong, Maria-Alexandra Resz and Monica Marian
Interdisciplinary Evaluation of the Săpânța River and Groundwater Quality: Linking Hydrological Data and Vegetative Bioindicators
Reprinted from: *Water* **2025**, *17*, 1975, https://doi.org/10.3390/w17131975 . . . . **22**

Dragana Milijašević Joksimović, Dejana Jakovljević and Tamara Jojić Glavonjić
Evaluation of Water Quality in the Tamiš River in Serbia Using the Water Pollution Index: Key Pollutants and Their Sources
Reprinted from: *Water* **2025**, *17*, 1024, https://doi.org/10.3390/w17071024 . . . . **51**

Simona Elena Avram, Lucian Barbu Tudoran, Gheorghe Borodi and Ioan Petean
Microstructural Comparison of the Mineralization Within Borsec and Tusnad Public Springs
Reprinted from: *Water* **2025**, *17*, 2892, https://doi.org/10.3390/w17192892 . . . . **68**

Chendong Yang, Xiujuan Feng, Liangfu Jiang, Haotong Guo, Benqing He, Ziqi Wang, et al.
Concentration Characteristics, Source Apportionment, and Health Risk Assessment of Groundwater Contaminants in a Karst Coal Mining Area (Guizhou Province, China)
Reprinted from: *Water* **2026**, *18*, 351, https://doi.org/10.3390/w18030351 . . . . **87**

Marta Szostak, Krzysztof Szoszkiewicz, Krzysztof Achtenberg and Dariusz Drożdżyński
Behavioral Responses of *Unio tumidus* Freshwater Mussels to Neonicotinoid Pesticide Contamination
Reprinted from: *Water* **2025**, *17*, 289, https://doi.org/10.3390/w17030289 . . . . **104**

Simona Elena Avram, Bianca Violeta Birle, Lucian Barbu Tudoran, Gheorghe Borodi and Ioan Petean
Investigation of Used Water Sediments from Ceramic Tile Fabrication
Reprinted from: *Water* **2024**, *16*, 1027, https://doi.org/10.3390/w16071027 . . . . **121**

Catalina Topa, Gabriel Murariu, Valentina Calmuc, Madalina Calmuc, Maxim Arseni, Cecila Serban, et al.
A Spatial–Seasonal Study on the Danube River in the Adjacent Danube Delta Area: Case Study—Monitored Heavy Metals
Reprinted from: *Water* **2024**, *16*, 2490, https://doi.org/10.3390/w16172490 . . . . **138**

About the Editor

Thomas Dippong

Thomas Dippong (Professor at the Technical University of Cluj-Napoca) is a chemical engineer with a Ph.D. and habilitation in chemistry. His current research activities are related to the characterization of nanoparticles for various applications as part of an ongoing research project in partnership with the Technical University of Cluj-Napoca within the field of ferrite embedded in silica matrix. He is an expert in analytical chemistry, organic/inorganic chemistry, thermal treatment, instrumental analysis, synthesis of magnetic nanomaterials and nanocomposites, characterization techniques, food thermal behaviour, volatile profile, and water quality. Prof Dippong has published 166 peer-reviewed publications (110 papers in high-ranked scientific ISI-Thomson journals (65 Q1, 20 Q2, and 25 Q3) and 56 in other national and international journals), with his work having received 3715 citations. He currently has an h-index of 49 (WoS). He has given 37 lectures at international conferences, as well as having published 2 books with international publishing houses and 15 books with national publishing houses. In the last 5 consecutive years (2020–2024), Prof Thomas Dippong was included in the prestigious list of the world's top 2% of researchers. He has been a contract manager for many projects and is currently an active member of four projects. Prof. Dippong has reviewed 800 scientific articles for 150 ISI-Thomson journals. He has been the Guest Editor of 10 Special Issues published by three prestigious Q1 ISI-Thomson journals.

Preface

This Special Issue focuses on water and sediment quality, with emphasis on source identification, temporal and spatial distribution, speciation, pollution indices, exposure pathways, bioaccumulation, toxicological effects, environmental risk assessment, human health risks, and the transformation of heavy metals and emerging pollutants in aquatic environments. The Issue also includes computational and theoretical investigations of reaction mechanisms, kinetics, and thermodynamics underlying water remediation pathways.

The presence of organic pollutants can induce bacterial growth, making the interaction between microbial contamination and chemical pollutants an important aspect of water quality and the implementation of novel detection methods for emerging contaminants. Therefore, these studies address bacteriological contamination in relation to organic pollutants in aquatic environments and conduct analyses of pollution assessment tools (indices, diagrams, modelling). Such pollution largely results from anthropogenic activities, industrial processes, and the extensive use of metals and chemicals.

This Special Issue comprises works with a focus on the assessment of human health risks linked to contaminated water, along with response strategies for both groundwater and surface water resources and the examination of the impact on biota in surface water and the transfer of pollutants from aquatic environments to biota. It also highlights the development of advanced technologies aimed at preventing groundwater pollution and restoring water quality, including photocatalysis, adsorption, biodegradation, and membrane separation.

The quality of surface and groundwater is often a reflection of human impacts on the environment, as heavy metals, nutrients, and organic compounds increasingly alter aquatic ecosystems. This Special Issue presents technologies for preventing groundwater pollution and restoring groundwater quality, including photocatalysis, adsorption, biodegradation, and membrane separation, along with analyses of risks and response strategies for the sustainable management of groundwater resources.

Another emergent theme in this Special Issue is the development of advanced, sustainable water-treatment solutions that integrate physicochemical methods, membrane separation, advanced oxidation, and biological processes, supported by economic analyses of pollutant-removal technologies. The Issue also addresses pollution sources, impacts, control, and management measures, and the implementation of innovative, non-toxic materials for decontaminating water.

Thomas Dippong
Guest Editor

water MDPI

Article

Evaluating Groundwater Metal and Arsenic Content in Piatra, North-West of Romania

Thomas Dippong [1], Cristina Mihali [1] and Alexandra Avram [2,*]

[1] Faculty of Sciences, Technical University of Cluj-Napoca, 76 Victoriei Street, 430122 Baia Mare, Romania; dippong.thomas@yahoo.ro
[2] Faculty of Chemistry and Chemical Engineering, Babes-Bolyai University, 11 Arany Janos Street, 400028 Cluj-Napoca, Romania
* Correspondence: alexandra.avram@ubbcluj.ro

Abstract: The present study introduces a monitoring initiative focused on the quality of groundwater in the Piatra locality, situated in the North-West region of Romania. This paper employs an evaluation of 21 physico-chemical parameters, encompassing factors such as electrical conductivity, pH, chemical oxygen demand, turbidity, total hardness, NH_4^+, NO_3^-, Cl^-, PO_4^{3-}, Li, Na, K, Ca, Mg, Ba, Sr, Al, Fe, Mn, Sn, and Ti. Additionally, it examines five heavy metals (Cr, Cu, Ni, Pb, and Zn) and arsenic in water sourced from six distinct private wells. Each well, with its characteristics, serves as a unique drinking water source. The assessment encompassed the evaluation of pollution levels, quality status, and risk factors for all drinking water sources, utilizing pollution, quality, and risk indices. The aim of this study was to establish the level of toxicity in water, assess its impact on human health, and disseminate information to the public about the appropriate utilization of individual water sources. The results indicated a general contamination with chloride, ammonium, manganese, chromium, and iron. Human health risk assessment indices revealed that the consumption of studied waters presented non-carcinogenic risks associated with Cr for adults and with Cr, As, Pb, and Cu for children for some of the groundwater sources. The water quality index (WQI) categorizes the samples as possessing excellent and good quality. This research represents one of the initial endeavors to assess the groundwater source quality in connection with the potential human health risks posed by the metals studied within the protected area of the Tisa River Basin.

Keywords: groundwater; heavy metals; human health risk; drinking water sources

1. Introduction

Groundwater, found underground in soil, sand, and rock spaces known as aquifers, represents 97% of the world's drinking water [1]. Worldwide, approximately 2.5 billion people depend solely on groundwater to meet their daily drinking requirements [2]. According to the UNESCO Water Report 2023, over the last 40 years, a 1% water consumption increase per year has been estimated [3]. A projection by Burek et al. [4] states that this trend could eventually lead to a 20–30% increase by the year 2050.

In the case of Romania, according to the National Administration of Romanian Waters, there is around 9.6 billion m^3 of water beneath the surface, with almost half as part of the water table and the rest as part of deep-sea reserves [5]. Groundwater's advantage lies in its natural renewal through recharge, where precipitation infiltrates the soil to refill aquifers, ensuring a sustainable and dependable supply [6]. However, continental-scale projections for Europe suggest a probable 5–20% decrease in annual mean precipitation in southern Europe and the Mediterranean from 2071 to 2100, with precipitations for Romania already having decreased to 30 mm/decade [5]. In numerous rural regions of Romania, the primary concern is not merely the availability of water but rather the accessibility of potable water. The 2030 Agenda prioritizes developing water and sanitation infrastructure for an improved quality of life and health in Romania. Even with this being a strategic priority, in 2016, only 65.2% of the population had water access, making Romania the least-developed EU country in this aspect, with significant rural disparities and limited national infrastructure and investment contributing to low wastewater collection (63.46%) and sewage treatment (56.71%) rates [7]. Despite the European Union's (EU) drinking water legislation, the quality of drinking water from both private and numerous public wells remains unregulated and frequently lacks protection against contaminants. The 98/83/EC Council Directive exempts drinking water systems catering to fewer than 50 individuals or producing less than 10 m^3 per day [8]. In rural areas, individual well digging risks water quality and quantity due to expertise and design issues, especially during high-demand or drought periods [9]. The Piatra area, where the groundwater quality study was conducted, lies within the Upper Tisa region near the Ukrainian border and is intersected by the Tisa River and its tributary, Șugătag. This region, including the Upper Tisa River and Tisa meadow, is designated as a Natura 2000 protected area due to its rich biodiversity, housing various species such as amphibians, fish, and migratory birds like wild ducks, storks, and swans. The primary occupations of Piatra residents include agriculture, animal husbandry, and wood processing. The area offers picturesque landscapes and opportunities for recreational fishing. Geologically, it features Quaternary formations from the Pleistocene and Holocene periods, influenced by alluvium from flood periods. The protected area spans low terrace formations of the Tisa River, characterized by alluvial/proluvial materials like gravels, boulders, and sands [10]. Land use in Piatra includes forests, meadows, pastures, arable lands, and residential areas, with predominant alluvial, well-drained soils supporting diverse natural vegetation. The area experiences a temperate-continental climate, with increased humidity and cold temperatures influenced by the Baltic and Scandinavian regions [11].

According to soil taxonomic classifications, the area possesses diverse soil types: alluvisols dominate the meadows along the Tisa River with well-drained properties and a pH range from moderately acidic to weakly alkaline, while Regosols are found along the Săpânța River on gravel/boulder deposits with slightly acidic pH levels; Gleysols occur in depressions between Teceu and Remeți and Săpânța and Câmpulung la Tisa, influenced by a prolonged water table presence and poor-quality humus; and Luvisols range from gelic luvosols to stagno-gleic luvisols, present in terraces and settlement areas, with varying pH levels from strongly acidic to moderately acidic [12–14]. Shallow dug wells pose contamination risks due to their wide diameter, shallow depth, and inadequate regulation, elevating their vulnerability to environmental shifts and fostering well deterioration. All the soils in the area exhibit unique behaviors from a pedological perspective [15]. Alluvisols, though well-drained, pose risks if surface runoff or contamination seeps through soil layers, which can be mitigated by proper well construction. Regosols, which are slightly acidic, may lead to heavy metal leaching if not lined adequately. Gleysols, which are waterlogged, risk contamination from stagnant water and humus, necessitating well construction vigilance. Luvisols, with varying acidity, can affect water purity and surface contamination if not properly sealed or protected.

Thus, the assessment of groundwater quality considers biological, hydrological, and physico-chemical factors. Prolonged environmental interactions influence its chemical composition, with agricultural waste like fertilizers, insecticides, and pesticides threatening quality through infiltration [16]. Despite purification through evaporation and precipitation, water needs rely on limited rainwater and groundwater, facing persistent pollution driven by rapid resource degradation [17]. Contaminants impacting natural water sources comprise heavy metals (Cu, Ag, Zn, Cd, Hg, Cr, Mo, Mn, Co, and Ni), inorganic pollutants (PO_4^{2-}, NO_3^-, and NH_4^+), metalloids (B, Si, and Te), organic pollutants (benzene, phenol, toluene, chloroaniline, and methylene blue), and microorganisms (*S. aureus*, *E. coli*) [18]. Certain heavy metals, such as Cr, Pb, Ni, and Cd, carry potential risks due to their toxic, genotoxic, and carcinogenic effects, while others like Se, Mo, Mn, Co, Cu, and Fe are essential micronutrients for various biological functions [18–27]. However, excessive intake of these metals can result in neurotoxic effects. Additionally, it has been reported that approximately 20% of global cancer cases result from improper water consumption, and in 2022, 829,000 deaths worldwide were attributed to inadequate sanitation and various human activities near groundwater sources [28]. Evaluating the degree of heavy metal pollution through pollution indices—PI (pollution index) and HEI (heavy metal evaluation index)—is essential for safe water consumption. The overall quality of water can be assessed using chronic daily intake (CDI). The hazard quotient (HQ) and the hazard index (HI) are employed for the evaluation of non-carcinogenic health risks based on the consumption of water contaminated with metals [29]. Heavy metals, pesticides, and detergents, due to their accumulation in vital organs, are major pollutants, impacting key human body systems [30]. Human health risk assessment is vital for estimating the harmful effects and likelihood of health risks to ecosystems and human health [31]. In this process, spatial modeling of heavy metals in groundwater plays a crucial role, considering factors such as chronic daily intake, hazard quotient, and hazard index [32,33]. To prevent potential health risks, it is essential to implement risk factors for water sources characterized by high concentrations of contaminants [34]. Recent investigations in Nigeria, Algeria, Iran, Kenya, Thailand, Bulgaria, Slovakia, and Romania have shown high levels of non-carcinogenic outcomes for chemical elements including Ni, Hg, Cd, Pb, As, and nitrogen pollutants (NO_3^-, NO_2^-, and NH_4^+), which are prevalent in groundwater due to activities like household, agricultural, and industrial processes, carrying the potential to trigger cancer or methemoglobinemia [35].

The primary objectives of this investigation centered on assessing the quality and pollution levels of groundwater drawn from dug wells, specifically used for drinking purposes, aiming to ascertain its suitability for consumption and its potential impact on human health. Another objective is to identify the extent of the anthropogenic pressures and, implicitly, to elaborate measurements in order to reduce the negative anthropogenic influences. What sets this study apart is its comprehensive approach to evaluating groundwater quality in Piatra town, situated within the Plain of Tisa River. This method involves diverse methodologies, including statistical analysis of key physicochemical indicators, cluster analysis identifying wells influenced by similar pollution sources, human risk assessment, and the calculation and interpretation of water quality indices. By introducing innovative perspectives and methodologies, this study contributes to existing literature, aiming to enhance knowledge in the field. The findings are expected to provide insights into how groundwater responds to stress from human activities, offering valuable recommendations for population health risk prevention and sustainable practices.

2. Materials and Methods

2.1. Water Sampling

The samples of groundwater were gathered from six distinct locations within the town of Piatra, Romania, throughout all seasons in the year 2023. Figure 1 illustrates the locations of all six sampling points comprised exclusively of dug wells. The open dug wells, situated on residents' properties, are excavated to a depth of 4–10 m with an 80 cm diameter. They are constructed using cement/asbestos or rock tubes and equipped with a pulley system for water extraction. Sampling points were chosen randomly to ensure a comprehensive coverage of the town's entire area, encompassing potential sources of contaminated groundwater. Physicochemical parameter analysis was conducted within 48 h of sampling. Adhering to SR ISO 5667-3/2018 [36] guidelines, collected water samples were stored in clean polyethylene containers and refrigerated until analyzed in triplicate [37].

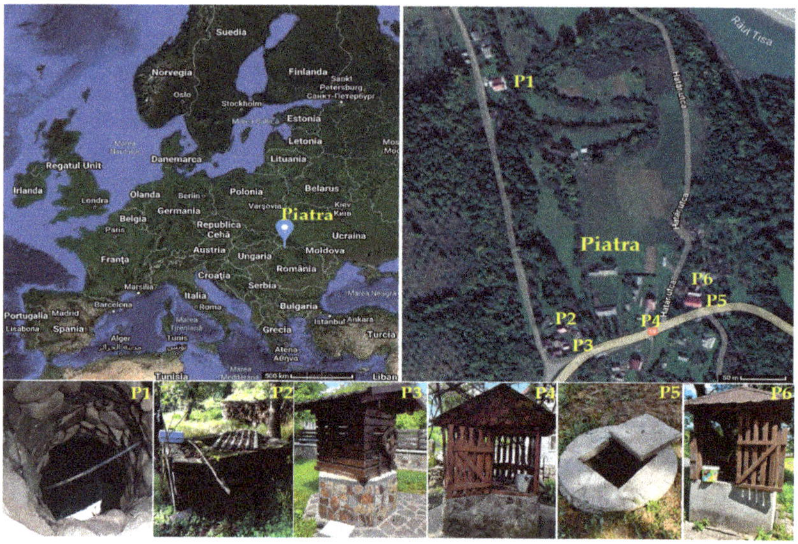

Figure 1. Sampling point (P1–P6) in the studied area.

2.2. Methods of Analysis

Field investigations and sample analysis were conducted over 2022 using portable devices. Electrical conductivity (EC) and pH were measured with a WTW INOLAB 740 conductometer (WTW, Weilheim in Oberbayern, Germany) and an HI 253 Hanna Instruments pH meter (Hanna Instruments, Woonsocket, RI, USA), respectively, following SR EN ISO 7888/1985 [38] and SR EN ISO 10523/2012 [39] standards [37]. Dissolved oxygen, chloride concentration, total hardness (ht), and aluminum were assessed using specified methods and instruments [37]. The measurement of Al^{3+} was conducted using a Specord 50 UV–VIS spectrophotometer (Analytik Jena, Jena, Germany) [37]. Ammonium, nitrates, phosphates, dissolved iron, and water turbidity were also measured following relevant standards and using appropriate devices, such as the WTW 355 IR portable turbidimeter (WTW, Weilheim in Oberbayern, Germany) [37]. The portable devices underwent calibration before each measurement [37]. The preservation of water samples involved the addition of nitric acid (HNO_3, 65%, Merck, Darmstadt, Germany) in a 1:1 ratio until reaching an acidic pH range of 1–2. Following this, the samples underwent digestion with concentrated HNO_3 (70%, Merck, Germany) and H_2O_2 (30%, Merck, Germany), then were dissolved in concentrated HNO_3, diluted with filtered ultrapure water, further diluted to a volume of 100 mL using an acid solution, and stored at 4 °C until the analysis [37]. All reagents utilized were analytical-grade (PA), with deionized water employed for their preparation.

In the case of samples investigated using the graphite furnace, preparation involved a filtration step followed by an acidification with concentrated HNO_3, 0.5 mL for 100 mL of the water sample, excluding the mineralization stage [37]. Elements such as Mg, Ca, Mn, Fe, and Zn at parts per million (ppm) levels were examined through flame absorption atomic spectrometry (FAAS) using a Perkin Elmer NexION 300S spectrophotometer equipped with flame and graphite furnace atomizers (Perkin Elmer, Norwalk, CT, USA). An in-depth analysis was conducted on trace metals (μg/L), including Cu and Ni, through the utilization of graphite furnace atomic absorption spectrometry (GFAAS) employing a pyrolytic platform [37]. The spectrophotometer, featuring hollow cathode lamps tailored to individual metals and a continuous background correction system, employed an air-acetylene flame for samples at the ppm level and a graphite furnace for samples at the trace level [37]. Calibration, quality control of data, and assurance adhered to established standard operating procedures, which included conducting triplicate analyses and measuring blanks in parallel. The findings were presented as the average values derived from three distinct replicates, employing analytical-grade reagents and certified 1000 ppm standard metal solutions for both calibration and standard preparation [37].

2.3. Statistical Analysis

Descriptive statistics of the registered data were accomplished by computing the mean, standard deviation, coefficients of variation, standardized skewness, and kurtosis for each water parameter with the help of Excel (Microsoft Office 2021, Microsoft Corporation, Washington, DC, USA). Cluster analysis, a classification method applied to environmental data, was employed to illustrate the similarity or dissimilarity of observation points or analyzed parameters using the Statgraphic program (version Centurion 19, 2023, The Plains, VA, USA).

2.4. Heavy Metal Evaluation Index (HEI)

The heavy metal evaluation index method was employed to investigate the overall quality of water based on the concentrations of heavy metals. The *HEI* values were calculated for the elements Zn, Pb, Ni, Cu, Cr, As, Mn, Fe, and Sr. Only the heavy metals with a maximum allowable limit were considered. As was also considered due to its high toxicity. *HEI* was calculated according to Equation (1) [40,41].

$$HEI = \sum_{t=1}^{n} \frac{C_{M_i}}{C_{(MAC)_i}} \quad (1)$$

where C_{Mi} is the measured concentration of the "*i*" considered a heavy metal and $C_{(MAC)i}$ is the maximum admissible concentration in drinking water, according to Romanian legislation or another standard for drinking water quality. By using *HEI*, the water samples can be classified into the following pollution degrees: low pollution degree when $HEI < 40$, medium degree of pollution for $40 < HEI < 80$, and high degree of pollution when $HEI > 80$.

2.5. Water Quality Index (WQI)

The water quality index (*WQI*) is an approach to assessing the quality of diverse water samples, consolidating various indicators of water composition into a singular representative value that conveys comprehensive information about water quality [29,40,42–46]. The *WQI* method was used to assess the overall quality of groundwater samples based on a weighted arithmetic technique, considering 23 indicators of water quality: EC, pH, turbidity, NH_4^+, NO_3^-, PO_4^{3-}, and the metals: Li, Na, K, Ca, Mg, Ba, Sr, Al, Fe, Mn, Cr, Cu, Ni, Pb, Zn, and As. For each water quality parameter, a weight (w_i) and a relative weight (W_i) were assigned based on Equations (2) and (3):

$$w_i = \frac{1}{S_i} \quad (2)$$

where S_i is the maximum allowable concentration in potable water (MAC), according to Law 311/2004 458/2002 Law M.O. No. 552/29 July 2002—law on the quality of drinking water, as shown in Table 1.

$$W_i = \frac{w_i}{\sum_{i=1}^{n} w_i} \quad (3)$$

where *n* is the number of the assessed water indicators; *n* was 23 in the present study.

Table 1. Physico-chemical composition of groundwater sources in Piatra locality, a Natura 2000 protected area, and standards (maximum allowable concentration—MAC) in accordance with Council Directive 98/83/EC [8].

Element Content in Water, μg/L	P1	P2	P3	P4	P5	P6	Mean	SD	CV, %	Stnd. Skewness	Stnd. Kurtosis	MAC
EC (μS/cm)	763	398	172	1205	192	252	497	410	82.49	1.29	0.34	2500
pH	7.08	7.12	7.55	7.18	6.98	7.61	7.25	0.26	3.61	0.72	−0.84	6.5–9.5
DO (mg/L)	6.79	5.65	9.06	5.33	9.77	9.55	7.69	2.01	26.13	−0.16	−1.34	
T (NTU)	1.31	3.92	**8.75**	1.14	2.56	**5.48**	3.86	2.90	75.20	1.01	0.22	5
NH_4^+ (mg/L)	**0.88**	0.48	0.28	**1.09**	0.04	**0.77**	0.59	0.39	66.82	−0.21	−0.64	0.5
NO_3^- (mg/L)	4.21	2.38	1.28	2.23	0.15	5.32	2.60	1.89	72.95	0.33	−0.38	50
h_t (°g)	24.7	10.2	2.75	25.6	4.42	2.16	11.64	10.85	93.22	0.71	−0.99	min 5
Cl^- (mg/L)	41.5	66.5	39.5	**565.2**	62.8	29.2	134.12	211.67	157.83	2.42	2.95	250
PO_4^{3-} (mg/L)	**0.86**	0.13	0.15	0.19	0.17	0.14	0.27	0.29	105.44	2.42	2.94	0.4

Notes: The maximum allowable concentration (MAC) for drinking water, as stipulated by Law 311/2004 and Law M.O. No. 458/2002, in accordance with the law on the quality of drinking water (552/29 July 2002). SD represents the standard deviation; *CV* denotes the coefficient of variance; and Stnd signifies standardized. The values highlighted in bold surpassed the MAC.

WQI were computed following Equations (4) and (5) according to the literature [29,47–53]:

$$WQI = \frac{\sum_{i=1}^{n} Q_i \times W_i}{\sum_{i=1}^{n} W_i} \quad (4)$$

Q_i is the quality rating of each physico-chemical indicator determined following Equation (4):

$$Q_i = \left(\frac{C_i - V_i}{S_i - V_i}\right) \times 100 \quad (5)$$

In this context, C_i represents the measured value of each physico-chemical parameter or metal concentration, while V_i stands for the ideal value of the chemical indicator. The ideal value, V_i was zero for the considered indicators, with the exception of pH. The optimal value for pH was considered to be 7 [47,52,53].

2.6. Human Health Risk Assessment

Human health risk assessment by groundwater consumption was established according to the method described in the literature [46,54–56], by computing the chronic daily intake of water contaminants, *CDI* (mg/kg·day), hazard quotients (*HQ*) for each water contaminant, and the hazard index (*HI*). Nutrients such as NH_4^+ and NO_3^-, metals (Al, Sr, Ba, Fe, Mn, Cr, Cu, Ni, Pb, and Zn), and As were considered to compute the hazard index, *HI*. The human health risk due to groundwater consumption was evaluated by calculating the chronic daily intake of water, *CDI* (mg/kg·day), hazard quotients (*HQ*), and the hazard index (*HI*), according to Equations (6) and (7) for adults (male and female) and children.

$$CDI = \frac{C \times IR \times EF \times ED}{BW \times AT} \quad (6)$$

where C is the measured pollutant concentration in the groundwater sample (mg/L); *IR* is the ingestion rate per unit time (1.5 L/day for a child and 2 L/day for an adult); *EF*: exposure frequency (350 days/year considering 15 days of holidays or visits); *ED*: exposure duration (6 years in the case of children; 30 years in the case of adults); *BW*: body weight; the average body weight for Romanian males is 85 kg, while for females, it is 72 kg [57], and an average body weight of 15 kg was considered for children [55].

The hazard quotients, HQ_i, for the considered pollutants in water were calculated according to Equation (7):

$$HQ_i = \frac{CDI_i}{RfD_i} \quad (7)$$

where RfD_i is the chronic oral reference dose, an estimate of the daily oral exposure level for the population, and the highest acceptable daily intake level for a specific element or pollutant that does not lead to adverse health effects [47,52,53].

The *RfD* values expressed in mg/kg/day are 0.97 NH_4^+; 1.60 NO_3^-; 0.6 Sr; 0.7 Fe; 0.14 Mn; 0.0003 As; 0.0003 Cr; 0.0005 Cu; 0.0054 Ni; 0.0003 Pb; 0.3 Zn; 7 Al; and 0.2 Ba [58–61].

The hazard risk, *HI*, which aims to assess the human health risk through more toxic elements, was computed as the sum of all *HQ*, calculated for each particular pollutant according to Equation (8). *HI* values greater than one, $HI \geq 1$, are associated with considerable non-carcinogenic health hazards [47,53,62].

$$HI = \sum_{i=1}^{n} HQ \tag{8}$$

3. Results and Discussion

3.1. Physico-Chemical Characteristics of Groundwater Samples

The results for the physico-chemical content of the examined water samples (P1–P6) are presented in Table 1. As seen from Table 1, the electrical conductivity (EC) values for wells P1 to P6 exhibit a notable variation, suggesting diverse levels of ion concentration in the respective groundwater sources. Well P4 stands out with the highest conductivity at 1205 μS/cm, potentially indicating a higher concentration of dissolved ions or minerals. In contrast, wells P3 and P6 demonstrate the lowest conductivity values at 172 μS/cm and 252 μS/cm, respectively, suggesting a lower presence of dissolved ions. Wells P2, P5, and P1 fall in between, showcasing moderate electrical conductivity values of 398 μS/cm, 192 μS/cm, and 736 μS/cm, respectively. These variations could be attributed to geological factors, such as differing soil compositions or proximity to potential contamination sources, influencing the conductivity of the groundwater in each well. A previous study [46] on another well within the region registered much higher values (1575–2480 μS/cm), attributed to the interaction with aquifer rocks. EC serves as evidence of salt content in water, and the ingestion of water with elevated EC levels may contribute to a range of health issues, including but not limited to cancer, diarrhea, hepatitis, and gastroenteritis, impacting vital organs such as the heart, kidneys, and stomach [48].

Dissolved oxygen (DO) indicates water stratification and the contamination degree [49]. In this study, DO varied between 5.33 and 9.77 mg/L. Wells P3, P5, and P6 have relatively high DO concentrations, measuring 9.06 mg/L, 9.77 mg/L, and 9.55 mg/L, respectively. These elevated levels suggest favorable conditions for aerobic organisms and may indicate well-oxygenated water. Conversely, wells P2 and P4 show lower DO values at 5.65 mg/L and 5.33 mg/L, possibly indicating reduced oxygen availability in these wells. Well P1 falls in between, with a dissolved oxygen level of 6.79 mg/L. Discrepancies in these values may be ascribed to a variety of factors, including groundwater flow patterns, biological activities, and environmental conditions. The low amounts of DO (P2 and P4) lead to a lack of freshness, a fad taste, and unfriendly conditions of microorganisms that make the water not potable [50]. High amounts of DO (P3, P5, and P6) increase the organic suspended matter rich in pathogens [49,50].

pH ranged between 6.98 and 7.61, within the thresholds established for drinking water, indicating a weak acidic to weak basic character, which can indicate a low pollution level with organic and inorganic compounds [51]. Due to rock interaction and rich amounts of carbonates, pH has a basic character and changes the taste of water, possibly leading to skin and eye rashes [52]. Turbidity ranged between 1.14 and 8.75 NTU. Such variations in the turbidity levels across the six wells indicate differences in water clarity. Well P3 exhibits the highest turbidity at 8.75 NTU, suggesting a higher concentration of suspended particles or sediments in the water. P6 follows closely with a turbidity of 5.48 NTU, indicating moderately cloudy water. Wells P2 and P5 demonstrate moderate turbidity levels at 3.92 NTU and 2.56 NTU, respectively. P1 and P4 exhibit lower turbidity levels at

1.31 NTU and 1.14 NTU, suggesting relatively clear water. These variations in turbidity may be influenced by factors such as land use, soil composition, and human activities in the vicinity of each well. High turbidity levels can indicate potential contamination or natural sedimentation processes. Higher values may be caused by the presence of bacteria, plankton, iron and aluminum hydroxide, sludge, and colloidal matter. Water characterized by elevated turbidity posed an epidemiological threat because it facilitated the suspension of particles, serving as a medium for pathogens [40]. The consumption of water with high turbidity levels may result in health issues, particularly intestinal diseases, and can also induce distortion in aquatic ecosystems [53].

With the exception of P1, P4, and P6, all other samples are highly rich in NH_4^+, almost two times the MAC established for the drinking water. The high amount is related to anthropogenic activities, for example, the intense use of fertilizers or fecal deposits. If consumed, water with NH_4^+ in high amounts might cause convulsion, hepatic encephalopathy, coma, and death. Such water contaminated with NH_4^+ can be treated through energetic chlorination and filtration process [63]. The NO_3^- concentrations are below the MAC (50 mg/L), between 1.28 and 4.21 mg/L (Table 1). The sources of NO_3^- include intensive agricultural practices, sewage and septic tank leakage, manure and contaminated sludge deposits, as well as microbial decomposition. Groundwater contamination with NO_3^- is influenced by the geological and hydrogeological structure. Consuming water rich in NO_3^- is linked to the onset of illnesses like cancer and methemoglobinemia [62]. As for chlorides, all wells are within the allowable limit, except for well P4, which has more than double the maximum allowable limit of 250 mg/L. The anomalous presence of heightened chlorine levels in a singular well, as opposed to others within the same town, underscores the complexity of localized water quality dynamics. Chlorides (salts of metals with hydrochloric acid), indicative of water salinity, are essential for the body's electrolyte balance, but excessive salinity renders water unsuitable for drinking due to potential chemical aggressiveness [35]. PO_4^{3-} are lower than 0.4 mg/L, with potential sources represented by the intense use of fertilizers and detergents, but also with geogenic origin. An elevated phosphate concentration can change both the flavor and color of water [64]. Variations in the levels of ammonium, chlorides, nitrates, and phosphates among wells in the same close region or town can be attributed to diverse factors, including differences in geological formations, land use practices, and proximity to pollution sources. Additionally, variations in well construction, depth, and maintenance practices may influence the vulnerability of wells to contamination. These local factors influence the interaction of water with surrounding soils, rocks, and contaminants, contributing to the unique composition of each well and the resulting differences in water quality parameters. For example, one study [65] focuses on 10 wells from Remeti locality in very close proximity to Piatra town (the locale housing the six wells analyzed in the present paper). The ammonium levels reported in Remeti wells were up to more than four times higher than the MAC, presenting a maximum of 2.38 mg/L compared to the maximum in Piatra, which is P4 = 1.09 mg/L; also, in Remeti, phosphates were within admissible limits except for one well (0.78 mg/L) comparable to the ones in this study. Nitrate amounts were also excessively higher in wells from Remeti than in the P1–P6 samples in this study. This could be explained by more intense agricultural practices in the former compared to the latter, coupled also with the heavy practice in Romanian countryside involving the application of animal manure. Teceu locality is also in close proximity to Piatra town, and water analysis on a well in Teceu [46] revealed values for ammonium, chlorides, nitrates, and phosphates within the legal limits for potable water. However, it should be stated that [46] only analyzed one well in that specific town, which would be a limitation for stating the pollution levels for wells in the area.

3.2. Metal and as Content in Groundwater Samples

The concentrations of major metals (Ca, Mg, Na, and K), of metals found at lower concentrations (Li, Al, Ba, and Sr), of heavy metals (Fe, Mn, Sn, Ti, Cr, Cu, Ni, Pb, and Zn), and As are shown in Table 2.

Table 2. The metal and As content of groundwater sources in Piatra locality, a Natura 2000 protected area and standards (maximum allowable concentration—MAC) in accordance with Council Directive 98/83/EC [8].

Element Content in Water, μg/L	P1	P2	P3	P4	P5	P6	Mean	SD	CV, %	Stnd. Skewness	Stnd. Kurtosis	MAC
Li	2.3	3.1	2.5	6.1	2.4	4.7	3.52	1.55	44.07	1.17	−0.01	50
Na	18,345	16,889	25,455	59,745	24,332	11,045	25,969	17,361	66.85	1.95	2.12	200,000
K	2077	2045	2105	7245	2726	3044	3207	2020	62.99	2.23	2.57	10,000
Ca	1890	3558	1345	8921	2432	3779	3654	2746	75.15	1.83	1.86	100,000
Mg	10,124	20,144	8142	45,224	6885	12,784	17,217	14,505	84.25	1.93	1.90	50,000
Ba	17.2	20.1	15.5	80.2	31.6	27.3	31.93	24.41	76.31	2.12	2.35	700
Sr	88	126	75	**446**	132	**204**	178.5	138.6	77.64	1.92	1.92	200
Al	27.3	18.3	88.3	4.3	19.1	17.5	29.13	29.91	102.68	2.11	2.42	200
Fe	175	195	**225**	175	132	147	174.83	33.27	19.03	0.28	−0.13	200
Mn	2.8	3.2	8.5	1.4	1.7	3.1	3.45	2.58	74.89	1.99	2.20	50
Sn	22.1	94.5	18.3	68.9	57.4	11.3	45.42	33.31	73.34	0.49	−0.76	-
Ti	14.3	19.4	10.3	48.6	31.0	41.2	27.47	15.37	55.95	0.35	−0.89	-
As	5.4	8.5	0.22	9.8	6.2	0.55	5.11	3.99	77.99	−0.33	−0.88	10
Cr	7.6	12.6	**55.8**	3.4	2.6	1.7	13.95	20.90	149.81	2.25	2.60	50
Cu	2.5	4.1	3.3	2.5	5.1	6.4	3.98	1.55	38.84	0.70	−0.34	100
Ni	3.1	7.7	12.4	3.6	2.2	5.6	5.77	3.80	65.94	1.22	0.53	20
Pb	3.7	9.3	4.1	4.1	8.3	7.1	6.10	2.44	40.05	0.30	−1.21	10
Zn	4.2	5.3	21.3	15.3	1.2	1.8	8.18	8.20	100.23	1.04	−0.30	5000

Notes: The maximum allowable concentration (MAC) for drinking water, as stipulated by Law 311/2004 and Law M.O. No. 458/2002, in accordance with the law on the quality of drinking water (552/29 July 2002). SD represents standard deviation; CV denotes the coefficient of variance; and Stnd signifies standardized. The values highlighted in bold surpassed the MAC.

The average values for the metal concentrations in the groundwater samples were as follows: Na > Mg > Ca > K > Sr > Fe > Sn > Ba > Al > Ti > Cr > Zn > Pb > Ni > As > Cu > Li > Mn.

Calcium and magnesium are essential elements for the human body; thus, their presence in water is not typically a cause for concern, unless there are underlining medical conditions (i.e., kidneys). While their presence in high amounts leads to hard water, which is organoleptically changed, the World Health Organization (WHO) acknowledges that hard water has no known adverse health effects and may even contribute to the intake of essential minerals. However, the WHO does report that the taste sensitivity threshold for Ca^{2+} within the range of 100–300 mg/L, contingent on the accompanying anion, and it is likely that the taste threshold for magnesium is lower than that for calcium. In certain cases, consumers are known to tolerate water hardness exceeding 500 mg/L [66]. Here, the highest magnesium value is found in fountain P4, while the lowest value is found in well P3. All values fall within the maximum allowable limit of 50,000 μg/L. The presence of such ions, the presence of calcium- and magnesium-rich minerals in rocks and soils, and the characteristics of aquifers through which water flows. Also derived from geological source is potassium, with the highest level being observed in the water from fountain P4, while the lowest value is in the water from fountain P2. However, all values fall within the maximum allowable limit of 10,000 μg/L.

Aluminum (Al) stands out as the most prevalent metallic element found in the Earth's crust. While poorly absorbed in the gastrointestinal tract, its bioavailability increases in drinking water, potentially impacting human health. According to Law 311/2004, the legal limit for aluminum content is 200 μg/L. One potential origin of aluminum is the application of $Al_2(SO_4)_3$ during the water treatment procedure [29]. Recognized as a neurotoxin on a broad scale, exposure to aluminum has been associated with neurodegenerative conditions, including Parkinson's, Alzheimer's, and multiple sclerosis. Studies indicate cognitive decline at Al intake $\geq$0.1 mg/day from water, with elevated levels posing an increased risk of cognitive impairment in the elderly, leading to heightened vulnerability to hip fractures and adverse health effects [67].

Manganese is also naturally abundant, and water rich in Mn is characterized by an unpleasant metallic taste and a muddy odor. The presence of manganese in the water distribution system forms deposits that can settle as a black precipitate [29]. According to Water Law 311/2004, the permitted limit for manganese content should not exceed 50 μg/L. The values obtained in the studied wells are low, ranging from 1.4 to 8.5 μg/L. Manganese originates from diverse sources, encompassing industrial practices like alkaline battery and cleaning product manufacturing, agricultural activities involving the application of fungicides, fertilizers, and pesticides, as well as mining operations.

Barium is a naturally occurring element that can be found in groundwater, including well water. While low levels of barium are generally considered to be naturally present and not harmful, elevated concentrations can pose health risks. Barium can enter groundwater through the weathering of certain rocks and minerals. Barium can deposit in muscles, lungs, and bones because it resembles calcium but is absorbed more quickly [41]. The highest barium value was observed in the water from fountain P4, while the lowest value was observed in the water from fountain P3. All values fell within the permissible limit, which is 7000 μg/L.

Arsenic is a toxic contaminant that is colorless, odorless, and tasteless and is commonly found in high concentrations in groundwater. Arsenic found in groundwater is susceptible to abrupt variations [29]. For this study, the measured values in groundwater vary between 0.22 and 9.8 μg/L. Wells P1, P3, and P6 exhibit arsenic concentrations well below the permissible limit at 5.4 μg/L, 0.22 μg/L, and 0.55 μg/L, respectively. Wells P2 and P5 approach the threshold with values of 8.5 μg/L and 6.3 μg/L, suggesting a moderate risk of arsenic exposure. P4 records the highest arsenic level at 9.8 μg/L, nearing the maximum limit and indicating a potential health concern. Communities relying on well water for drinking, particularly in rural areas, face heightened vulnerability to arsenic contamination. Geological conditions, often prevalent in rural regions, may lead to naturally occurring arsenic in aquifers. Limited resources for water monitoring and regulatory oversight contribute to the unknowing consumption of arsenic-contaminated water. Rural populations, dependent on wells and lacking alternative water sources, are at direct risk of health issues associated with arsenic exposure. Challenges in accessing clean water alternatives, agricultural practices, and a lack of awareness further compound the issue. Ingesting water containing a substantial amount of arsenic can result in significant immediate and/or prolonged health issues, including but not limited to vomiting, diabetes, heart diseases, cancer, spontaneous abortion, childhood cancer, and potential fatality [68]. Following ingestion, arsenic is swiftly absorbed by the gastrointestinal tract and undergoes metabolism. As for chromium, according to Water Law 311/2004, the legal limit is 50 μg/L. In this study, the obtained values varied between 1.7 and 55.8 μg/L. Sample P3 exceeds the legal limit of 50 μg/L, highlighting the need for immediate attention and remediation to ensure compliance with established water quality standards. The other samples, while below the legal limit, may still warrant monitoring and preventive measures to maintain water quality within acceptable levels. The variation in chromium levels, with higher concentrations in sample P3 compared to others in the same village, may stem from localized geological, anthropogenic, or hydrogeological factors. Chromium, a highly toxic heavy metal, is linked to health risks such as cancer, DNA damage, and oxidative stress. It is present in water in hexavalent (Cr(VI)) and trivalent (Cr(III)) forms, and Cr(VI) is especially toxic for individuals with respiratory issues [69].

In compliance with Water Law 311/2004, the permissible limit for nickel content stands at 20 μg/L. Recorded nickel values, ranging from 2.2 to 12.4 μg/L, are influenced by factors like pH, soil composition, and depth [29]. Acknowledged as a significant contributor to groundwater pollution by the U.S. Environmental Protection Agency, nickel's potential toxic and health risks necessitate a comprehensive understanding of public safety [70]. Despite being a crucial element for the good functioning of enzymes, blood, the endocrine system, and gene synthesis, various studies highlight nickel's disruptive impact on glucose metabolism and insulin secretion through biological mechanisms [70]. The presence of nickel in drinking water from wells can exert various influences on the surrounding environment. Nickel concentrations, when used for irrigation, may result in soil contamination, impacting plant health and altering the overall ecosystem in the well area. Interactions with the aquifer and subsurface geology can lead to the leaching of nickel into the well water from surrounding rock formations or anthropogenic sources. The overall groundwater quality in the well area is affected, posing concerns for both human consumption and agricultural use. Persistent contact with heightened nickel concentrations in drinking water poses health concerns, while the potential for corrosion or deterioration of well infrastructure adds another layer of concern.

According to Water Law 311/2004, the legal limit for lead content is 10 μg/L. The obtained lead content values ranged between 3.7 and 9.3 μg/L. Lead is one of the 275 priority-controlled pollutants by the U.S. and Chinese Environmental Protection Agencies [68]. Lead and its compounds can enter groundwater through mining activities. Lead is challenging to eliminate after its accumulation in the human or animal body because it can cause a diverse array of physical and mental issues [71,72]. The accumulation of lead in the human body causes damage to all organs, including the central nervous system, affects the liver, thyroid, and bones, and causes brain injuries and infertility [29,72]. Due to the physiology of the body, children, being more vulnerable than adults to lead contamination, suffer from constant brain damage, with approximately 18 million affected by lead poisoning [72]. Controlling lead pollution in drinking water is of vital importance [71].

As for copper, according to Water Law 311/2004, the legal limit is 100 μg/L. The highest value was found in well P6 (6.4 μg/L). Increased copper levels can arise from natural processes, like the breakdown of rocks, and human activities, including mining, industry, and agriculture. Drinking water containing elevated copper levels may result in stomach and headache discomfort, along with irritation of the eyes and nose [73]. Hence, monitoring the copper content is essential.

According to Water Law 311/2004, the legal limit for zinc content is 5000 μg/L. The zinc content at the sampling points is low. Sampling point P3 recorded double the zinc content (21.3 μg/L) compared to other points, probably due to a higher interaction of groundwater with rocks. Zinc, a naturally occurring element, undergoes slow enrichment in groundwater through interactions with rocks, influenced by inorganic carbon content and pH, and it is recognized for its significant mobility within water systems. Zinc exhibits opalescence and an astringent taste [29]. The mobility of Zn in water is predominantly affected by pH, with other factors like clay content, phosphorus availability, concentration of organic matter, and redox conditions also contributing to its behavior [74].

The strontium content in well water samples falls within the admissible limits of 200 μg/L for four of the six samples, namely P1, P2, P3, and P5. Wells P4 and P6 notably surpass the permissible limit, with sample P4 reading almost twice the legal amount, raising concerns about potential health risks associated with elevated strontium levels. The variations in strontium concentrations among wells in the same village could be attributed to geological factors, such as the composition of the subsurface rocks, which may contain higher concentrations of strontium. As this element is essential for general human health, it may be neglected in the overall water assessment. However, Sr has the potential to substitute for calcium and magnesium in bone, potentially impacting bone growth and strength. One study [75] emphasizes the noteworthy inverse correlation between the incidence of rickets in children and the Ca/Sr ratio in the potable water available for public consumption in China. A lower Ca/Sr ratio is associated with a higher incidence of rickets, suggesting a potential link between the calcium-to-strontium ratio and bone health in children. This finding underscores the importance of considering the Ca/Sr ratio in assessing the overall effect of strontium in the water supply for public consumption, particularly in regions with high strontium concentrations, to ensure adequate management and supervision of water quality.

Iron is the most problematic element in water. The water samples from 5 of the 6 wells analyzed fall below the legal limit of 200 μg/L, with values ranging from 132 (P5) to 195 (P2) μg/L. Sample P3, however, registers iron values of 225 μg/L, thus surpassing the allowed amount. These values indicate a potential issue with iron contamination in the well water. Elevated iron levels in drinking water can lead to several problems, including unpleasant taste, discoloration, and staining of plumbing fixtures. Additionally, high iron concentrations may indicate the presence of other contaminants or geological conditions that contribute to the contamination. The presence of higher iron levels could likely be attributed to the historical mining activities in Maramures County (different metalliferous resources including iron, copper, and manganese [76]), home to Piatra town and the six analyzed wells.

3.3. Cluster Analysis of Water Samples

Figure 2 shows the cluster analysis in the Q mode for groundwater sampling points based on the metal and As concentrations. The pairs of wells P5–P6 and P1–P2 showed the highest similarity, while the highest dissimilarity was found for well P4, one of the deepest wells (740 cm). The water from the P4 well exhibited the highest concentrations of several elements, including Na, K, Ca, Mg, Ba, Sr, Ti, As, and Zn, while the level of Ni was found to be the lowest. P3 also showed high dissimilarity from the samples P1, P2, P5, and P6. P3 registered the highest concentrations of Cr, Al, and Zn compared to the average values of the other groundwater sources. In the case of the pair of wells P5 and P6, the lowest levels of Cu, Al, Pb, Ti, and Mn were measured. P1 and P2 are wells with low depths (210 and 200 cm) while the depths of P5 and P6 are medium (550 cm and 400 cm).

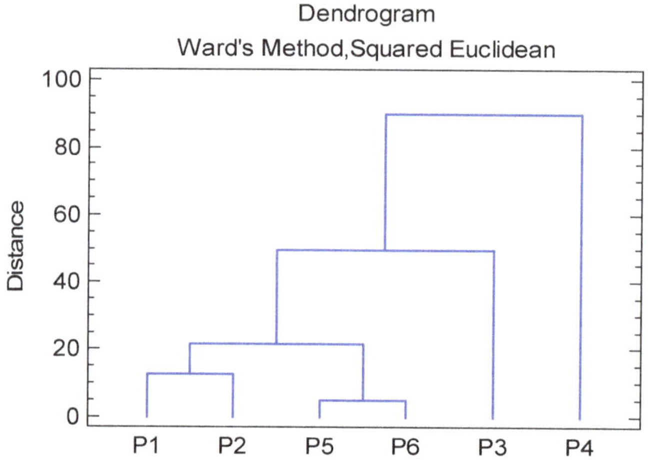

Figure 2. Cluster analysis in Q mode for groundwater sampling points based on the metal and As concentrations.

Cluster analysis was also performed in the R mode (Figure 3), showing the similarities or dissimilarities between the metals and As concentrations in groundwater samples.

The R mode cluster analysis reveals two primary clusters, namely C1 and C2. C1 contains the group of Al-Mn-Cr linked to Ni and also to the Fe-Zn pair of heavy metals. This element is derived from the alumino-silicates that include Al, Mn, Cr, Fe, and Zn. Cluster C2 comprises two subclusters: C2a and C2b. C2a includes the pair As-Sn and a group of eight alkaline and alkaline earth metals (K, Na, Li, Ba, Ca, Sr, and Mg), along with Ti, the element ranking as the Earth's ninth most prevalent [77,78]. Conversely, C2b encompasses the heavy metals Cu and Pb, nonferrous heavy metals naturally occurring in volcanic rocks.

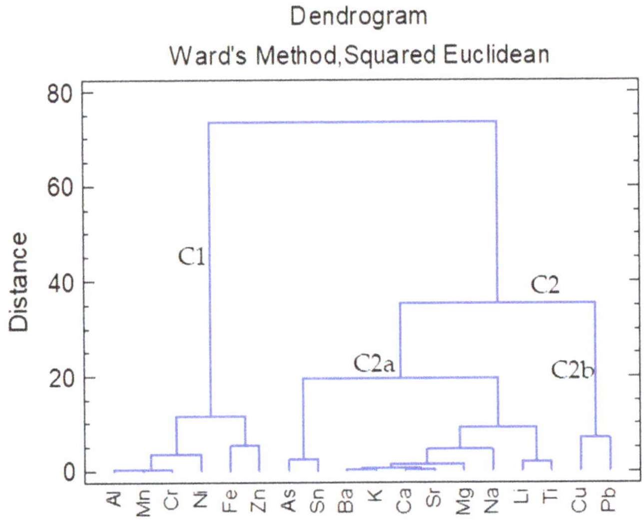

Figure 3. Cluster analysis in R mode for groundwater sampling points based on metal and As concentrations.

3.4. Heavy Metals Evaluation Index (HEI)

The *HEI* values depicted in Figure 4 varied in the range of 2.59 and 4.77 and can be classified as having a low pollution degree. The highest *HEI* value was calculated for P4 with 4.77, followed by P2 with 4.09, and the well with the lowest value of *HEI* was P1 (2.59). In the case of P4, Sr contributed the most to the *HEI* value (2.23), followed by As (0.98), Fe (0.88), and Pb (0.41).

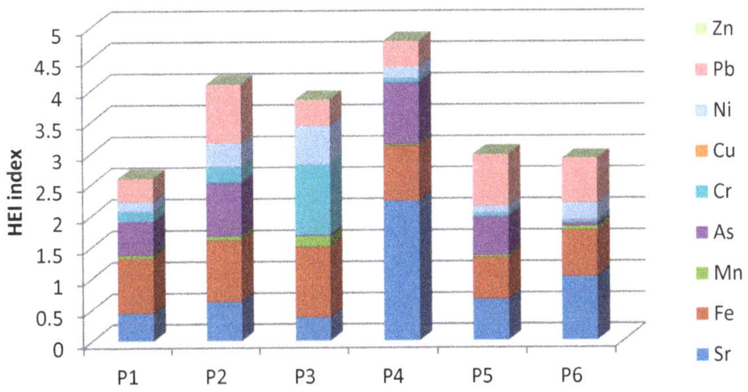

Figure 4. *HEI* values of the groundwater sources in Piatra locality based on As and metal concentrations (Sr, Fe, Mn, Cr, Cu, Ni, Pb, and Zn).

The highest values of elemental *HEI* values were due to Sr (P4, P6), As (P4, P2), Fe (P1–P4), Pb (P2, P5), and Cr (P3).

An investigation into the extent of heavy metal contamination of dug and drilled wells in Seini town, NW Romania, reported higher *HEI* values than in the present study, ranging values between 1.5 and 14, still below 40, showing a low degree of pollution due to anthropic influences [40].

In areas impacted by industrial operations, the health risk index (*HEI*) demonstrated significantly elevated values. For instance, in a plain located in western Iran, influenced by an industrial town, the *HEI* values varied from 21.4 to 133.3.

The computation of HEI provides a quick evaluation of the comprehensive quality of drinking water [79].

3.5. Water Quality Assessment by WQI

Table 3 displays the physico-chemical parameters employed in WQI computation, along with the corresponding parametric values, weights, relative weights, and the Q_i range.

Table 3. The physico-chemical parameters employed in WQI computation; the values and weights used; and the variation of Q_i.

Physico-Chemical Parameter, Measure Units	S_i * Value (p_{vi})	Weight, w_i	Relative Weight, W_i	Variation of Q
EC (μS/cm)	2500	4×10^{-4}	0.17×10^{-6}	6.88–48.2
pH (pH units)	9.5	0.11	0.000322	0.8–24.4
Turbidity, NTU	5	0.2	0.000585	22.8–175
NH_4^+ mg/L	0.5	2	0.00585	8–218
NO_3^-, mg/L	50	0.02	5.85×10^{-5}	0.3–10.64
Cl^-, mg/L	250	0.04	1.17×10^{-5}	11.68–226.08
PO_4^{3-}, mg/L	0.4	2.5	0.007319	32.5–215
Li, mg/L	0.05	20	0.058548	4.6–12.2
Na, mg/L	200	0.005	1.46×10^{-5}	5.52–29.87
K, mg/L	10	0.1	0.000293	20.45–72.45
Ca, mg/L	100	0.01	2.93×10^{-5}	1.35–8.92
Mg, mg/L	50	0.02	5.85×10^{-5}	13.77–90.45
Ba, mg/L	0.7	1.429	0.004182	2.21–11.46
Sr, mg/L	0.2	5	0.014637	37.5–223.0
Al, mg/L	0.2	5	0.014637	2.15–44.15
Fe, mg/L	0.2	5	0.014637	66.0–112.5
Mn, mg/L	0.05	20	0.058548	2.8–17.0
As, mg/L	0.01	100	0.292742	2.2–98
Cr, mg/L	0.05	20	0.058548	3.4–111.6
Cu, mg/L	0.1	10	0.029274	2.5–6.4
Ni, mg/L	0.02	50	0.146371	11–62
Pb, mg/L	0.01	100	0.292742	37–93
Zn, mg/L	5	0.2	0.000585	0.024–0.426

Notes: * according to Law 311/2004 458/2002; Council Directive. 98/83/EC; WHO, 2011 [8,66,80].

Figure 5 illustrates the assessment of groundwater quality for wells P1–P6 using the WQI method. WQI scores ranged from 31.75 to 63.43, with a mean value of 43.68 ± 12.50. P1, P3, P5, and P6 were classified as having excellent water quality. P4, having a 50.97 score, was very close to being classified as excellent quality, and P2 water was in the good quality class.

The main contributors to the WQI scores were As with a score of WQI_{As} in the range of 0.64–24.88 with the highest value found for the P2 sample, Ni (1.6–9.08) with the highest value for P3, and Pb (10.83–27.23) with the highest value for P2. The high value of WQI for Cr was calculated for P3 (6.53) while P4 registered a high value for WQI_{Sr} (3.26). One study [46] stated that WQI indices, utilizing physico-chemical parameters of water, were employed to evaluate the water quality evolution of Teceu Lake. This lake is situated in the Upper Tisa protected area in the northwest of Romania, along with a groundwater source in close proximity. The assessment was conducted using WQI indices for the period spanning January to December 2022, presenting WQI in the range of 17.71–37.94 for groundwater source (excellent quality) while the WQI score of Teceu Lake water was between 22.95 and 146.31 and indicated excellent-quality, good-quality, and poor-quality water depending on the month in which the water samples were collected [46]. The WQI values showed an increasing trend during the month of the year, with maximum values in October and November, especially due to nutrient content, ammonium, and phosphates. In another investigation [40], focusing on Seini town in the northwest of Romania, *WQI* indices were derived from sixteen chemical indicators, encompassing key physico-chemical parameters, Al, and heavy metals. The findings revealed that 65% of the groundwater samples exhibited excellent quality, while the remaining samples demonstrated poor and very poor quality. This was attributed to elevated concentrations of NH_4^+, NO_3^-, Fe, Cu, and Pb surpassing the maximum allowable concentrations (MAC). Globally, *WQI* values ranging from 51.84 to 159.41 indicate that the water quality of four lakes situated in the Bangalore Urban district, the most densely populated district in the Indian state of Karnataka, falls within poor, very poor, and unsuitable categories. This assessment is based on the consideration of 10 parameters: pH, turbidity, total alkalinity, total acidity, total phosphorus, chemical oxygen demand (COD), biochemical oxygen demand (BOD), dissolved oxygen (DO), nitrates, and total nitrogen [81]. Moreover, in Cameroon, the *WQI* scores of groundwater in the Ngoua watershed, the primary water supply source for Douala city, located at the shore of the Atlantic Ocean, vary between 2.12 and 187.21. In the computation of the *WQI*, the main physico-chemical indicators of water were considered: pH, turbidity, EC, total dissolved solids, salinity, and the concentrations of major cations and anions. The main contaminants in groundwater in the area were sulfates and nitrates [82].

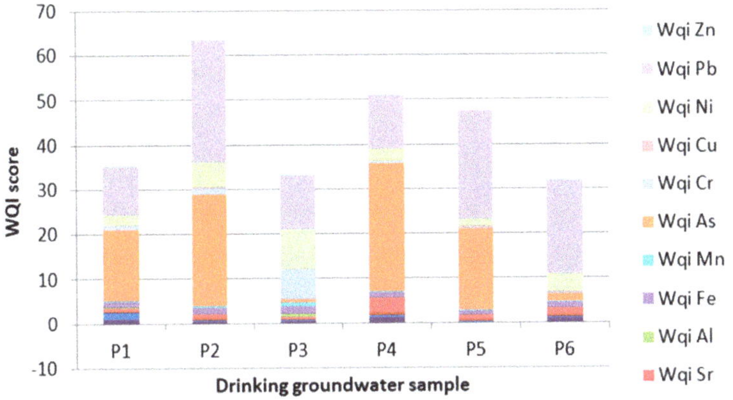

Figure 5. The water quality indices (*WQI*) of groundwater samples P1–P6 are shown as a stacked graphic, showing the main contribution to the *WQI* score.

3.6. Human Health Risk Assessment

The human health risk associated with P1–P6 groundwater consumption was assessed for adults (men and women) and children. The average *CDI* results of nitrogen compounds (NO_3^- and NH_4^+), metals, and As were obtained in the following order: NO_3^- > NH_4^+ > Sr > Fe > Cr > Pb > Ni > As > Cu. The highest values of *CDI* were found for NO_3^- in P6, with 0.12 mg/kg·day for adults (men and women) and 0.51 mg/ kg·day for children. High *CDI* values were calculated for NH_4^+ in P4: 0.0245 mg/kg·day for men and women and in P1 (0.0198 mg/kg·day) in the case of men and women, while for children the *CDI* was 0.105 mg/kg·day. Between metals and As, the highest *CDI* was obtained for Sr in P4, with 0.01 mg/kg·day for adults and 0.042 mg/kg·day for children. *CDI* for Cr in P3 was 0.0013 mg/kg·day for men and women and 0.005 mg/kg·day for children. A high value of *CDI* was obtained for As in P4: 0.00022 and 0.00094 mg/kg·day for adults and children, and in P2: 0.00019 and 0.00081 mg/kg·day for adults and children. The hazard quotients and hazard index due to groundwater intake are shown in Table 4.

Table 4. Hazard quotients, *HQ*, and hazard index, *HI*.

Groundwater Sample/Pollutant	P1	P2	P3	P4	P5	P6
			NH_4^+			
Male	0.0204	0.01116	0.0065	0.0254	0.0009	0.0179
Female	0.0206	0.0112	0.0065	0.0255	0.0009	0.0180
Children	0.0870	0.0475	0.0277	0.1078	0.0040	0.0761
			NO_3^-			
Male	0.0594	0.0336	0.0180	0.0314	0.0021	0.0750
Female	0.0596	0.0337	0.0181	0.0316	0.0021	0.0754
Children	0.2523	0.1426	0.0767	0.1336	0.0090	0.3188
			Sr			
Male	0.0033	0.0047	0.0028	0.0168	0.0050	0.0077
Female	0.0033	0.0048	0.0028	0.0168	0.0050	0.0077
Children	0.0055	0.0079	0.0047	0.0279	0.0083	0.0128
			Fe			
Male	0.0056	0.0063	0.0073	0.0056	0.0043	0.0047
Female	0.0057	0.0063	0.0073	0.0057	0.0043	0.0048
Children	0.0240	0.0267	0.0308	0.0240	0.0181	0.0201
			As			
Male	0.4061	0.6392	0.0165	0.7370	0.4662	0.0414
Female	0.4080	0.6421	0.0166	0.7404	0.4684	0.0416
Children	1.7260	2.7169	0.0703	3.1324	1.9817	0.1758
			Cr			
Male	0.5715	0.9475	4.1962	0.2557	0.1955	0.1278
Female	0.5742	0.9519	4.2155	0.2569	0.1964	0.1284
Children	2.4292	4.0274	17.8355	1.0868	0.8310	0.5434
			Cu			
Male	0.1128	0.1850	0.1489	0.1128	0.2301	0.2888
Female	0.1133	0.1858	0.1496	0.1133	0.2312	0.2901
Children	0.4795	0.7863	0.6329	0.4795	0.9781	1.2274
			Ni			
Male	0.0130	0.0322	0.0518	0.0150	0.0092	0.0234
Female	0.0130	0.0323	0.0520	0.0151	0.0092	0.0235
Children	0.0550	0.1367	0.2202	0.0639	0.0391	0.0994
			Pb			
Male	0.2782	0.6994	0.3083	0.3083	0.6242	0.5339
Female	0.2795	0.7026	0.3097	0.3097	0.6270	0.5364
Children	1.1826	2.9726	1.3105	1.3105	2.6530	2.2694
			HI			
Male	1.38	2.50	4.72	1.43	1.53	1.02
Female	1.39	2.51	4.74	1.44	1.53	1.02
Children	6.24	10.86	20.21	6.37	6.52	4.74

The risk indices were applied for NH_4^+, NO_3^-, Sr, Fe, As, Cr, Cu, Ni, and Pb, for which the studied groundwater sources indicated high concentrations of these constituents around the MACs. Hazard quotient (*HQ*) values exceeded one for chromium in well P3 for both adults and children, as well as for children in groundwater sources P1–P4. High *HQ* Pb values for children showed that they are exposed to health risks due to water ingestion from all the groundwater sources investigated. The lowest *HI* values were assessed in the case of P6 (1.02 for adults and 4.74 for children), while the highest values were calculated for P3 groundwater. In the case of children, *HQ* values higher than one were computed for Cr in P1–P4, for Pb in all the groundwater sources, and for As in P1, P2, P4, and P5, and for Cu in P6. Other research indicated significantly higher *HQ* values for children compared to those calculated for adults [56,83,84].

Conversely, the health risk assessment of the groundwater sample in Seini town, NW of Romania, indicated $HQ > 1$ for NO_3^- in some of the samples but low values of *HQ* for Cu, Pb, and Mn [40]. Napo et al., 2021 [56], assessed the health risk associated with oral exposure to groundwater in Togo's coastal sedimentary basin for males, women, and children, with a mean value of *HQ* of 0.963, 1.226, and 2.098, especially due to manganese, nitrate, arsenic, fluoride, and cadmium. The groundwater sources were affected by seawater intrusion, evaporite dissolution, and anthropogenic contamination. Moreover, groundwater in the Xinzhou Basin, situated in the semiarid region of central-eastern Shanxi Province in North China [83], was evaluated by computing the health risk posed by the contaminants NO_3^-, NO_2^-, and F^-, which exceeded the standard limits in some of the samples and found *HQ* oral values of 0.02 to 2.14 for men, 0.02 to 2.72 for women, and 0.04 to 4.66 for children. Health indices provide a precise evaluation of water quality risks for human consumption. Examining multiple parameters enhances the accuracy of identifying contaminants and health concerns. Utilizing these indices guides effective mitigation strategies, facilitates informed decision-making, and ultimately safeguards the well-being of communities relying on well water sources for multiple purposes.

4. Conclusions

This study delves into the assessment of water quality using pollution and quality indices, with a specific emphasis on evaluating potential human health risks linked to water consumption. The emphasis is on toxicological aspects that may impact human health, examining the correlation among chemical indicators that collectively influence well-being. The results indicate that groundwater specimens sourced from the safeguarded Tisa River Basin exhibit elevated levels of NH_4^+, Fe, Cl^-, and Mn, surpassing maximum allowable concentrations (MACs). Additionally, Cr concentrations exceed the MACs. Notably, ammonium concentrations were highest in P1, P4, and P6, resulting from the melting of snow, seepage with organic content, and the breakdown of vegetation and phytoplankton. The *HEI* values showed low levels of pollution. The *WQI* method indicated P1, P2, P4, and P6 as excellent-quality water, while the groundwater in P4 and P2 was classified as good quality. However, health risk assessment indicated a risk associated with water consumption higher than one in the case of all the groundwater sources, but the highest *HQ* value was computed for Cr in the P3 groundwater source. *HI* values were greater than 1 for all groundwater samples whose ingestion poses a health risk to consumers. The highest health risk assessed was computed for children, who are more vulnerable to the water pollutant due to their low mass. The findings underscore the need for sustainable and protective policies to mitigate potential adverse effects on human health. This research serves as a valuable foundation for future studies, particularly in areas with industrial and agricultural activities. Furthermore, it provides crucial information for water treatment specialists to design centralized water supply systems, ensuring that residents are informed about water quality and the potential health implications associated with their consumption.

Author Contributions: Conceptualization, T.D.; methodology, T.D. and A.A.; software, C.M.; validation, T.D. and C.M.; formal analysis, T.D., C.M. and A.A.; investigation, T.D., C.M. and A.A.; resources, T.D.; data curation, T.D. and A.A.; writing—original draft preparation, T.D. C.M. and A.A.; writing—review and editing, T.D., C.M. and A.A.; visualization, T.D., C.M. and A.A.; supervision, T.D. All authors have read and agreed to the published version of the manuscript.

Funding: This research received no external funding.

Data Availability Statement: Data is contained within the article.

Conflicts of Interest: The authors declare no conflicts of interest.

References

1. Abanyie, S.K.; Apea, O.B.; Abagale, S.A.; Amuah, E.E.Y.; Sunkari, E.D. Sources and factors influencing groundwater quality and associated health implication: A review. *Emerg. Contam.* **2023**, *9*, 100207. [CrossRef]
2. Grönwall, J.; Danert, K. Regarding Groundwater and DrinkingWater Access through A Human Rights Lens: Self-Supply as A Norm. *Water* **2020**, *12*, 419. [CrossRef]
3. UNESCO World Water Assessment Programme. The United Nations World Water Development Report. 2023. Available online: https://unesdoc.unesco.org/ark:/48223/pf0000384659 (accessed on 3 December 2023).
4. Burek, P.; Satoh, Y.; Fischer, G.; Kahil, M.T.; Scherzer, A.; Tramberend, S.; Nava, L.F.; Wada, Y.; Eisner, S.; Flörke, M.; et al. *Water Futures and Solution: Fast Track Initiative (Final Report)*; IIASA Working Paper; International Institute for Applied Systems Analysis (IIASA): Laxenburg, Austria, 2016; Available online: https://pure.iiasa.ac.at/id/eprint/13008/ (accessed on 3 December 2023).
5. Romania ClimThe World Bank Europe and Central Asia region. Romania Climate Change and Low Carbon Green Growth Program, Component B Sector Report, Integrated Water Resources Rapid Assessment, January 2014. Available online: https://documents1.worldbank.org/curated/en/675851468094456585/pdf/842640WP0P14660Box0382136B00PUBLIC0.pdf (accessed on 3 December 2023).
6. Hussain, F.; Hussain, R.; Wu, R.-S.; Abbas, T. Rainwater Harvesting Potential and Utilization for Artificial Recharge of Groundwater Using Recharge Wells. *Processes* **2019**, *7*, 623. [CrossRef]
7. Firoiu, D.; Ionescu, G.H.; Bandoi, A.; Florea, N.M.; Jianu, E. Achieving Sustainable Development Goals (SDG): Implementation of the 2030 Agenda in Romania. *Sustainability* **2019**, *11*, 2156. [CrossRef]
8. Council Directive. 98/83/EC—On the Quality of Water Intended for Human Consumption. Available online: https://eur-lex.europa.eu/legal-content/EN/TXT/?uri=celex:31998L0083 (accessed on 20 December 2023).
9. Schuteima, G.; Hooks, T.; McDermott, F. Water quality perceptions and private well management: The role of perceived risks, worry and control. *J. Environ. Manag.* **2020**, *267*, 110654. [CrossRef]
10. Sandulescu, M. *Geotectonics of Romania*; Editura Tehnică: București, Romania, 1984. (In Romanian)
11. Cioruta, B.-V.; Coman, M. Definition, Role, and Functions of Soil Related to the Knowledge Society and the Somes,-Tisa Hydrographic Area (Romania). *Sustainability* **2022**, *14*, 8688. [CrossRef]
12. Florea, N.; Munteanu, I. *Sistemul Român de Taxonomie a Solurilor (SRTS)*; Institutul Naţional de Cercetare-Dezvoltare Pentru Pedologie, Agrochimie şi Protecţia Mediului–ICPA: Bucureşti, Romania, 2012. (In Romanian)
13. Toth, G.; Montanarella, L.; Stolbovoy, V.; Mate, F.; Bodis, K.; Jones, A.; Panagos, P.; Liedekerke, M.V. Soils of the European Union. JCR–Scientific and Technical Reports, European Commision Joint Research Centre Institute for Environment and Sustainability. 2012. Available online: https://data.europa.eu/doi/10.2788/87029 (accessed on 2 December 2023).
14. Stanila, A.-L.; Dumitru, M. Soil zones in Romania and pedogenetic processes. *Agric. Agric. Sci. Procedia* **2016**, *10*, 135–139. [CrossRef]
15. Oprea, R. *Compendiu de Pedologie*, 2nd ed.; Editura Universitara: Bucharest, Romania, 2013; ISBN 978-606-591-832-0. (In Romanian)
16. Todd, D.K.; Mays, L.W. A Review Study on "Quality Assessment of Groundwater Resource in Centre India Region". *Int. J. Innov. Res. Technol.* **2004**, *4*, 968–972.
17. Arulnangai, R.; Sihabudeen, M.M.; Vivekanand, P.A.; Kamaraj, P. Influence of physico chemical parameters on potability of groundwater water in ariyalur of Tamil Nadu, India. *Mater. Today Proc.* **2021**, *36*, 923–928. [CrossRef]
18. Velarde, L.; Nabavi, M.S.; Escalera, E.; Antti, M.L.; Akhtar, F. Adsorption of heavy metals on natural zeolites: A review. *Chemosphere* **2023**, *328*, 138508. [CrossRef] [PubMed]
19. Hussain, T.S.; Faisal, A.A.H. Nanoparticles of (calcium/aluminium/CTAB) layered double hydroxide immobilization onto iron slag for removing of cadmium ions from aqueous environment. *Arab. J. Chem.* **2023**, *16*, 105031. [CrossRef]
20. Khan, Z.I.; Ahmed, K.; Ahmed, T.; Zafar, A.; Alrefaei, A.F.; Ashfaq, A.; Akhtar, S.; Mahpara, S.; Mehmood, N.; Ugulu, I. Evaluation of nickel tixicity and potential health implications of agriculturally diversely irrigated wheat crop varieties. *Arab. J. Chem.* **2023**, *16*, 104934. [CrossRef]
21. Senze, M.; Kowalska-Goralska, M.; Czyz, K. Availability of aluminum in river water supplying dam reservoirs in Lower Silesia considering the hydrochemical conditions. *Environ. Nanotechnol. Monit. Manag.* **2021**, *16*, 100535. [CrossRef]
22. Mohanty, S.; Benya, A.; Hota, S.; Kumar, M.S.; Singh, S. Eco-toxicity of hexavalent chromium and its adverse impact on environment and human health in Sukinda Valley of India: A review on pollution and prevention strategies. *Environ. Chem. Ecotox.* **2023**, *5*, 46–54. [CrossRef]
23. Siraj, K.; Kitte, S.A. Analysis of Copper, Zinc and Lead using Atomic Absorption Spectrophotometer in ground water of Jimma town of Southwestern Ethiopia. *Int. J. Chem. Anal. Sci.* **2013**, *4*, 201–204. [CrossRef]
24. Hossain, M.; Bhattacharya, P.; Frape, S.K.; Ahmed, K.M.; Jacks, G.; Hasan, M.A.; Bromssen, M.; Shahiruzzaman, M.; Morth, C.M. A potential source of low-manganese, arsenic-safe drinking water from Intermediate Deep Aquifers (IDA), Bangladesh. *Groundw. Sustain. Dev.* **2023**, *21*, 100906. [CrossRef]
25. Markeb, A.A.; Moral-Vico, J.; Sanchez, A.; Font, X. Optimization of lead (II) removal from water and wastewater using a novel magnetic nanocomposite of aminopropyl triethoxysilane coated with carboxymethyl cellulose cross-linked with shitosan nanoparticles. *Arab. J. Chem.* **2023**, *16*, 105022. [CrossRef]

26. Wei, Z.; Gu, H.; Le, Q.V.; Peng, W.; Lam, S.S.; Yang, Y.; Li, C.; Sonne, C. Perspectives on phytoremediation of zinc pollution in air, water and soil. *Sustain. Chem. Pharm.* **2021**, *24*, 100550. [CrossRef]
27. Beinabaj, S.M.H.; Heydariyan, H.; Aleii, H.M.; Hosseinzadeh, A. Concentration of heavy metals in leachate, soil, and plants in Teheran's landfill: Investigation of the effect of landfill age on the intesity of polluation. *Heliyon* **2023**, *9*, e13017. [CrossRef]
28. Chorol, L.; Gupta, S.K. Evalution of groudwater metal pollution index through analytical hierarchy process and its health risk assement via Monte Carlo simulation. *Process Saf. Environ. Prot.* **2023**, *170*, 855–864. [CrossRef]
29. Rupias, O.J.B.; Pereira, S.Y.; Silva de Abreu, A.E. Hydrogeochemistry and groundwater quality assessment using the water quality index and heavy-metal pollution index in the alluvial plain of Atibaia river—Campinas/SP, Brazil. *Groundw. Sustain. Dev.* **2020**, *15*, 100661. [CrossRef]
30. Alengebawy, A.; Abdelkhalek, S.T.; Qureshi, S.R.; Wang, M.-Q. Heavy metals and pesticides toxicity in agricultural soil and plants: Ecological risks and human health implications. *Toxics* **2021**, *9*, 42. [CrossRef] [PubMed]
31. Zhang, S.; Han, Z.; Peng, J.; Chen, Y.; Zhan, L.; Li, J. Human health risk assessment for contaminated sites: A retrospective review. *Environ. Int.* **2023**, *171*, 107700. [CrossRef] [PubMed]
32. Badeenezhad, A.; Soleimani, H.; Shahsavani, S.; Parseh, I.; Mohammadpour, A.; Azadbakht, O.; Javanmardi, P.; Faraji, H.; Nalosi, K.B. Comprehensive health risk analysis of heavy metal pollution using water quality indices and Monte Carlo simulation in R software. *Sci. Rep.* **2023**, *13*, 15817. [CrossRef] [PubMed]
33. Munene, E.N.; Hashim, N.O.; Ambusso, W.N. Human health risk assessment of heavy metal concentration in surface water of Sosian river, Eldoret town, Uasin-Gishu County Kenya. *MothodsX* **2023**, *11*, 102298. [CrossRef] [PubMed]
34. Li, P.; Karunanidhi, D.; Subramani, T.; Srinivasamoorthy, K. Sources and consequences of groundwater contamination. *Arch. Environ. Contam. Toxicol.* **2021**, *80*, 1–10. [CrossRef] [PubMed]
35. Huang, G.; Hou, Q.; Han, D.; Liu, R.; Song, J. Large scale occurrence of aluminium-rich shallow groundwater in the Pearl River Delta after the rapid urbanization: Co-effects of anthropogenic and geogenic factors. *J. Contam. Hydrol.* **2023**, *254*, 104130. [CrossRef] [PubMed]
36. *SR EN ISO 5667-3:2018*; Calitatea apei. Prelevare. Partea 3: Conservarea ¸si Manevrarea Probelor de Apa. ISO: Geneva, Switzerland, 2018. (In Romanian)
37. Dippong, T.; Mihali, C.; Cical, E. *Metode de Determinare a Proprietatilor Fizico-Chimice ale Alimentelor*; Risoprint: Cluj-Napoca, Romania, 2016; ISBN 978-973-53-1773-7. (In Romanian)
38. *ISO 7888:1985*; Water Quality—Determination of Electrical Conductivity. ISO: Geneva, Switzerland, 1985.
39. *SR EN ISO 10523:2012*; Calitatea apei. Determinarea pH-ului. ISO: Geneva, Switzerland, 2012. (In Romanian)
40. Dippong, T.; Mihali, C.; Hoaghia, A.M.; Cical, E.; Cosma, A. Chemical modeling of groundwater quality in the aquifer of Seini town-Somes Plain, Northwestern Romania. *Ecotoxicol. Environ. Saf.* **2019**, *168*, 88–101. [CrossRef] [PubMed]
41. Bhuiyan, M.A.H.; Islam, M.A.; Dampare, S.B.; Parvez, L.; Suzuki, S. Evaluation of hazardous metal pollution in irrigation and drinking water systems in the vicinity of a coal mine area of northwestern Bangladesh. *J. Hazard. Mater.* **2010**, *179*, 1065–1077. [CrossRef]
42. Varol, M.; Gökot, B.; Bekleyem, A. Dissolved heavy metals in the Tigris River (Turkey): Spatial and temporal variations. *Environ. Sci. Pollut. Res.* **2013**, *20*, 6096–6108. [CrossRef]
43. González, S.O.; Almeida, C.; Calderón, M.; Mallea, M.A.; González, P. Assessment of the water self-purification capacity on a river affected by organic pollution: Application of chemometrics in spatial and temporal variations. *Environ. Sci. Pollut. Res.* **2014**, *21*, 10583–10593. [CrossRef]
44. Uddin, M.G.; Nash, S.; Agnieszka, I.; Olbert, A.I. A review of water quality index models and their use for assessing surface water quality. *Ecol. Ind.* **2021**, *122*, 107218. [CrossRef]
45. Mare Roșca, O.; Dippong, T.; Marian, M.; Mihali, C.; Mihalescu, L.; Hoaghia, M.A.; Jelea, M. Impact of anthropogenic activities on water quality parameters of glacial lakes from Rodnei mountains, Romania. *Environ. Res.* **2020**, *182*, 109136. [CrossRef] [PubMed]
46. Dippong, T.; Mihali, C.; Avram, A. Water Physico-Chemical Indicators and Metal Assessment of Teceu Lake and the Adjacent Groundwater Located in a Natura 2000 Protected Area, NW of Romania. *Water* **2023**, *15*, 3996. [CrossRef]
47. Manoj, K.; Ghosh, S.; Padhy, P.K. Characterization and classification of hydrochemistry using multivariate graphical and hydrostatistical techniques. *Res. Chem. Sci.* **2013**, *3*, 32–42.
48. Harter, T.; Davis, H.; Mathews, M.C.; Meyer, R.D. Shallow groundwater quality on dairy farms with irrigated forage crops. *J. Contam. Hydrol.* **2022**, *55*, 287–315. [CrossRef] [PubMed]
49. Chen, M.; Al-Maktoumi, A.; Al-Mamari, H.; Izady, A.; Nikoo, M.R.; Al-Busaidi, H. Oxygenation of aquifers with fluctuating water table: A laboratory and modeling study. *J. Hydrol.* **2020**, *590*, 125261. [CrossRef]
50. Dimri, D.; Daverey, A.; Kuman, A.; Sharma, A. Monitoring water quality of River Ganga using multivariate techniques and WQI (Water Quality Index) in Western Himalayan region of Uttarakhand, India. *Environ. Nanotechnol. Monit. Manag.* **2021**, *15*, 100375. [CrossRef]
51. Lu, B.T.; Luo, J.L.; Norton, P.R. Environmentally assisted cracking mechanism of pipeline steel in near-neutral pH groundwater. *Corros. Sci.* **2010**, *52*, 1787–1795. [CrossRef]
52. Wilkin, R.T.; DiGiulio, D.C. Geochemical impacts to groundwater from geologic carbon sequestration: Controls on pH and inorganic carbon concentrations from reaction path and kinetic modelling. *Environ. Sci. Technol.* **2010**, *44*, 4821–4827. [CrossRef]
53. Mohammed, I.; Al-Khalaf, S.K.H.; Alwan, H.H.; Naje, A.S. Environmental assessment of Karbala water treatment plant using water quality index (WQI). *Mater. Today Proc.* **2022**, *60*, 1554–1560. [CrossRef]

54. Nawaz, R.; Nasim, I.; Irfan, A.; Islam, A.; Naeem, A.; Ghani, N.; Irshad, M.A.; Latif, M.; Un Nisa, B.; Ullah, R. Water Quality Index and Human Health Risk Assessment of Drinking Water in Selected Urban Areas of a Mega City. *Toxics* **2023**, *11*, 577. [CrossRef] [PubMed]
55. Emmanuel, U.C.; Chukwudi, M.I.; Monday, S.S.; Anthony, A.I. Human health risk assessment of heavy metals in drinking water sources in three senatorial districts of Anambra State, Nigeria. *Toxicol. Rep.* **2022**, *9*, 869–875. [CrossRef]
56. Napo, G.; Akpataku, K.V.; Seyf-Laye, A.S.M.; Gnazou, M.D.T.; Bawa, L.M.; Djaneye-Boundjou, G. Assessment of Shallow Groundwater Quality Using Water Quality Index and Human Risk Assessment in the Vogan-Attitogon Plateau, Southeastern (Togo). *J. Environ. Poll. Hum. Health* **2021**, *9*, 50–63. [CrossRef]
57. Average Height and Weight by Country. Available online: https://www.worlddata.info/average-bodyheight.php (accessed on 2 December 2023).
58. United States Environmental Protection Agency. *IRIS (Integrated Risk Information System). A–Z List of Substances*; United States Environmental Protection Agency: Washington, DC, USA, 1987. Available online: https://www.epa.gov/iris (accessed on 3 December 2023).
59. Qaiyum, M.S.; Shaharudin, M.S.; Syazwan, A.I.; Muhaimin, A. Health Risk Assessment after Exposure to Aluminium in Drinking Water between Two Different Villages. *J. Water Resour. Prot.* **2011**, *3*, 268–274. [CrossRef]
60. Tsuji, S.J.; Perez, V.; Garry, M.R.; Alexander, D.D. Association of low-level arsenic exposure in drinking water with cardiovas-cular disease: A systematic review and risk assessment. *Toxicology* **2014**, *323*, 78–94. [CrossRef]
61. Mawari, G.; Kumar, N.; Sarkar, S.; Frank, A.L.; Daga, M.K.; Singh, M.M.; Joshi, T.K.; Singh, I. Human Health Risk Assessment due to Heavy Metals in Ground and Surface Water and Association of Diseases With Drinking Water Sources: A Study from Maharashtra, India. *Environ. Health Insights* **2022**, *16*, 1–11. [CrossRef]
62. Rivett, M.O.; Buss, S.R.; Morgan, P.; Smith, J.W.; Bemment, C.D. Nitrate attenuation in groundwater: A review of biogeochemical controlling processes. *Water Res.* **2008**, *42*, 4215–4232. [CrossRef]
63. Laha, F.; Gashi, F.; Cadraku, H. Variations in the physico-chemical parameters of Grinaja catchment groundwater. *IFAC-PapersOnLine* **2019**, *49*, 200–205.
64. Adesakin, T.A.; Oyewale, A.T.; Bayero, U.; Mohammed, A.N.; Aduwo, I.A.; Ahmed, P.Z.; Abubakar, N.D.; Barje, I.B. Assessment of bacteriological quality and physico-chemical parameters of domestic water sources in Samaru community, Zaria, Northwest Nigeria. *Heliyon* **2020**, *6*, e04773. [CrossRef]
65. Dippong, T.; Mihali, C.; Mare Rosca, O.; Marian, M.; Payer, M.M.; Tîbîrnac, M.; Bud, S.; Coman, G.; Kovacs, E.; Hoaghia, M.-A. Physico-chemical characterization of water wells from Remeti, Maramures. *Stud. Univ. Babes-Bolyai Chem.* **2021**, *66*, 197–211. [CrossRef]
66. World Health Organization. *Guidelines for Drinking-Water Quality*, 4th ed.; WHO: Geneva, Switzerland, 2011; pp. 1–541. ISBN 978-92-4-154815-1.
67. Weisner, M.L.; Harris, M.S.; Mitsova, D.; Liu, W. Drinking water disparities and aluminum concentrations: Assessing socio-spatial dimensions across an urban landscape. *Soc. Sci. Humanit. Open* **2023**, *8*, 100536. [CrossRef]
68. Nguyen, P.M.; Mulligan, G.N. Study on the influence of water composition on iron nail corrosion and arsenic removal performace of the Kanchan arsenic filter (KAF). *J. Hazard. Mater. Adv.* **2023**, *10*, 100285. [CrossRef]
69. Khan, M.; Din, I.; Aziz, F.; Qureshi, I.U.; Zahid, M.; Mustafa, G.; Sher, A.; Hakim, S. Chromium adsorption from water using mesoporous magnetic iron oxide-aluminum silicate absorbent: An investigation of adsorbtion isotherms and kinetics. *Curr. Res. Green Sustain. Chem.* **2023**, *7*, 100368. [CrossRef]
70. Qu, Y.; Ji, S.; Sun, Q.; Zhao, F.; Li, Z.; Zhang, M.; Li, Y.; Zheng, L.; Song, H.; Zhang, W.; et al. Association of urinary nickel levels with diabetes and fasting blood glucose levels: A nationwide Chine population-based study. *Ecotox. Environ. Saf.* **2023**, *252*, 114601. [CrossRef] [PubMed]
71. Li, L.; Sun, F.; Liu, Q.; Zhao, X.; Song, K. Development of regional water quality criteria of lead for protecting aquatic organism in Taihu Lake, China. *Ecotox. Environ. Saf.* **2021**, *222*, 112479. [CrossRef] [PubMed]
72. Abbaszade, G.; Tserendorj, D.; Salazar-Yanez, N.; Zachary, D.; Volgyesi, P.; Toth, E. Lead and stable lead isotopes as tracers of soil pollution and human health risk assessment in former industrail citisen of Hungary. *Appl. Geochem.* **2022**, *145*, 105397. [CrossRef]
73. Obasi, P.N.; Akudinobi, B.B. Potential health risk and levels of heavy metals in water resources of lead–zinc mining communities of Abakaliki, southeast Nigeria. *Appl. Water Sci.* **2020**, *10*, 184. [CrossRef]
74. Krok, B.; Mohammadian, S.; Noll, H.M.; Surau, C.; Markwort, S.; Fritzsche, A.; Nachev, M.; Sures, B.; Meckenstock, R.U. Remediation of zinc-contaminated groundwater by iron oxide in situ adsorption barriers—From lab to the field. *Sci. Total Environ.* **2022**, *897*, 151066. [CrossRef]
75. Peng, H.; Yao, F.; Xiong, S.; Wu, Z.; Niu, G.; Lu, T. Strontium in public drinking water and associated public health risks in Chinese cities. *Environ. Sci. Pollut. Res.* **2021**, *28*, 23048–23059. [CrossRef]
76. Manea, A.; Dumitru, M.; Vrinceanu, N.; Eftene, A.; Anghel, A.; Vrinceanu, A.; Ignat, P.; Dumitru, S.; Mocanu, V. Soil heavy metal status from Maramures county, Romania. In Proceedings of the GLOREP 2108 Conference, Timisoara, Romania, 15–17 November 2018.
77. La Maestra, S.; D'Agostini, F.; Sanguineti, E.; González, A.Y.; Annis, S.; Militello, G.M.; Giovanni Parisi, G.; Scuderi, A.; Gaggero, L. Dispersion of Natural Airborne TiO2 Fibres in Excavation Activity as a Potential Environmental and Human Health Risk. *Int. J. Environ. Res. Public Health* **2021**, *18*, 6587. [CrossRef]
78. Shi, H.; Magaye, R.; Castranova, V.; Zhao, J. Titanium dioxide nanoparticles: A review of current toxicological data. *Part. Fibre Toxicol.* **2013**, *10*, 15. [CrossRef]

79. Lorestani, B.; Merrikhpour, H.; Cheraghi, M. Assessment of heavy metals concentration in groundwater and their associated health risks near an industrial area. *Environ. Health Eng. Manag.* **2020**, *7*, 67–77. [CrossRef]
80. Law 311 from 6rd June 2004 That Improves and Complements Law 458 from 29 July 2002 Regarding the Quality of Drinking Water. Official Gazette. Part I, no. 582/30.06.2004. (In Romanian). Available online: https://www.lege-online.ro/lr-LEGE-311%20-2004-(53106)-(1).html (accessed on 3 December 2023).
81. Mishra, M.; Singhal, A.; Sriniva, R. Effect of urbanization on the urban lake water quality by using water quality index (WQI). *Mater. Today Proc.* **2023**. [CrossRef]
82. Billong, P.T.N.; Feumba, R.; Ndjigui, P.-D. Hydrogeochemical appraisal of groundwater quality in Ngoua watershed (Doua-la-Cameroon): Implication for domestic purposes. *Sci. Afr.* **2023**, *22*, e01910. [CrossRef]
83. Chen, F.; Yao, L.; Mei, G.; Shang, Y.; Xiong, F.; Ding, Z. Groundwater Quality and Potential Human Health Risk Assessment for Drinking and Irrigation Purposes: A Case Study in the Semiarid Region of North China. *Water* **2021**, *13*, 783. [CrossRef]
84. Hopenhayn, C.; Ferreccio, C.; Browning, S.R.; Huang, B.; Peralta, C.; Gibb, H.; Hertz-Picciotto, I. Arsenic Exposure from Drinking Water and Birth Weight. *Epidemiology* **2003**, *14*, 593–602. [CrossRef] [PubMed]

Disclaimer/Publisher's Note: The statements, opinions and data contained in all publications are solely those of the individual author(s) and contributor(s) and not of MDPI and/or the editor(s). MDPI and/or the editor(s) disclaim responsibility for any injury to people or property resulting from any ideas, methods, instructions or products referred to in the content.

water

MDPI

Article

Interdisciplinary Evaluation of the Săpânța River and Groundwater Quality: Linking Hydrological Data and Vegetative Bioindicators

Ovidiu Nasca [1], Thomas Dippong [1], Maria-Alexandra Resz [2,*] and Monica Marian [1]

[1] Faculty of Science, Technical University of Cluj-Napoca, 76 Victoriei Street, 430122 Baia Mare, Romania; ovidiu_nasca@yahoo.com (O.N.); thomas.dippong@cb.utcluj.ro (T.D.); monicamarian17@gmail.com (M.M.)
[2] SC Research Institute for Auxiliary Organics Products SA, ICPAO, 8 Carpati Street, 551022 Medias, Romania
* Correspondence: m.alexandraresz@gmail.com

Abstract

This study was carried out to fill the present research gap in the study area by assessing water chemistry, potential heavy metal contamination, and the associated health risk evaluation that goes along with it in surface water bodies and groundwater in the NE of Maramureș County, near the Tisa River. The main methods we applied were Piper, Ficklin–Caboi, and Gibbs diagrams for determining the water typology and chemistry, the Overall Water Quality Index (OWQI) and vegetation cover to determine the water quality, a contamination index for analyzing the contamination degree, and a human health risk assessment through water ingestion after exposure of children and adults. This article's main findings specify that waters were characterized and classified into the $CaMgHCO_3^-$ dominant category of water type, with precipitation, agricultural, and domestic inputs, related to the Cl^- (mean ranging between 1.01–5.65 mg/L) and NO_3^- (mean ranging between 2.23–5.52 mg/L) content. The OWQI scores indicated excellent quality, below the critical value, ranging between 0.70 and 6.57. The applied risk assessment indicated that the daily intake of toxins is higher in the case of children than in adults, up to four and five times. The hazard quotient scores, ranging between 0.00093 and 0.248 for adults and between 0.0039 and 1.040 for children, indicated that if consumed, the studied waters can pose potential negative effects on children.

Keywords: cover vegetation; heavy metals; human health risk assessment; water typology

1. Introduction

Water is among the most fundamental elements of life, being the most prevalent on the surface of the globe. Surface waters are defined as units characterized by biotic and abiotic elements. The biotic elements comprise aquatic fauna and flora, which are strongly influenced by the natural conditions in which they live. The abiotic elements encompass morphological, geological, hydrological, and climatological factors [1]. Water, found in different states, is a limited resource. River systems have played a pivotal role in the development of human societies and the enhancement of the human condition [2–7]. Human well-being is impacted by water quality through direct human uses like drinking and irrigation in agriculture and industry, as well as indirect effects like its impact on ecosystem services like aquaculture and fisheries productivity and wetland biodiversity [3–5]. Rapid urbanization and other human activities, such as hydrogeochemical processes, water–rock interaction, and evaporation, have an important impact on the geochemical evolution of groundwater [5–9]. According to other studies, contaminants like nitrates, heavy metals, and microbiological infections are common and present major health dangers to the local populace. The rapid expanding human population and hydrological processes (recharge characteristics, possible recharge zones, groundwater origin, pollution sources, and surface water interactions in areas like wetlands, as well as water-stable isotopes) favor the water contamination process [3–9]. The equilibrium in natural ecosystems may be impacted or even destroyed by human selfishness regarding nature and the living environment, which are exploited carelessly and without supervision [10]. Because water bodies receive untreated wastewater from households, municipal farms, industrial activities, and other sources, most aquatic ecosystems (ponds, lakes, rivers, and streams) are of low quality because of human activity [11]. Investigating natural ecosystems includes looking into the main risks to biodiversity, such as overuse of natural resources, water pollution, land fragmentation, habitat destruction or degradation, and non-native species incursions [12,13].

Aquatic habitats are characterized by biodiversity with higher sensitivity related to the terrestrial ecosystem [13–15]. Two amphibian species (*Triturus cristatus* and *Bombina variegata*) and one reptilian species (*Emys orbicularis*) of conservation significance round out the faunal list [15]. The composition and geographic distribution of amphibians (vital and therefore protected species in Romania) echo the negative effects of human pressure on humid environments (pollution, habitat deterioration) [14]. The water level, water flow velocity, and untreated wastewater discharge from draining water with suspended sediment all affect the nutrient concentrations in moving water bodies. For example, the release of phosphorus particles has been demonstrated to induce degradation in the quality of water through the process of sedimentation [16].

Efficient tools, specifically contamination and health risk indices, are based on the monitored and measured chemical parameters in waters and the standard values or reference doses established by authorities through regulations and guidelines at national and international levels, correlated with the water quality and its purpose, applied in order to determine the potential contamination degree and the human health risks. The most common and successful indices used at the global level are represented by the contamination index (CI), heavy metal evaluation index (HMEI), chronic daily index (CDI), hazard quotient (HQ), or the hazard index (HI) [1,2]. They are mandatory to be applied in case of potential contamination. In this framework, the present study area was selected based on the use of water sources and the contamination sources situated near the water sources. The main objective of the current research was to measure and examine the chemical content (13 chemical indicators, 10 trace metals, and 4 major cations) related to water sources sampled from Natural Area Tisza, used with drinking purpose. Based on the obtained concentrations, the water typology was analyzed (Piper, TIS (Total Ionic Salinity), Gibbs, Ficklin–Caboi charts), and the degree of contamination with heavy metals was calculated based on the contamination indices (PI and HMEI). The overall water quality was determined by applying the WQI (Water Quality Index). Likewise, a human health risk evaluation with non-carcinogenic implication was implemented in two scenarios (children and adults) with the help of three risk indices (CDI, HQ, and HI). The obtained results are significant and can be used by policy makers and local authorities in preventing and reducing the contamination and potential risk induced for human health.

The major research question challenged in the current study implies the chemical composition of two different water types, surface water and groundwater, potentially influenced by anthropogenic activities. The site-specific context of this study implies a protected area from the NE part of Maramureș County, Romania, where several sampling points were selected on the Săpânța River, a tributary of the Tisa River, and a groundwater source.

The specific objectives of this study include (*a*) the determination of the physico-chemical marker distribution and sources concerning the Săpânța River and groundwater sampling points; (*b*) a water typology analysis based on the Piper and Total Ionic Salinity (TIS) plots; (*c*) the assessment of the potential metal contamination degree of surface waters and groundwaters, using diverse contamination indices, as well as changes in the quality of waters, which were analyzed by using vegetation parameters as an indicator; (*d*) the evaluation of the ecological and biological variety related to the Săpânța River and groundwater sampling points, with the aim of implementing conservation regulations, and (*e*) a human health risk assessment of metals through water ingestion in the cases of children and adults.

The significance of this work implies the assessment of water quality (drinking water source) and public health, emphasizing the scientific contribution of the obtained results given by its interdisciplinary use in water and human protection. Similarly, the findings of this study, also implying the identification of various hazardous metals, are significant in improving the application of environmental risk management in the studied area, aiming for contamination mitigation.

2. Materials and Methods

2.1. Study Area and Sampling

Protected natural areas are designated for the conservation of species and their habitats. Natura 2000 sites are unique in that they encompass extensive surface areas, thereby concentrating both biodiversity values and human communities engaged in diverse activities. The Tisa and Săpânța Rivers are part of the protected area, encompassing a diverse array of habitats and species of common interest. The Săpânța River plays a pivotal role in enhancing biodiversity in a critical section of the protected area. It is situated in the Gutâi Mountains amidst natural beech forests, ensuring optimal water quality. However, the rivers traverse the town of Săpânța, which has a population of approximately 3000, as well as a substantial influx of tourists due to its nationally and internationally recognized objectives, namely the Merry Cemetery and the Peri Monastery. The presence of an anthropogenic disturbance, in the form of a micro-hydropower plant on the Săpânța meadow, in conjunction with the impact of agriculture in the Tisa meadow, has given rise to inquiries regarding the quality of the river and the waters connected to it, including groundwater and the Tisa River.

From a geological point of view, the area belongs to the Maramures Depression [17]. The formations that are included in the Maramures Depression are attributed to the Badenian, Sarmatian, and Pliocene ages. The geological structure consists of Pannonian marls as a substrate of sediments from the Tisa month. The formations attributed to the Pleistocene are represented by the low terrace deposits of the Tisa River. These are formed by alluvial–proluvial material represented by gravel, boulders, and sand. Petrographically, in the mobile deposits, eruptive formations predominate in association with metamorphic and sedimentary rocks, which are responsible for a part of minerals and ions, with the potential to increase conductivity in relation to the surface water of the river.

The Săpânța River is localized in the Săpânța locality, in the NE part of Maramureș County near the Tisa River (Figure 1). The sampling points for all four (4) studied water bodies are presented in Figure 1, which also indicates the land use (mainly agricultural crops, a touristic area, and forest). The sampling points had the following coordinates: SR1 (47°59′16″ N, 23°41′46″ E), SR2 (47°59′3″ N, 23o41′51″ E), SR3 (47°58′55″ N, 23°41′52″ E), and GW (47°58′54″ N, 23°42′3″ E). The sampling strategy was applied after standard procedures, and several physico-chemical analyses were performed on-site.

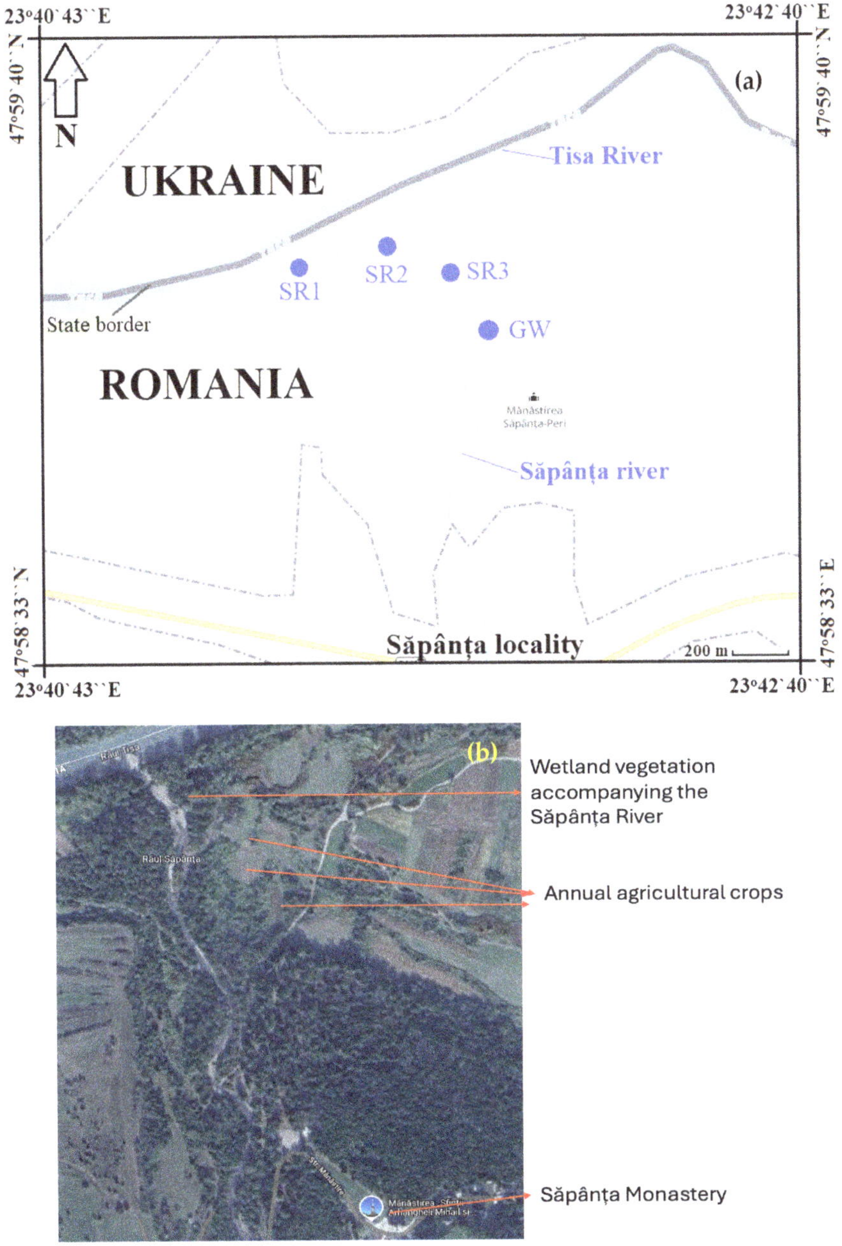

Figure 1. Localization of the study area (**a**). Map of the studied area—mosaic of natural vegetation and annual agricultural crops (**b**).

2.2. Water Quality Analysis

Analysis of the physico-chemical parameters was applied 48 h after the water samples were collected. Samples were collected in sterile polyethylene flasks to avoid contamination, and they were refrigerated until analysis and were based on standard procedures [18]. Samples were collected in triplicate, as monthly measurements in 2024. A conductometer model, INOLAB 740 (WTW, Weilheim in Oberbayern, Germany), was used to measure the EC, and an HI 253 pH meter (Hanna Instruments, Woonsocket, RI, USA) with a mixed pH electrode was used for the pH determinations, in accordance with standard guidelines [19,20]. Turbidity was determined applying titrimetric methods and using a WTW 355 IR portable turbidimeter (WTW, Weilheim in Oberbayern, Germany) [21].

Subsequent to the incorporation of oxalic acid (99.0% purity, purchased by Sigma-Aldrich, St. Louis, MO, USA), the dissolved oxygen (DO) concentration was determined through back titration employing potassium permanganate (99.0% purity, purchased by Merck Darmstadt, Darmstadt, Germany) [22]. In agreement with standard procedures, total alkalinity was measured by acidic titration, while precipitation titration using silver nitrate (≥99.0% purity, purchased by Merck Darmstadt, Darmstadt, Germany) was used to quantify the concentration of Cl^- [23,24]. The EDTA titrimetric procedure was applied to determine total hardness (h_t) [25]. The analysis of Al, NH_4^+, and NO_3^- was conducted using a UV–VIS spectrophotometer, model Specord 50 (Analytik Jena, Jena, Germany), which conforms to standard methodologies [26–28]. The PO_4^{3-} was determined using the SR EN ISO 6878:2005 method [29]. The SR ISO 6332/1996 protocol was employed to analyze the dissolved Fe content [30], and the SR ISO 8662/2/1997 protocol was used to analyze the Mn content [31].

A 1:1 HNO_3 (65%, supplied by Merck, Darmstadt, Germany) solution was employed to preserve the water samples until an acid pH of 1–2 was attained. The samples were subjected to digestion using a mixture of 30% hydrogen peroxide and 70% concentrated HNO_3 (both supplied by Merck, Darmstadt, Germany). The digested samples were stored at 4 °C after being dissolved in HNO_3. Filtration and acidification with 0.5 mL of 70% HNO_3 per 100 mL of each sample, excluding the mineralization step, were required for samples that were to be examined using the graphite furnace [32,33].

The analysis was conducted using graphite flame absorption atomic spectrometry (GFAAS) on a spectrophotometer, the Analyst 800 model (PerkinElmer, Norwalk, CT, USA), determined using standard procedures [32,33]. Details related to the equipment and calibration are presented in the Supplementary Material.

Based on the physico-chemical and chemical analysis conducted on the water samples, variate plots were constructed, serving different purposes in analyzing the studied water samples.

The total salinity of water was determined based on the Total Ionic Salinity in the form of a TIS plot. TIS shows the salinity of waters expressed in meq/L. It is determined with the help of the SO_4^{2-}, Cl^-, and HCO_3^- content. The relation between the total metal load and pH determined in the present study was represented with the help of a Ficklin–Caboi diagram. This plot shows the metal load and pH type, classifying the waters into one of six categories (I. high acid, extreme metal; II. acid, extreme metal; III. near-neutral, extreme metal; IV. high acid, high metal; V. acid, high metal; and VI. near-neutral, high metal). TIS and Ficklin–Caboi plots were obtained with the help of Microsoft Excel 2016 MSO, version 2503 Build.

2.3. Water Typology

The water typology was determined based on the Piper plot. This plot was constructed with the free version 1.30.0.0 of GW_Chart, software approved by the U.S. Geological Survey. A Piper plot contains three different plots (two trilinear plots: one based on the cation content and one on the anions content, as well as one diamond plot which combines the results of both trilinear plots) and indicates the water typology based on the cation (Na, K, and Mg) and anion (CO_3^{2-}, HCO_3^-, SO_4^{2-}, and Cl^-) content measured in the water samples.

Water sources were identified with the help of Gibbs plots and the relation between NO_3^-/Cl^- and Cl^- content, using Microsoft Excel 2016 MSO, version 2503 Build. Gibbs plots include two plots, one based on the ratio among the TDS content (mg/L) and the anion ratio ($Cl^-/(Cl^- + HCO_3^-)$) content and one based on the ratio between the TDS content (mg/L) and the cation ratio ((K + Na)/(Ca + Na + K)). According to the representation of samples related to the plots, it classifies its origin in one of two categories (water–rock interaction or precipitation dominance). Agricultural inputs and domestic effluents are two other sources into which the waters are classified, based on the relation between the Cl^- content and ratio between NO_3^- and Cl^- content (mmol/L).

2.4. Potential Contamination Degree Evaluation

The overall water quality of waters was determined and studied by applying the OWQI (Overall Water Quality Index) approach. The OWQI converts quantitative data into qualitative data, identifying the degree of quality. The OQWI is based on the measured data, implying the physico-chemical and chemical indicator results (major cations, anions, trace metals), the unit weight, the maximum allowable concentration (MAC), and the ideal value of each indicator. This index is comprised with the help of Equation (1), as indicated by Das et al. [34]:

$$OWQI = \frac{\left[100 \times \left(\sum \frac{MD_i - ID}{MAC_i - ID}\right)\right]}{W_i}, \quad (1)$$

MD_i signifies the average results related to each chemical indicator determined in the water samples, ID is the ideal value, and W_i is the weight of each chemical parameter considered for calculating this tool. After applying Equation (1), the water sample is classified into one of five (5) quality categories (unsuitable for drinking if the OWQI score is below 100; very poor (OWQI <76), poor ($51 < OWQI \leq 75$), good ($26 < OWQI < 50$), and excellent ($OWQI \leq 25$)), characterizing the studied water sample [34].

The potential contamination in relation to heavy metals and As content was determined and evaluated by applying the contamination index (CI) and the heavy metal evaluation index (HMEI). Both index approaches consider the total quality of the water related to the heavy metal and As content. The CI is based on the weighted arithmetic quality average of each studied chemical, while the HMEI represents the ratio between the measured concentration and the MAC of each chemical. Equations (2) and (3) indicate the calculation methods for CI and for HMEI, respectively [35]:

$$CI = \frac{\sum_{i=1}^{n} W_i \times q_i}{\sum_{i=1}^{n} W_i}, \quad (2)$$

$$HMEI = \sum_{i=1}^{i} \frac{MD_i}{MAC_i}, \quad (3)$$

Element q_i represents the sub-index of the ith indicator, calculated as the ratio between the obtained result and MAC concerning heavy metal, multiplied by 100. The obtained CI score classifies the water as contaminated (CI >100) or not contaminated ($CI \leq 100$) with the studied heavy metals and As, compared to the critical contamination index value (100) [35]. The HMEI, otherwise, offers an evaluation of the overall quality of water, implying the impact of toxin implications on human health. Based on the HMEI scores, waters are characterized by one of the six quality classes: severely affected (HMEI >6.0), heavily affected ($4.0 \leq HMEI \leq 6.0$), moderately affected ($2.0 \leq HMEI < 4.0$), partly affected ($1.0 \leq HMEI < 2.0$), clean ($0.3 < HMEI < 1.0$), and very clean ($HMEI \leq 0.3$) [36].

2.5. Human Health Risk Assessment of Toxins

To quantify the exposure to heavy metals through water ingestion, the lifetime mean daily dose of ingested toxins, the Chronic Daily Index (CDI), was applied and analyzed. In this framework, Equation (4) was applied:

$$CDI = {}^{MD \times ED \times EF}/_{AT \times BW} \times IR, \quad (4)$$

where ED ($ED_{children}$ = 6 years, ED_{adults} = 70 years) and EF ($EF_{children}$ and EF_{adults} = 365 days) are the exposure duration and exposure frequency, and AT ($AT_{children}$ = 2190 days, AT_{adults} = 25550 days) and BW ($BW_{children}$ = 15 kg, BW_{adults} = 70 kg) are the average time exposure and body weight. Element IR represents the daily ingestion rate ($IR_{children}$ = 1.8 L/day, IR_{adults} = 2.0 L/day) [36].

Non-carcinogenic concerns include teratogenic and genetic consequences and are mostly related to exposure. Two different scenarios were applied in the present study, health risk affecting children and adults, through the ingestion pathway related to heavy metals present in water.

Furthermore, the non-cancer risk of heavy metals, implying potential negative effects, was assessed by calculating two health risk indices: the hazard quotient (HQ) and the hazard index (HI). For their determination, Equations (5) and (6) were applied as follows:

$$HQ = \frac{CDI}{RfD}, \quad (5)$$

$$HI = \sum HQ_{HM1} + HQ_{HM2} + \ldots + HQ_{HMn} \quad (6)$$

RfD represents the chronic oral reference dose, including the daily oral exposure level of the general population plus a sensitive subgroup that is most likely not at high risk of long-term negative effects. It is specific for each toxin used in the calculation of the CDI. The chronic hazard index (HI), defined as the total of all HQs determined for each heavy metal, was employed to quantify the potential human health risks associated with exposure at multiple HMs. HQ and HI are expressed in mg/kg/day. HQ and HI scores higher than one (HQ and HI $\geq$1.0) indicate non-carcinogenic risk, while scores lower than one imply no significant non-cancer risk from heavy metals ingested through water consumption [36,37].

The risk assessment methods were applied considering the Sr, Mn, Fe, and Zn concentrations, as well as four specific metals, including Pb, Ni, Cu, and As, which pose significant health risks to humans, especially when exposure occurs over extended periods due to their toxicity at low concentrations, carcinogenicity, and persistency.

3. Results and Discussion

3.1. Analysis of Water Chemistry

The results related to the physico-chemical and chemical indicators measured in the water samples are presented in Tables 1 and 2 as annual median values.

The electrical conductivity of the Săpânța River varied between 43.3 and 50.0 µS/cm. In comparison, the groundwater, as represented by a well near the river, exhibited a conductivity of 255 µS/cm. This value is well below the legal limit of 2500 µS/cm. The low conductivity values are attributable to two factors. Firstly, there is a low level of dissolved inorganic substances in ionized form [38–41]. Secondly, there is a dilution effect during atmospheric precipitation, which is also influenced by temperature [38–41]. The electrical conductivity of water is contingent upon the Cl^- and $HCO_3{}^-$ levels, depending on the water's composition. Evaporation and flooding influence the electrical conductivity of water as well [38–41].

Table 1. The annual median values of the physico-chemical parameters determined in water samples (SR1–SR3) and groundwater near Săpânța River (GW). The results are expressed as the mean ± standard deviation (n = 12 (monthly measurement in 2024)).

	SI	SR1	SR2	SR3	GW	Mean	Median	Min	Max	SD	Errors	MAC *
EC	μS/cm	50.0 ± 4.8	46.2 ± 4.5	43.3 ± 4.1	255 ± 23	98.6	48.1	43.3	255	104	±8.3	250
pH	-	7.95 ± 0.77	7.98 ± 0.78	7.75 ± 0.63	7.38 ± 0.21	7.77	7.85	7.99	7.98	0.28	±1.15	6.5–9.5
DO	mg/L	9.26 ± 0.89	8.69 ± 0.72	8.55 ± 0.81	7.99 ± 0.68	8.63	8.62	1.09	9.26	0.52	±1.10	-
T	NTU	2.15 ± 0.19	5.52 ± 0.47	3.16 ± 0.30	1.09 ± 0.09	2.98	2.66	1.86	5.52	1.89	±0.25	<5
H_t	°g	2.45 ± 0.15	1.86 ± 0.12	2.78 ± 0.22	11.55 ± 0.96	4.66	2.62	1.01	11.55	4.61	±0.19	>5
Cl^-	mg/L	1.01 ± 0.88	2.75 ± 0.16	1.88 ± 0.16	5.65 ± 0.53	2.82	2.32	0.045	5.65	2.01	±0.08	250
NH_4^+	mg/L	0.045 ± 0.004	0.057 ± 0.005	0.053 ± 0.005	0.066 ± 0.006	0.055	0.055	2.23	0.066	0.009	±0.002	0.5
NO_3^-	mg/L	2.23 ± 0.17	2.76 ± 0.20	2.88 ± 0.24	5.52 ± 0.49	3.35	2.82	3.84	5.52	1.48	±0.39	50
SO_4^{2-}	mg/L	3.84 ± 0.29	4.68 ± 0.33	4.89 ± 0.41	5.45 ± 0.26	4.72	4.79	3.84	5.45	0.67	±0.42	250
PO_4^{3-}	mg/L	0.02 ± 0.001	0.04 ± 0.002	0.03 ± 0.001	0.03 ± 0.002	0.03	0.03	0.02	0.04	0.008	±0.001	0.4
TDS	mg/L	44.2 ± 4.2	47.2 ± 3.9	53.5 ± 5.1	62.9 ± 5.9	52.0	50.6	44.2	62.9	8.3	±4.2	-
A_t	mmol/L	0.35 ± 0.02	1.22 ± 0.11	1.38 ± 0.09	2.65 ± 0.19	1.40	1.30	0.35	2.65	0.95	±0.05	-

Notes: * MAC based on international regulations, WHO guidelines [42], and national law [43] related to quality of water used as drinking water source.

The pH exhibited fluctuations within the permissible limits, ranging from 7.75 to 7.95 (Table 1), both in the Săpânța River and in the groundwater (7.38). The decrease in water pH is attributable to an increase in the content of carbon dioxide or other gases with acidic reactions, due to organic matter accumulation. An increase in water pH is associated with the alkalinity of the soil type, while significant fluctuations present stress for aquatic species [44,45].

The range of dissolved oxygen (DO) levels in the Săpânța River was between 8.55 and 9.26 mg/L, with a mean value of 8.7 mg/L. The DO level in groundwater (7.99 mg/L) is a function of its oxygen content, which is determined by the degree of organic substances present. The presence of pathogenic microflora in groundwater is a consequence of the water conditions, as well as the atmospheric pressure, water temperature, microorganisms in hydrogeochemical processes, and oxidizable substances. The geographical area of the basin, climate, altitude, temperature, flora, leakages of fertilizers applied on agricultural terrains, soil characteristics of the riverbed, and biological and chemical properties of the water are also factors that influence the oxygen content of groundwater [39,40,44]. The lower value of DO in the groundwater compared to the Săpânța River water may be indicative of an overabundance of bacteria capable of consuming the dissolved oxygen in the water [11,39,40,44]. The levels of DO in water are influenced by various factors, including water temperature, air pressure, aquatic vegetation, phytoplankton, and the presence of oxidizable materials and bacteria [43]. During the monitoring period, DO concentrations increased, concurrently with a decline in the electrical conductivity of the water [44]. A state of deoxygenation induced by environmental eutrophication, characterized by elevated PO_4^{3-} and NH_4^+ amounts and augmented algae activity, has been associated with diminished DO levels. This phenomenon can be attributed to anaerobic processes, as evidenced by the accumulation of deposits on pond bottoms [11,46].

Turbidity is attributed to suspended particulate matter, including but not limited to inorganic and organic compounds, clay, dissolved substances, and microorganisms. The elevated turbidity level (5.52 NTU) obtained in the present study is attributable to the presence of soil particles derived from runoff, dust, and mineral particulates resulting from mining activities. Additional contributing factors include geological characteristics of the river course and its sedimentary formations, the presence of cyanobacteria and other algae in the Săpânța River water, and the occurrence of in-bloom conditions [39,40]. Water turbidity is associated with a variety of factors, including clay, sludge, fine and inorganic organic matter, algae, soluble colored organic compounds and plankton, and their composition. Turbidity is also related to agricultural and irrigation activities, mainly to the intense use of chemical fertilizers and agricultural pesticides [11].

The water samples exhibited a characteristic low total hardness (H_t), with the groundwater registering 11.55 °g (classified within the hard water range) and the Săpânța River water (soft water) displaying 1.86–2.78 °g (soft water), indicative of low total alkalinity. The composition of the soil in the sampling locations exerted a significant influence on the overall hardness. The low level of total hardness, varying between 1.86 and 2.78 °g (soft water) in Săpânța River, indicated that the geology of the riverbed and human runoff are the primary sources of natural alkalinity with a low content of carbonate and bicarbonate [11–16]. Mg and Ca salts are correlated with the total, while the atmospheric pressure, temperature, and pH of water are responsible for the total hardness variation [39,47,48].

Anthropogenic pollution or interaction with rocks and soil can release Cl^- (1.01–5.65 mg/L) into water bodies. Water's salinity is correlated with the presence of Cl^-, which is salt [49]. The geological structure (soil and rock formations) of the aquifer contains minerals rich in Cl^- [48]. Elevated amounts of Cl^- can be introduced into groundwater by human activities including industrial discharge, the intense application of specific fertilizers, the use of road salt for deicing that includes chloride compounds, and water–rock interaction [49].

The NH_4^+ content is very low in all the studied points (0.045–0.066 mg/L), being well below the permissible limit of 0.5 mg/L. This is related to the runoff of organic matter from the soils around the Săpânța River and due to the decomposition of vegetation and phytoplankton fallen in the water [39–41,44]. Since NH_3 ions are linked to Mg and Fe and their oxidizing effect is reduced during reduction treatment, their presence may be caused by the level of pollution, contamination with excreta, and incomplete dissolution of nitrogen-based organic substates in the soil. They can be eliminated by a chlorination process. High temperatures and pH levels encourage the production of NH_4^+ in natural waterways [44].

The NO_3^- levels varied between 2.23 and 2.88 mg/L in the river Săpânța and 5.52 mg/L in the groundwater, being well below the MAC (50 mg/L), due to the oxidation process of NH_4^+ and organic substances with NO_3^- formation and to the reduced uptake of NO_3^- by aquatic plants [1,49]. The presence of NO_3^- in water can be attributed to soil composition, the mineralization of organic pollutants of a proteinaceous nature, and the application of nitrogen-containing fertilizers and pesticides (nitrogen-based fertilizers) in agricultural practices [1]. Low NO_3^- levels during warm months could be caused by aquatic plants and phytoplankton absorbing large amounts of nitrogen compounds [1,49].

The concentrations of SO_4^{2-} in the Săpânța River (3.84–4.89 mg/L) and a groundwater point (5.52 mg/L) are related to rocks with high concentrations of sulfur compounds, industrial waste, domestic wastewater discharges, polluted domestic and industrial waters, and water–soil interactions, as well as mineral leaching [49–51]. The primary sources of SO_4^{2-} include sulfide dissolution from vegetation and animal decomposition, pyrite, sedimentary rocks, organic matter, minerals in water, gypsum and lignite in soil, and the combustion of organic matter due to acid rain and volcanic activity [49–51].

The low PO_4^{3-} concentration (0.02–0.04 mg/L) is caused by inorganic fertilizer leakage from agricultural practices, which promotes algae growth and alters the water's color and flavor. Eutrophication decreases the DO content of water and poses a threat to the biodiversity of aquatic biota, and it is caused by PO_4^{3-} in water, which is derived from household products containing detergents based on P, industrial processes, and P-containing fertilizers intensively applied in agricultural practices [52].

The level of total dissolved solids (TDS), which ranges from 44.2–53.5 mg/L in the Săpânța River and 62.9 mg/L in the groundwaters, quantifies the total dissolved mineral level in water, and pH specifies the ability of water to react with alkalis or acids [51]. While high levels of TDS and alkalinity are related to strong industrial and agricultural activities, organic matter leaking into the water, and precipitation processes, the low values of both indicators in the water systems are associated with the catchment geology [53].

Temperature, dissolved salts, dissolved cations, and pH can all affect total alkalinity. Weak acids and their conjugate bases provide this essential buffering function, and high alkalinity safeguards the environment, especially against hazards like acid rain and sewage that can change pH levels. Low alkalinity, which is frequently brought on by human activity and geologic influences, indicates low levels of CO_3^{2-}, OH^-, and HCO_3^- [39,48].

By dissolving and infiltrating silica salts created by human activity, interchangeable cations among clay and water were found, with Na and K ranging between 1.22 and 3.82 mg/L and 0.76–2.63 mg/L (Table 2). Natural erosion of salt deposits, the dissolution of Na-containing rocks, irrigation and pluvial streams that interact with Na-rich soils, wastewater infiltration, and sewage system effluents are the primary sources of Na [54,55]. Human-caused activities, such as the use of animal manure as fertilizer and human waste, are the source of K, a basic element in aquatic vegetation and fish population nutrition that enters water through mineral dissolution. Because K is less soluble and more resistant to weathering and clay formation, its concentration in water is lower than that of Na [54,55].

Table 2. The annual median values of metals and As determined in water samples (SR1–SR3) and groundwater near Săpânța River (GW). The results are expressed as the mean ± standard deviation (n = 12 (monthly measurement in 2024)).

Metals	SI	SR1	SR2	SR3	GW	Mean	Median	Min	Max	SD	Errors	MAC *
Na	mg/L	1.22 ± 0.11	1.45 ± 0.14	1.56 ± 0.15	3.82 ± 0.35	2.01	1.51	1.22	3.82	1.21	0.17	200
K	mg/L	0.76 ± 0.05	1.63 ± 0.12	1.83 ± 0.16	2.63 ± 0.19	1.71	1.73	0.76	2.63	0.77	0.14	10
Ca	mg/L	5.91 ± 0.48	5.47 ± 0.51	5.33 ± 0.46	4.22 ± 0.39	5.23	5.40	4.22	5.91	0.72	0.50	100
Mg	mg/L	1.22 ± 0.10	0.94 ± 0.77	1.44 ± 0.12	3.89 ± 0.27	1.87	1.33	0.94	3.89	1.36	0.16	50
Fe	µg/L	5.52 ± 0.46	4.22 ± 0.38	5.15 ± 0.47	5.92 ± 0.52	5.20	5.34	4.22	5.92	0.73	0.52	20
Mn	µg/L	7.71 ± 0.68	7.16 ± 0.70	7.03 ± 0.66	6.88 ± 0.61	7.20	7.10	6.88	7.71	0.36	0.68	50
Al	µg/L	16.8 ± 0.12	17.3 ± 0.08	17.9 ± 0.12	18.5 ± 0.08	17.6	17.6	16.8	18.5	0.74	1.69	200
Sr	µg/L	19.5 ± 0.15	23.2 ± 0.20	27.5 ± 0.21	31.1 ± 0.26	25.3	25.4	19.5	31.1	5.05	2.49	7000
Zn	µg/L	11.1 ± 0.09	14.3 ± 0.08	13.5 ± 0.07	34.7 ± 0.15	18.4	13.9	11.1	34.7	10.9	1.80	5000
Cr	µg/L	0.88 ± 0.06	0.95 ± 0.07	0.99 ± 0.08	1.23 ± 0.09	1.01	0.97	0.88	1.23	0.15	0.09	50
Cu	µg/L	2.71 ± 0.12	3.41 ± 0.28	3.22 ± 0.25	5.22 ± 0.43	3.64	3.32	2.71	5.22	1.09	0.35	100
Ni	µg/L	0.88 ± 0.05	1.07 ± 0.07	1.25 ± 0.05	1.37 ± 0.02	1.14	1.16	0.88	1.37	0.21	0.10	20
Pb	µg/L	1.11 ± 0.02	1.37 ± 0.05	1.55 ± 0.09	1.84 ± 0.06	1.47	1.46	1.11	1.84	0.31	0.14	10
As	µg/L	0.39 ± 0.012	0.52 ± 0.02	0.49 ± 0.03	0.22 ± 0.01	0.41	0.44	0.22	0.52	0.14	0.04	10

Notes: * MAC based on international regulations, WHO guidelines [45], and national law [42] related to quality of water used as drinking water source.

Ca varies in the studied waters from 4.22 mg/L to 5.33 mg/L, while Mg varies from 0.94 mg/L to 3.89 mg/L (Table 2). Both originated from a geological origin, the same as K [1,49,52]. Since both Mg and Ca are necessary for human health, their presence in water is usually not a problem, depending on the anion that is present. Cation exchange and geogenic processes that indicate igneous rocks (amphibole, pyroxene, and feldspars) are the sources of Ca in aquifers. The presence of Mg in samples is related to natural processes including mineral alteration processes (olivine, micas, or amphiboles) [56,57].

Fe levels varied between 4.22 µg/L and 5.92 µg/L (Table 2), below the MAC of 200 µg/L, related to the geological conditions and potential contamination sources identified in the study area, namely the historical mining activities (Mn, Fe, Cu ore extraction) and industrial pollution and waste-related effluents from Maramures County [49,58]. Iron is naturally present in most aquatic vegetation, gravel, and soils (weathered formations causing a high organic matter content). It also takes part in photosynthesis as part of the reactive redox Fe–sulfur center, which converts N to NH_3^- [49,58–60].

The presence of elevated Mn levels (6.88–7.71 µg/L) is becoming increasingly prevalent, and the subsequent accumulation of this substance can exert deleterious effects on the central nervous system [58]. Potential sources are related to mining activities, intense agricultural activities, and industrial processes (cleaning products and alkaline battery production). Anthropogenic activities (such as wasteyard effluents and the use of fungicides) and geological abundance (such as igneous rock minerals, limestone, and metamorphic and dolomitic rocks) may be the source of Mn in water. Microbial activity and DO have an impact on the presence of Mn in various oxidation states in water [58,59].

In the present study, Al ranged between 16.8 and 18.5 μg/L, and it dissolves only in alkaline or acidic waters. Potential sources of Al are the use of $Al_2(SO_4)_3$ in the water treatment process and the use of aluminum sulfate-based agents in water treatment [59,60].

Geological variables, such as the composition of the subsurface rocks, are related to the Sr level in the studied samples, below the permissible limit of 200 μg/L, reaching 19.5–31.1 μg/L. To maintain proper management and supervision of water quality, this research emphasizes how crucial it is to take the Ca/Sr ratio into account when evaluating the overall impact of Sr in water sources used for consumption [55].

Zn concentrations measured in the samples range from 11.14 to 14.3 μg/L in Săpânța and 34.7 μg/L in groundwater, which is below the maximum acceptable concentration (MAC) of 5000 μg/L (Table 2). In the long run, Zn may rise to levels that are harmful to species living in the sediment, and clay content, phosphorus availability, redox conditions, and organic matter concentration all have an impact [61]. Zinc is a naturally occurring metal that slowly enriches groundwater through interactions with rocks. It is controlled by pH and inorganic carbon content, and it is known to be very mobile in water systems [61,62].

The very toxic heavy metal Cr (0.88–1.23 μg/L) is associated with oxidative stress, cancer, and DNA damage. Anthropogenic sources involve the use of Cr in chemical industries, refractory, metallurgy, leather tanning, and refining, and Cr is a naturally occurring element that usually occurs in volcanic emissions, soil, and erosion [63,64].

The relatively elevated Cu content (2.71–3.22 μg/L in Săpânța and 5.22 μg/L in the groundwater) may be attributed to anthropogenic activities applied in the study area (e.g., agriculture, mining, industry) or natural processes (e.g., breakdown of rocks and rock degradation) [49,50,65]. The geological characteristics of aquifer bedrock and copper–sulfide ores, such as bornite and chalcocite, are indicative of chemical reactions that result in the liberation of copper into aquifers [65]. Cu–sulphide ores and aquifer bedrock's geological characteristics imply that chemical reactions could take place, releasing Cu into water systems [49,50].

Ni concentrations in samples have been found to reach 0.88–1.37 μg/L (Table 2) at pH values below 7.38–7.75. Soil, depth, and pH can all affect the Ni content. When utilized for irrigation, soil contamination with Ni occurs, affecting plant health and altering the well-functioning of the environment. Rock formations based on Ni and its anthropogenic sources (tannery effluent, mafic and ultramafic rocks) may leak into the groundwater because of interactions with the aquifer and underlying geology [50].

According to Regulation no. 7, 2023 [42], the MAC is 10 μg/L, although the measured Pb concentration values ranged from 1.11 to 1.84 μg/L. Pb, and its compounds may enter groundwater as a result of mining activities. Both natural processes (carbonate rock mineralization, weathering processes) and those generated by humans (oil as fuel, metal industry, burning coal, the application of pesticides and fertilizers in agricultural activities, electronic waste, and mining exploitation) are the primary sources of Pb [9,50].

As, a non-volatile, odorless, and tasteless substance, is a potentially toxic pollutant that often occurs in elevated concentrations in groundwater. Groundwater containing As has been observed to be susceptible to abrupt alterations [50,65]. The measured levels in the Săpânța River for the present study range from 0.39 to 0.52 μg/L, whereas the groundwater sources have values of 0.22 μg/L. Geological factors, prevalent in rural areas, can be a contributing factor to the presence of naturally occurring arsenic in aquifers. The issue is exacerbated by agricultural practices, a lack of awareness, and the challenges associated with accessing clean water alternatives [50,65].

It was observed that the electrical conductivity, total hardness (Săpânța River falls into the soft water category), Cl^-, NH_4^+, NO_3^-, SO_4^{2-}, TDS, total alkalinity, and metal content (Na, K, Mg, Fe, Al, Sr, Zn, Cr, Cu, Ni, Pb) were lower in the river samples compared to the groundwater samples. On the other hand, pH, dissolved oxygen, turbidity, and some metals (Ca, Mn, and As) were higher in the river samples compared to the groundwater samples.

Based on the chemical and physico-chemical results, the acidity, metal load, and Total Ionic Salinity were determined by applying a Ficklin–Caboi plot (Figure 2a) and a TIS chart (Figure 2b).

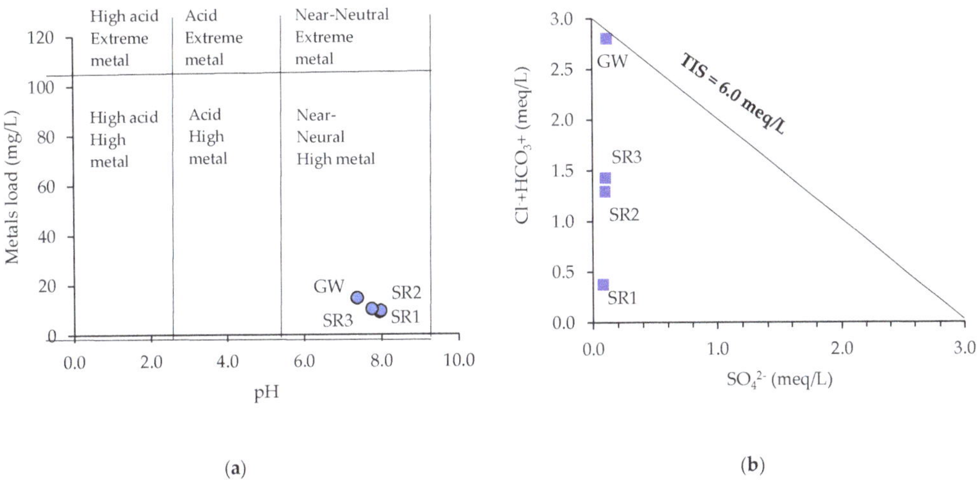

Figure 2. The metal load, acidity, and salinity of waters represented by the (**a**) Ficklin–Caboi chart, and (**b**) TIS diagram, both applied for all water samples (SR1–SR3, GW).

The metal load chart or the Ficklin–Caboi diagram indicates the total metal load content (Ca, Na, Mg, K, Fe, Al, As, Cr, Cu, Mn, Ni, Pb, Zn, and Sr) related to the pH determined in the water samples. The metal load, expressed in mg/L, was calculated as the sum of all studied metals. According to the Ficklin–Caboi diagram, most waters were characterized by near-neutral pH and a high metal load (Figure 2a). The acidity of waters can be related to fertilizer run-off and the direct discharge of wastewater.

All samples were characterized by a salinity lower than 6.0 meq/L, according to the TIS diagram (Figure 2b). It was noticed that the groundwater sampling point (GW) has the highest position in the graphical representation concerning the Cl^- + HCO_3^- content, while the river sampling points (SR1–SR3) were visibly lower. The difference is correlated with the water origin and specific chemistry, as well as natural processes.

The chemical equilibrium between the anionic (HCO_3^-, CO_3^{2-}, Cl^-, and SO_4^{2-}) and cationic (K, Mg, Ca, and Na) content was represented by a Piper plot (Figure 3). Based on the trilinear Piper chart, waters were characterized and classified into the $CaMgHCO_3^-$ dominant category of water type. As it can be observed, many samples are mainly influenced by CO_3^{2-}, HCO_3^-, and Ca ions. According to the cation trilinear diagram represented in Figure 3, the groundwater (GW) sampling point is observed to be slightly apart from the river samples, which cluster together. This fact is observed specifically in the cationic trilinear chart, concerning the Mg, Na, and K concentrations. Generally, groundwaters are rich in Mg, compared to rivers, due to long-term rock–water interaction, the geological composition of aquifers, and longer residence time. The dissolution of

Mg-bearing minerals comes from magnetite, dolomite, and ultramafic rocks, as well as olivine, micas, or amphiboles [57]. On the other hand, river water tends to be lower in Mg due to short residence time and dilution by rainfall. Generally, surface processes like runoff and atmospheric deposition have an impact on river water and their chemical composition. Ca is higher in rivers compared to groundwater due to the direct contact of agricultural runoff and urban waste and their frequent interaction with carbonate rocks [1,54].

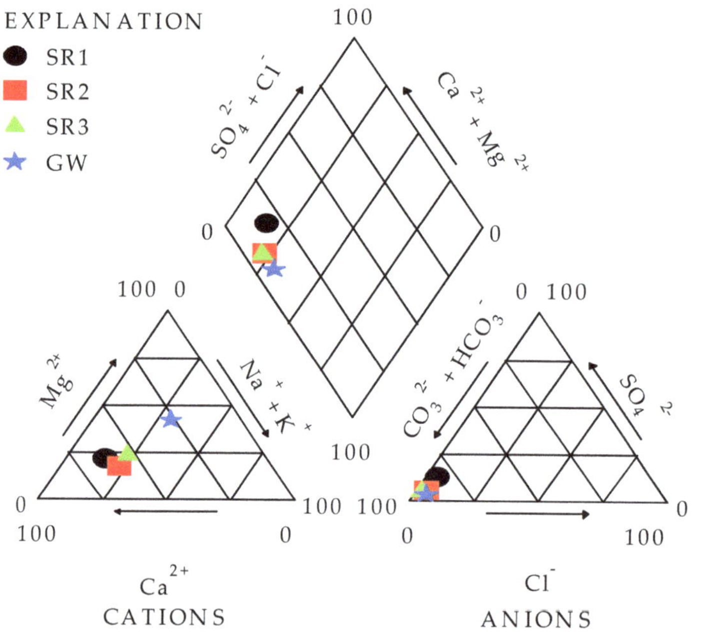

Figure 3. Piper plot applied of the water samples (SR1-SR3, GW).

The development of surface water chemistry is summed up in the original Gibbs [66] article. The primary controlling mechanisms indicated int the chart are water–rock interaction, precipitation, and evaporation. The scatter plot, which plots sodium/(Ca + Na) ratios (x-axis) against TDS on the y-axis, clearly illustrates the impact of these processes. Because they amalgamate waters with different transit periods, rivers are usually in a condition of dynamic disequilibrium with basin sediments and bedrock geochemistry. They also frequently have higher concentrations of Ca and HCO_3^- than Na and Cl, respectively. Based on these data, and with the help of Gibbs plots, represented by Figure 4, the potential sources of waters are visually shown. It was observed that, according to the anionic ratio and the cationic ratio plots, generally, samples were characterized by precipitation dominance and not water–rock interaction dominance. This characterization is associated with the low TDS values, as well as the low content of cation and anions, which is related to a geogenic origin, associated with water–rock interaction. The precipitation dominance is reflected by the chemical composition, which is mainly influenced by environmental parameters or processes, specifically atmospheric deposition and rainfall (intensity and amount), with minimal influence from rock interaction or evaporation. Precipitation dominance is maintained when high rainfall dilutes solute concentrations and contributes to low TDS, as well as when a consistent freshwater input exists. Anthropogenic sources, specifically urbanization, mining activities, industrial emissions, and agricultural practices, which occur in the present study, influence the precipitation dominance characterizing the studied water samples. Although impervious surfaces may decrease water–rock interaction by boosting runoff and decreasing infiltration, they can increase the inflow of pollutants via surface contaminants and atmospheric deposition. Even when precipitation predominates, an increase in acid rain alters the ion composition of rainwater, changing the water's chemical signature.

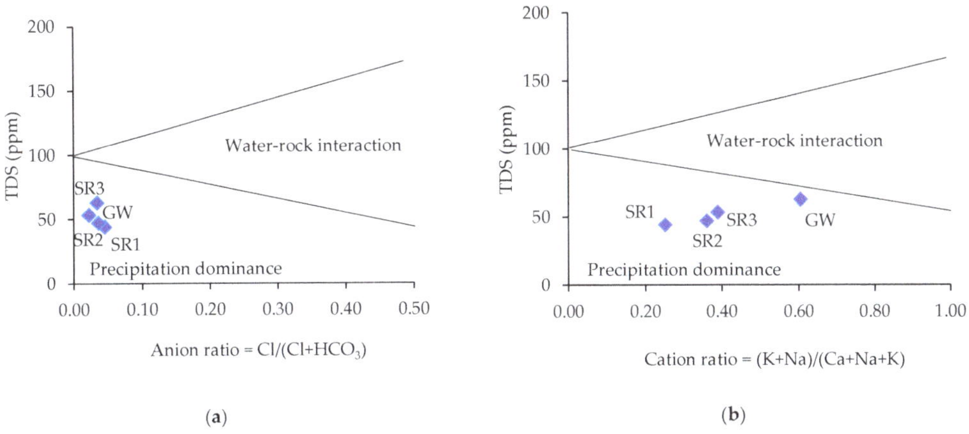

Figure 4. Gibbs chart applied to water samples: (**a**) anion ratio ($Cl^-/(Cl^- + HCO_3^-)$; (**b**) cation ratio ((K + Na)/(Ca + Na + K)).

Numerous factors, such as the geochemical environment (EC and pH), the exchange processes—which may differ from those significant for surface environments—and the solubility and availability of minerals, all affect the concentration of chemical components in natural groundwater. Besides natural processes and phenomena, anthropogenic activities, implying the intense use of chemical substances, use of natural resources, discharge of wastewater, and emitted gases, change and alter all environmental factors and processes. It is imperative to determine the potential contamination sources and their existence and monitorization near the study area to prevent and mitigate the potential impact on the water system in this case. Furthermore, Figure 5 indicates the potential agricultural and domestic inputs, related to the Cl^- and NO_3^- content. Correlated with the Ficklin–Caboi plot, Figure 5 showed the run-offs associated with the agricultural inputs. This chart focuses mainly on both sources, due to the frequency of their practice in the study area, and in other studies implying water used with the drinking purpose. NO_3^- and Cl^- are the main and commonly contaminants occurred in high amounts after agricultural and nearby domestic households.

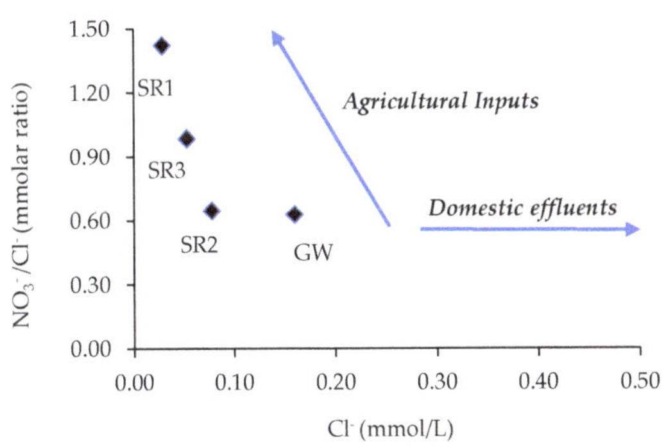

Figure 5. Potential sources of chemicals, based on the Cl^- and NO_3^- content.

3.2. *Analysis of Water Quality and Contamination Level*

The overall quality of studied waters was determined and analyzed with the help of the OWQI. In the present study, the OWQI was calculated based on two (2) physico-chemical indicators (pH, and electrical conductivity), and twelve (12) chemical indicators (NH_4^+, NO_3^-, Cl^-, Fe, Mn, Al, Cu, Zn, Pb, Cr, Ni, and As), and the standard values considered to be the maximum allowable concentrations (MAC) established by the international Regulation, WHO Guideline [42], and national Law [43] related to the quality of water used as drinking water source. The unit weight values (W_i) was the ration between one (1.0) and the MAC, and the ideal value (ID) was considered zero (ID = 0) for every chemical indicator, except for pH (ID_{pH} = 7.0) [34]. The obtained results were presented in Table 3, indicating that, generally, the studied water samples were characterized by excellent quality (OWQI scores < 7.00).

Table 3. Scores related to the overall water quality, and contamination level related to the heavy metals, and As content (indices calculated based on the mean values).

Sampling Points	Overall Water Quality Assessment			Contamination Level Assessment		
	$\Sigma[q \times W_i]$	OWQI	Water Quality	CI	HMEI	Contamination Level
SR1	446	1.11	Excellent	6.90	0.43	Low
SR2	280	0.70	Excellent	8.30	0.47	Low
SR3	467	1.16	Excellent	8.90	0.49	Low
GW	2643	6.57	Excellent	9.20	0.53	Low

The obtained scores indicated excellent quality. In the protected area, an excellent water quality, given by the OWQI scores aligns with the biodiversity conservation goals from at least three points of view, specifically: 1. Enhanced habitat quality, 2. Support for aquatic ecosystems, and 3. Indicator of effective management. Fish, amphibians, and invertebrates depend on clean water to survive and reproduce. Protected regions characterized by high OWQI values frequently have diverse aquatic communities, which enhances overall biodiversity. High-quality water is essential for maintaining the integrity of environments such as freshwater lakes, riparian zones, and wetlands. These ecosystems serve as protective barriers against environmental stressors and are essential habitats for a variety of species. Protected areas with successful conservation and management strategies often have excellent OWQI scores, showing that they are effectively reducing pollution and maintaining ecological balance [67].

The mean values of obtained OWQI scores ranged between 0.70 and 6.57. River sampling points SR1 and SR3 present a similar trend, while river sampling point SR2 was characterized by the lowest score, and the groundwater sampling point (GW) was characterized by the highest score, correlated with the highest content of Cl^-, NH_4^+, NO_3^-, SO_4^{2-}, Fe, Na, Pb, and Zn, which are correlated with the different geological origins, dissolution processes, and water source types.

Prior studies on alluvial aquifer samples also situated in the northern part of the country indicated the classification of studied waters into three quality levels, based on the WQI scores [3]. The index was based on the pH, NO_2^-, NO_3^-, NH_4^+, EC, Cl^-, total harness, turbidity, and major cations K, Na, Mg, and Ca. The obtained scores varied between 8.6 and 54, having a mean of 22. Based on these results, waters were classified into the good quality category or the poor and excellent quality categories [3]. These results are higher than the results obtained in the current study, up to eight times, mainly attributed to the higher NH_4^+ amounts characterizing the studied samples, based on Dippong et al. [3]. A different recent study implemented by Dippong et al. [9] implied the study of surface waters, from the northern part of Romania, by applying the WQI. The index was based on the pH, TDS, EC, total hardness, K, Ca, Mg, Na, Cl^-, NO_3^-, SO_4^{2-}, and NH_4^+. The obtained results exceeded the results obtained in the present study, by up to four times. The river samples were classified into two quality categories, good and excellent, with scores ranging between 16.0 and 25.2 [3].

Another different study, at the international level, in Bangladesh (Jamalpur Sadar area), studied surface water and groundwater that were characterized by different quality status. The scores obtained in the case of surface waters ranged between 9.90 and 327, indicating that 60% of samples were unsuitable for drinking and irrigation purposes, 19% had poor quality, 10% had very poor and good quality, and some presented excellent quality status. The high scores were correlated with high electrical conductivity (EC >2000 μS/cm), total dissolved solids (TDS >1300 mg/L), Cl^- (Cl^- >200 mg/L), and K (K >680 mg/L), as well as low pH (pH <5.0). The groundwater samples had lower quality compared to surface water, due to the higher standard values used for calculating the index [68]. Abdo et al. [69] studied surface water quality in Egypt (Kafr El Zayat and El-Kanater El-Khayria area) by applying this methodology. The obtained scores were below 125, indicating good and poor quality for the studied rivers, mainly due to the high amount of NH_4^+. The waters were rich in nutrients due to agricultural activities' runoffs, particularly those that are intensively cultivated and utilize substantial amounts of synthetic fertilizers and urban effluent [69].

The contamination degree with the studied heavy metals was determined and analyzed. In this framework, the CI and HMEI methods were applied. CI and HMEI were calculated for eight heavy metals (Mn, Ni, Zn, Pb, Cu, Cr, Fe, and Sr) and one metalloid (As). Standard values used in the methodology are represented by the WHO Guideline [42] and Romanian Regulation [43], both concerning the quality of water used as a drinking source. Tables S1 and S2 include the water quality classifications and contamination degrees based on the index scores, and the results concerning the contamination with heavy metals and As level characterizing the studied waters are indicated in Table 3. The obtained results indicated low contamination levels for heavy metals and As, their scores being lower than the critical limit of 100. It was observed that the river sampling point SR1 has the lower score, and the highest value was obtained in the case of the groundwater sampling point GW, correlated with the highest concentrations of Zn, Pb, Cu, and Cr determined in GW.

3.3. *Water Quality Based on Vegetation Cover near Săpânța River*

Water is one of the indispensable elements related to life, plants, and vegetation in particular. Water quality influences and selects plant species, and at the same time, plants, through root absorption processes, absorb water with nutrients, including pollutants, acting as living and active filters contributing to improving the quality of environmental factors, including water. The study of vegetation can contribute to the assessment of water quality, both through the composition and structure of the vegetation and through chemical analyses of the water. Wetlands with over 30% of the surface area covered by tree canopy are defined as forested wetlands (also known as "tree swamps") [47,56].

Forested wetlands provide critical habitats for a variety of plants and animals [47,56,70], support unique plant communities, and account for a large proportion of the plant diversity in forested landscapes [70–72]. They accumulate large amounts of plant biomass and can therefore be significant carbon stores. They also play an important role in regulating water flow, improving water quality, and recharging groundwater.

Forested wetlands are supplied with water in varying proportions from precipitation, groundwater, and surface water runoff, and their soils remain saturated for longer or shorter periods of time [56,73–75]. In seasonally saturated forested wetlands surrounded by upland areas, the amount and timing of saturation are largely controlled by precipitation and upland runoff [74]. A thorough understanding of vegetation–hydrology associations is essential for assessing, managing, and restoring the ecological processes that enhance the stability of wetland ecosystems, especially in disturbed or otherwise managed landscapes [75]. It also supports the sustainable management of forest and water resources and promotes the conservation of biodiversity in forested wetlands.

Studies mention the correlation between hydrology, other environmental factors, and plant diversity depending on the wetlands or vegetation types. It is important to gain a comprehensive understanding of the hydrological conditions that contribute to and promote biodiversity in wetland ecosystems. This understanding is crucial for the conservation of forested wetlands, but also for the assessment and maintenance of the quality of abiotic factors. In order to capture the correlation between the water factor and the forested wetlands on the banks of the Săpânța River, each water sampling point was studied through the five following phytocenological surveys: the type of vegetation (characteristic or not for the reference area), the anthropization degree of the vegetation cover expressed by the number of invasive species, the presence and abundance of nitrophilous species, and the presence/abundance of species indicative of chemical factors restrictive to vegetation.

In the four water sampling points, the vegetation is woody, characterized by fragments of warbler forests, which total 95 species of vascular plants. In Table 4, the first positions are occupied by the species characteristic of the plant associations Salicetum albae-fragilis, which achieve high dominance abundance indices that define the fen features. Among the 16 species of woody plants, trees, and shrubs, 11 species are typical fen species, totaling 60–90% of the total plant cover. The floristic composition and structure of the marginal stands of the Săpânța River, denote a favorable state of conservation, generally indicating an abiotic substrate with favorable parameters.

Table 4. Floristic composition and structure of the vegetation cover at the water sampling points.

	GW	**SR3**	**SR2**	**SR1**
Number of sample areas	5	5	5	5
Slope inclination (%)	20	5	5	5
Exposition	N-E	N	N	N
Tree height (m)	5–20	5–20	5–20	5–30
Tree diameter (cm)	10–20	10–30	10–30	10–40
Stratification	4	4–5	4–5	5–6
Grassy carpet covering (%)	70–80	80	75	80–100
Sample surface (mp)	1000	1000	1000	1000
Plant species	Ad m (dominance abundance index according to the Braun-Blanquet scale			
Salix alba	2	2	2	2
Salix purpurea	–	1	1	2
Salix fragilis	–	–	+	+
Alnus glutinosa	+	2	2	–
Populus alba	+	+	1	1
Populus nigra	–	+	–	1
Aegopodium podagraria	+	+	+	+
Humulus lupulus	–	–	+	+
Fraxinus excelsior	–	+	+	+
Matteucia struthiopteris	–	+	+	–
Rubus caesius	–	+	+	+
Sambucus nigra	+	–	+	+
Clematis vitalba	+	+	+	+
Portulaca oleracea	+	–	–	–

Table 4. *Cont.*

	GW	SR3	SR2	SR1
Leucojum vernum	+	–	–	+
Galium odoratum	+	+	+	+
Scilla bifolia	+	+	+	+
Viola reichenbachiana	+	+	+	+
Alliaria petiolata	+	+	–	+
Stellaria aquatica	+	+	–	–
Stellaria nemorum	+	–	–	+
Isopyrum thalictroides	+	+	+	+
Anemone ranunculoides	+	–	–	–
Eupatorium cannabinum	+	–	+	+
Quercus robur	1	–	–	–
Crataegus monogyna	1	–	–	–
Geranium robertianum	+	–	–	+
Ranunculus ficaria	1	1	1	1
Potentilla reptans	–	+	+	+
Veronica anagallis-aquatica	1	–	–	–
Ranunculus cassubicus	+	–	+	+
Ranunculus sardous	+	–	–	+
Chelidonium majus	+	–	–	+
Festuca pratensis	2	+	+	+
Erigeron annuus	–	–	–	±1
Geum urbanu	+	+	+	+
Fragaria vesca	+	+	+	+
Lolium perennum	1	–	–	–
Caltha palustris	1	–	–	–
Cruciata levipes	+	+	+	+
Telekia speciosa	+	+	+	+
Corylus avellana	+	–	+	–
Carpinus betulus	+	–	–	–
Hepatica nobilis	+	–	–	–
Glechoma hederacea	+	+	+	+
Galium molugo	+	+	+	+
Carex pillosa	1	–	–	+
Alisma plantago-aquatica	+	–	–	–
Phragmites australis	3	–	–	–
Cirsium vulgare	+	–	–	–
Rumex aquaticus	+	+	–	–
Urtica dioica	2	–	–	1
Corydalis cava	+	+	+	+
Mercurialis perennis	+	+	+	–
Trifolium repens	+	–	–	+
Rorippa austriaca	+	+	+	+
Crocus heuffelianus	+	+	+	+
Cardamine glanduligera	+	+	–	–
Anemone nemorosa	+	+	+	–
Juncus conglomeratus	+	–	–	+
Scrophularia nodosa	+	–	–	+
Scirpus sylvaticus	1	–	–	–
Chrysosplenium alternifolium	+	+	+	+
Asarum europaeum	+	+	+	–
Primula vulgaris	+	+	+	–
Lycopus europaeus	+	+	+	+
Bidens tripartitus	+	+	+	+

Table 4. *Cont.*

	GW	SR3	SR2	SR1
Lythrum salicaria	−	−	−	+
Artemisia vulgarsi	−	−	−	+
Polygonum hydropiper	+	+	+	+
Arctium lappa	−	−	−	+
Equisetum fluviatile	+	+	+	+
Solanum dulcamara	+	−	−	+
Myosotis nemorosa	+	+	+	+
Scutellaria hastifolia	−	−	−	+
Gratiola officinalis	−	+	+	+
Mentha aquatica	1	+	+	−
Stellaria media	+	−	−	+
Cucubalus baccifer	+	+	+	+
Agrostis stolonifera	+	+	+	+
Glyceria fluitans	+	+	−	+
Trifolium pratense	+	−	−	+
Silene alba	−	−	+	+
Chenopodium hybridum	+	−	−	+
Amaranthus retroflexus	−	−	+	+
Chenopodium murale	+	−	−	+
Sambucus ebulus	+	+	−	−
Calystegia sepium	+	−	−	+
Stachys palustris	−	−	+	+
Echinocystis lobata	+	−	−	+
Acer negundo	−	−	+	+
Reynoutria japonica	−	−	−	2
Solidago canadensis	−	−	+	1
Robinia pseudoacacia	+	−	−	1
Erigeron canadensis	−	−	−	+

The four water sampling points are distinguished by relatively homogeneous plant cover, with many common features. Since this study is focused on the analysis of water quality in the Săpânța River and groundwater, we use elements of the vegetation cover to confirm (or not) the physico-chemical analyses of the mentioned waters. Plants often tolerate increased concentrations of chemical elements depending on numerous factors: phenology, robustness, growth stage, chemical elements, and water chemistry [76].

Although there are no studies for precise quantifications, the absorption of chemical elements from water, with respect to sediments in plant roots depends on a series of factors such as pH, reducing potential, temperature, salinity, organic matter content, and element levels [77,78]. In addition to these factors, others such as seasonal physiological variations, accumulation, and compartmentalization capacity can influence the level of accumulation in plants [79,80]. Certainly, roots are the main route of uptake of chemical elements from the substrate; as a result, they tend to reflect the level of chemical elements in sediments, soil, or water [81].

According to the indicator species theory, there are species with preferences for precise intensities of abiotic factors. As a result, the presence of these species known as indicators can replace or converge with habitat analyses. Floristic changes may be identified through changes in species composition over time and the invasion of fast-growing nitrophytes (N-demanding species) [82,83].

One such category is that of nitrophilic species, faithful to defined concentrations of N salts in the substrate. The analysis of the preferences towards N of the species identified at the sampling points revealed the following situation presented in Table 5.

Table 5. Distribution of species in the sample plots according to their preferences for N.

Number of Species	N × 3	N4	N5	N6	N7	N8	N9	Nx *	Nec. *
95	3	3	12	8	27	19	9	10	2
GW	3	3	8	7	20	16	5	11	1
SR3	0	2	6	7	13	6	3	6	0
SR2	0	3	5	8	15	6	4	7	0
SR1	0	3	7	6	20	14	6	8	2

Notes: * Categories of indicator species for N; Nx—amphitolerant species; Nec.—species for which preferences towards N are not known.

A proportion of 55 species out of the total of 95 are indicator plants for N-rich substrates. The four analyzed points are grouped differently. Points SR1 and SR2 have a lower proportion of nitrophilic plants (Figure 6 and Table 6). The cluster analysis highlights a distance index of about 12% between the two groups of samples.

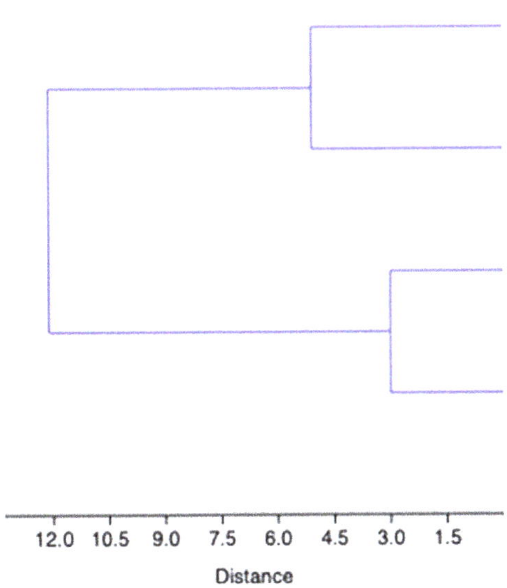

Figure 6. Cluster analysis based on Euclidean calculus.

Table 6. Similarity and distance indices between the four sampling points.

	SR1	SR2	SR3	GW
SR1	0	10.296	11.533	5.099
SR2	10.296	0	3.000	12.728
SR3	11.533	3.000	0	13.892
GW	5.099	12.727	13.892	0

According to this distribution, two of the points include several nitrophilic species: GW and SR3. The presence of these plants is due to an excess of N salts in the superficial horizon of the soils, correlated with the movement of domestic animals and the dumping of garbage, but also the fertilizers used on the agricultural lands in the immediate vicinity. Because N salts have a fertilizing effect on plant species, they absorb nitrates and ammonium salts at the root level, producing natural filtration and preventing the concentration of these salts in river waters, as well as in the groundwater.

In general, the banks of the Săpânța River are accompanied by plant associations characteristic of water edges, with robust phytocenoses, which, through their floristic composition and structure, denote good water quality, both in the river and in the groundwater. However, at the water sampling points, direct observations, respective to qualitative analysis of the component species from the perspective of the indicator species theory highlight subtle changes that occur progressively, especially due to anthropization. As in similar cases mentioned by the bibliographic resources [83], the presence of nitrophilic species in the grassy carpet, even if they have low values of the AD index (abundance dominance according to the Braun–Blanquet scale), indicates an input of N salts. The geological substrate and the soils, respectively, exclude the natural source of nitrates, the input being of anthropic nature, because all the water sampling points are in anthropized areas by the presence of roads traveled by domestic animals and people, by vehicles, but especially by agricultural lands cultivated with annual plants, which require significant quantities of fertilizers. Figure 7 indicates a greater number of nitrophilic species in the vegetation cover at points GW and SR1, the first being located in a marshy area with shallow surface water stagnation, with marshy micro-depression land and in the immediate vicinity of agricultural lands, and the second near the confluence of the Săpânța River into the Tisa, an intensely trafficked area, where waste is dumped and where invasive plant species, also having nitrophilic preferences, have free rein for expansion and are found in high proportions (Table 4, last six positions). Even if the number of nitrophilic plant species is increasing, for now they do not achieve high values of the dominance abundance indices, which signifies slow and transient increases in N salts, as well as a reversibility of the process, to the extent that the use of fertilizers and anthropization will be maintained at controlled levels.

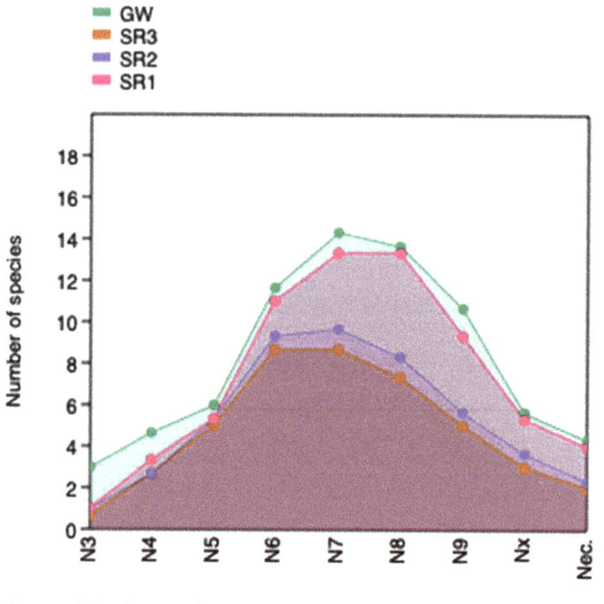

Figure 7. Distribution of species with different preferences for N in the water sample points.

Another aspect of the connection between water quality, environmental factors, and vegetation quality is the presence of indicator species for different heavy metals. Water and soil contamination is increasing in most ecosystems around the world [84–86]. Chemical elements can have both natural origin in geochemical processes and human activities (mining, metal processing, fossil fuel burning, use of fertilizers and pesticides) [87–89]. Studies indicate that several plant species, such as Trifolium pratense, Amaranthus retroflexus, and

Plantago lanceolata, can accumulate quantities of the heavy metals Zn, Ni, Cd, and Pb, respectively. Based on physiological characteristics and bioconcentration values, Trifolium pratense and Plantago lanceolata highlight qualities as bioindicators for heavy metals for numerous contaminants. Trifolium pratense demonstrates adaptability and absorption of Zn, Ni, and Cd, while Plantago lanceolata excels in the accumulation of Zn and Cd. In the vegetal carpet of the stands on the banks of the Săpânța River, such plants appear sporadically, meaning that the incidence of heavy metals is minimal. These results converge with the water analyses.

3.4. Health Risk Assessment for Toxins

All negative effects posed by ingesting toxins (heavy metals and As in the present study case) were considered through health risk assessment from the non-carcinogenic risk perspective. The evaluation was implemented with the help of three methods: the chronic daily intake (CDI), the hazard quotient (HQ), and the hazard index (HI). These methods were applied in two different study cases: exposure of children and exposure of adults to toxin ingestion through water consumption. Non-carcinogenic risk assessment was analyzed after a period of 6 years (children) and 70 years (adults) of water ingestion, with the water containing heavy metals and As. The obtained results are shown in Table 7.

Table 7. Non-carcinogenic risk for children and adults at each toxin analyzed in river (R1–R3) and groundwater (GW) sampling points.

HQ_{score} \ Sample Point		SR1	SR2	SR3	GW
As	children	0.780000	**1.04000** *	0.980000	0.440000
	adults	0.185714	0.247619	0.233333	0.104762
Cu	children	0.008130	0.010230	0.009660	0.015660
	adults	0.001936	0.002436	0.002300	0.003729
Fe	children	0.016560	0.012660	0.015450	0.017760
	adults	0.003943	0.003014	0.003679	0.004229
Ni	children	0.005280	0.006420	0.007500	0.008220
	adults	0.001257	0.001529	0.001786	0.001957
Pb	children	0.011100	0.013700	0.015500	0.018400
	adults	0.002643	0.003262	0.003690	0.004381
Sr	children	0.003900	0.004640	0.005500	0.006220
	adults	0.000929	0.001105	0.001310	0.001481
Zn	children	0.004456	0.005720	0.005400	0.013880
	adults	0.001061	0.001362	0.001286	0.003305
Mn	children	0.006609	0.006137	0.006026	0.005897
	adults	0.001573	0.001461	0.001435	0.001404
HI_{score}	children	0.83603	**1.09951**	**1.04504**	0.52604
	adults	0.19906	0.26179	0.24882	0.12525

Notes: * scores higher than the maximum admissible limit of 1.00 → presence of non-carcinogenic risk at studied toxins.

The increasing trend of ingested toxins identified through the obtained daily intake as average CDI scores is As < Ni < Pb < Cu < Fe < Mn < Zn < Sr.

Generally, children are more vulnerable than adults; the daily intake is higher in the case of children than in adults, up to four and five times. Children present an average intake (CDI) of As ranging from 2.64×10^{-5} to 6.24×10^{-5} mg/kg/day, followed by CDI_{Ni} varying between 1.06×10^{-4} and 1.64×10^{-4} mg/kg/day; CDI_{Pb}, 1.33×10^{-4}

and 2.21 × 10^{-4} mg/kg/day; CDI_{Cu}, 3.25 × 10^{-4} and 6.26 × 10^{-4} mg/kg/day; CDI_{Fe}, 5.06 × 10^{-4} and 7.10 × 10^{-4} mg/kg/day; CDI_{Mn}, 8.26 × 10^{-4} and 9.25 × 10^{-4} mg/kg/day; CDI_{Zn}, 1.34 × 10^{-3} and 4.16 × 10^{-3} mg/kg/day; and CDI_{Sr}, 2.34 × 10^{-3} and 3.73 × 10^{-3} mg/kg/day.

Adults present a lower daily intake of toxins through water ingestion than children, but with a similar trend: CDI_{As} ranges from 6.29 × 10^{-6} to 1.49 × 10^{-5} mg/kg/day; CDI_{Ni}, 2.51 × 10^{-5} to 3.91 × 10^{-5} mg/kg/day; CDI_{Pb}, 3.17 × 10^{-5} to 5.26 × 10^{-5} mg/kg/day; CDI_{Cu}, 7.74 × 10^{-5} to 1.49 × 10^{-4} mg/kg/day; CDI_{Fe}, 1.21 × 10^{-4} to 1.69 × 10^{-4} mg/kg/day; CDI_{Mn}, 1.96 × 10^{-4} to 2.20 × 10^{-4} mg/kg/day; CDI_{Zn}, 3.18 × 10^{-4} to 9.91 × 10^{-4} mg/kg/day; and CDI_{Sr}, 5.57 × 10^{-4} and 8.86 × 10^{-4} mg/kg/day.

Based on the CDI results, the hazard quotient (HQ) was calculated for each studied heavy metal and As. The obtained scores indicated that if consumed, the studied waters can pose potential negative effects on humans, especially on children. HQ scores were significantly high in the case of As, exceeding the critical unity in river sampling point SR2 and up to score HQ = 1.00 in the case of river sampling point SR3, followed by SR1 (Table 7). As is an element that generally occurs naturally in the Earth's crust. It can be introduced into surface and groundwater by natural processes and anthropogenic activities. The natural sources of As are represented by different processes, such as geological weathering, the presence of acid sulphate soils, or atmospheric deposition. Different minerals, specifically arsenolite, arsenopyrite, orpiment, or realgar contain As. Through the weathering of these minerals, As is mobilized, reaching and finally enriching the chemical composition of the water systems. In soils rich in FeS minerals, after draining or air exposure, oxidation occurs, favoring the mobilization of As to the surface water and groundwater. On the other hand, As has been observed to enter soil and implicitly water systems as a result of the intense use of herbicides, pesticides, and livestock feed additives containing As. Additionally, As enters into groundwater as a result of irrigation with water that is rich in As. Emissions of As generated by fossil fuel combustion and industrial activities, specifically steel and glass production, contribute to the contamination of air and implicitly water (after atmospheric deposition). Nevertheless, the extraction of different metals releases As as a by-product, which enters groundwater and surface water. A comprehensive understanding of these sources is imperative for evaluating the potential dangers to the environment and human health [69].

The average hazard scores ranged between 5.07 × 10^{-3} for Sr exposure and 8.10 × 10^{-1} for As exposure through ingestion in the case of children and 1.21 × 10^{-3} for Sr and 1.93 × 10^{-1} for As in the case of adults. Exposure to As through water ingestion, in the case of children, can pose an impact on health, according to the high HQ scores. The low reference dose and the relatively high amounts of As increased the scores. The average HQ scores obtained for children ranged between 0.44 and 1.04 for As exposure, 8.13 × 10^{-3} and 1.57 × 10^{-2} for Cu, 1.27 × 10^{-2} and 1.78 × 10^{-2} for Fe, 5.28 × 10^{-3} and 8.22 × 10^{-3} for Ni ingestion, 1.11 × 10^{-2} and 1.84 × 10^{-2} for Pb ingestion. They were between 3.90 × 10^{-3} and 6.22 × 10^{-3} and between 4.46 × 10^{-3} and 1.39 × 10^{-2} in the cases of Sr and Zn ingestion, while the hazard calculated for Mn exposure ranged between 5.89 × 10^{-3} and 6.61 × 10^{-3}.

The overall non-carcinogenic risk for all studied toxins was determined by applying the hazard index, HI. Results concerning HI for adults ranged between 0.125 and 0.2618, indicating that exposure at all studied heavy metals and As poses no negative effects if consumed through water ingestion. On the other hand, children are more vulnerable and can present non-carcinogenic risk if they consume the studied water sources. HI calculated for children indicated values exceeding the critical value in 50% of samples. HI is high due to the high HQ related to exposure to As.

In order to monitor, determine, analyze, and prevent contamination concerning waters, it is mandatory to follow several steps. A potential assessment model was proposed in this framework, presented with the help of Figure 8.

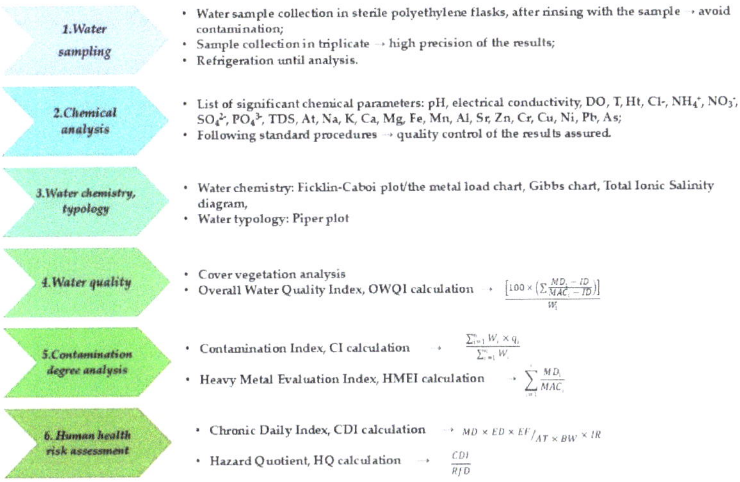

Figure 8. Research scheme aiming for determining the quality of water samples.

Being a natural protected area for the conservation of vegetation and flora and fauna species, anthropogenic activities have been limited. Agricultural land use has also been limited. However, during the surveys of the vegetation cover, it was repeatedly observed that local people repeatedly deviate from the rules of not using fertilizers within the perimeter of the protected natural area. Both natural fertilizers, animal manure, and synthetic fertilizers were found, as well as packaging and temporary storage on the edges of fields. At present, water quality analyses, both in the river and in the groundwater, do not show exceedances of nitrogen salts and heavy metals, respectively. Subtle nitrification is evidenced only by the species composition of the plant associations in the area. We propose that in prospective studies, soil, water, and vegetation analyses will be carried out in a network of points, both in the natural vegetation cover, cultivated plants, and in the river in order to determine the source of excess nitrogen, the trend in the concentration of elements in the substrate and in living organisms, and the direction of migration of these elements from abiotic to biotic. Understanding these mechanisms is necessary to inform more restrictive measures in the use of resources in natural areas in order to secure biodiversity resources and the health of inhabitants.

4. Conclusions

The purpose of this study was to assess the river water and groundwater chemistry and quality in Săpânța locality, situated in North Romania. The site-specific context of this study implies several sampling points in the Săpânța River, a tributary of the Tisa River, and a groundwater source. Most water exhibited a substantial metal load and an almost neutral pH level, based on the Ficklin–Caboi diagram. The Piper plot showed the chemical equilibrium between the cationic (K, Mg, Ca, and Na) and anionic (HCO_3^-, CO_3^2, Cl^-, and SO_4^{2-}) contents, indicating the typology of the water, which is $CaMgHCO_3^-$. A total of twelve (12) chemical indicators (NH_4^+, NO_3^-, Cl^-, Fe, Mn, Al, Cu, Zn, Pb, Cr, Ni, and As), two (2) physico-chemical indicators (pH and electrical conductivity), and the standard values deemed to be the maximum allowable concentrations pertaining to the quality of water used as a drinking water source were used to calculate the OWQI. Both the CI and HMEI approaches were used in this framework. Eight heavy metals (Mn, Ni, Zn, Pb, Cu, Cr, Fe, and Sr) and one metalloid (As) had their CI and HMEI determined. All scores were below the critical limit of 100, indicating lack of contamination with the studied toxins.

Following a period of consistent ingestion of 6 years for children and 70 years for adults, a comprehensive examination of As and eight heavy metals was conducted to ascertain their potential non-carcinogenicity in humans. Results indicated that in general, children are more susceptible than adults, and their daily intake might be up to four or five times more than that of adults. The average daily intake (CDI) of As in children ranges from 2.64×10^{-5} to 6.24×10^{-5} mg/kg As, whereas the CDI of As varies from 1.06×10^{-4} to 1.64×10^{-4} mg/kg. The increasing trend of ingested toxins identified through the obtained daily intake as average CDI scores is As < Ni < Pb < Cu < Fe < Mn < Zn < Sr. Adults ingest less pollutants through their water consumption each day than children, although the trend is similar. According to the hazard quotient, expressed as HQ scores, the sampled water may have harmful impacts on people, particularly children, if they are ingested. HQ scores were up to HQ = 1.00 in the case of river sampling points SR3 and SR1, respectively, and notably high in the case of As, above the critical unity in river sampling point SR2. Fundamental information about environmental risk management and water pollution control must be provided by applying a pollution and quality evaluation with the use of comprehensive indicators. Using the results, conservation restrictions will be implemented by evaluating the biological and ecological diversity of the river and groundwater inside the study area.

Vegetation analyses were used as a precursor to water analyses. The presence of indicator species, both nitrophilic and those that resist and accumulate heavy metals, constitute warning signals for changes occurring in the composition of the substrate. In the case of the Săpânța River, the interpretation of the floristic composition and the structure of the vegetation cover reflects and indicates a subtle involution from stands with a favorable conservation status, towards vegetation with a nitrophilic accent. In the short and medium term, vegetation, together with the clays in the substrate, acts as a filter for nitrogen salts, so that they are not found in the water. Monitoring indicator species can signal disturbances in soil and water chemistry from exogenous sources, signaling the deterioration of environmental quality, risks associated with human health, and the need to monitor the level of pollution of various kinds.

Based on this study's results, certain suggestions were made to improve the efficacy of future directions for research and mitigation efforts targeted at reducing metal contamination in drinking water sources. These consist of (1) determining and analyzing the temporal variation of chemical parameters; (2) being concerned with water purification solutions in order to enhance agricultural and livestock practices and acquire a dependable water source; (3) a comprehensive assessment of biological and ecological diversity to apply effective implementation of conservation measures in the Săpânța River and groundwater sampling points.

Supplementary Materials: The following supporting information can be downloaded at: https://www.mdpi.com/article/10.3390/w17131975/s1, Table S1: Ranges, formulas, and water quality classifications of related to the quality indices, applied in the current study for the water samples SR1–SR3 and GW. Table S2: Weight (Wi), ideal value (ID), and the maximum allowable concentration (MAC_i) used for calculating the OWQI, CI, and HMEI [34–36].

Author Contributions: Conceptualization, T.D.; methodology, T.D. and M.-A.R.; software, T.D. and M.-A.R.; validation, T.D., M.M. and M.-A.R.; formal analysis, T.D., O.N. and M.-A.R.; investigation, T.D. and M.-A.R.; resources, T.D., O.N. and M.-A.R.; data curation, T.D.; writing—original draft preparation, T.D., O.N. and M.-A.R.; writing—review and editing, T.D. and M.-A.R.; visualization, T.D., O.N., M.M. and M.-A.R.; supervision, T.D. All authors have read and agreed to the published version of the manuscript.

Funding: This research received no external funding.

Data Availability Statement: Data is contained within the article.

Conflicts of Interest: The authors declare no conflicts of interest.

References

1. Nidhi, J.; Yevatikar, R.; Raxamwar, S.T. Comparative study of physico-chemical parameters and water quality index of river. *Mater. Today Proc.* **2021**, *60*, 859–867.
2. Unech, R.; Bibi, C.; Welch, B.; Jamilson, H.; Sundaranar, K. Assesement of the bacteorogical quality and physico-chemical parameters of household water sources. *Int. J. Ecol. Sci. Environ. Engin.* **2020**, *6*, 14–17.
3. Dippong, T.; Hoaghia, M.A.; Senila, M. Appraisal of heavy metal pollution in alluvial aquifers. Study case on the protected area of Ronișoara Forest, Romania. *Ecol. Indic.* **2022**, *143*, 109347. [CrossRef]
4. Dippong, T.; Mihali, C.; Marian, M.; Mare Rosca, O.; Resz, M.A. Correlations between chemical, hydrological and biotic factors in rivers from the protected area of Tisa Superioară, Romania. *Process Saf. Environ. Prot.* **2023**, *176*, 40–55. [CrossRef]
5. Horvat, M.; Horvat, Z. Long term sediment transport simulaţion of the Danube, Sava and Tisa rivers. *Int. J. Sediment Res.* **2020**, *35*, 550–561. [CrossRef]
6. Dippong, T.; Senila, M.; Cadar, O.; Resz, M.A. Assessment of the heavy metal pollution degree and potential health risk implications in lakes and fish from northern Romania. *J. Environ. Chem. Eng.* **2024**, *12*, 112217. [CrossRef]
7. Sanja, M.; Sakan, M.; Ninod, M.; Dordevic, S. Trace element study in Tisa river and Danube alluvial sediment in Serbia. *Int. J. Sediment Res.* **2013**, *28*, 234–245.
8. Dippong, T.; Resz, M.A. Heavy metal contamination assessment and potential human health risk of water quality of lakes situated in the protected area of Tisa, Romania. *Heliyon* **2024**, *10*, e28860. [CrossRef]
9. Dippong, T.; Resz, M.A.; Tănăselia, C.; Cadar, O. Assessing microbiological and heavy metal pollution in surface waters associated with potential human health risk assessment at fish ingestion exposure. *J. Hazard. Mater.* **2024**, *476*, 135187. [CrossRef]
10. Roberto, C.G. Freshwater Biodiversity: A Review of Local and Global Threats. *Int. J. Environ. Stud.* **2016**, *73*, 887–904.
11. Geist, J. Integrative freshwater ecology and biodiversity conservation. *Ecol. Ind.* **2011**, *11*, 1507–1516.
12. Shatwell, T.; Thiery, W.; Kirillin, G. Future projections of temperature and mixing regime of European temperate lakes. *Hydrol. Earth Syst. Sci.* **2019**, *23*, 1533–1551. [CrossRef]
13. Houlahan, J.; Findlay, E.C.S.; Schmidt, B.R.; Meyer, A.H.; Kuzmin, S.L. Quantitative evidence for global amphibian population declines. *Nature* **2000**, *44*, 752–755. [CrossRef]
14. Stuart, N.S.; Chanson, J.S.; Cox, N.A.; Young, B.E.; Rodrigues, A.S.L.; Fischman, D.L.; Waller, R.W. Status and trends of amphibian declines and extinctions worldwide. *Science* **2004**, *3*, 1783–1785. [CrossRef] [PubMed]
15. Geng, M.; Wang, K.; Yang, N.; Li, F.; Zou, Y.; Chen, X.; Deng, Z.; Xie, Y. Spatiotemporal water quality variations and their relationship with hydrological conditions in Dongting Lake after the operation of the Three Gorges Dam, China. *J. Clean. Prod.* **2021**, *283*, 124644. [CrossRef]
16. Faithful, J.W. Physico-chemical changes in two northern headwater lakes in the Northwest Territories, Canada, during winter to spring seasonal transitions. *J. Great Lakes Res.* **2020**, *42*, 166–172. [CrossRef]
17. Mutihac, R.; Van Hulle, M. Comparison of principal component analysis and independedent component analysis for blind source separation. *Rom. Rep. Phys.* **2004**, *56*, 20–32.
18. *SR EN ISO 5667-3:2018*; Water Quality—Sampling—Part 3: Preservation and Handling of Water Samples. ISO: Geneva, Switzerland, 2018. (In Romanian)
19. *SR ISO 7888:1985*; Water Quality—Determination of Electrical Conductivity. ISO: Geneva, Switzerland, 1985. (In Romanian)
20. *SR EN ISO 10523:2012*; Water Quality—Determination of pH. ISO: Geneva, Switzerland, 2012. (In Romanian)
21. *SR EN ISO 7027:2001*; Water Quality—Determination of Turbidity. ISO: Geneva, Switzerland, 2001. (In Romanian)
22. *SR EN ISO 5814:2013*; Water Quality—Determination of Dissolved Oxygen Content—Electrochemical Probe Method. ISO: Geneva, Switzerland, 2013. (In Romanian)
23. *SR EN ISO 99693-1:2002*; Water Quality—Determination of Alkalinity—Part I: Determination of Total and Permanent Alkalinity. ISO: Geneva, Switzerland, 2002. (In Romanian)
24. *SR EN ISO 10304-1:2009*; Water Quality—Determination of Dissolved Anions by Liquid Phase Ion Chromatography—Part 1: Determination of Bromide, Chloride, Fluoride, Nitrate, Nitrite, Phosphate and Sulfate Ions. ISO: Geneva, Switzerland, 2009. (In Romanian)

25. *SR ISO 6059-2008*; Determining the of the Amount of Calcium and Magnesium Concentrations—EDTA Titrimetric Method. ISO: Geneva, Switzerland, 2008. (In Romanian)
26. *SR ISO 10566-2001*; Water Quality—Determination of Aluminum Content—Spectrometric Method with Purple Pyrocatechol Method. ISO: Geneva, Switzerland, 2001. (In Romanian)
27. *SR ISO 7150-1:2001*; Water Quality—Determination of Ammonium Content—Part 1: Manual Spectrometric Method. ISO: Geneva, Switzerland, 2001. (In Romanian)
28. *SR ISO 7890-3-2000*; Water Quality—Determination of Nitrate Content—Method by Molecular Absorption Spectrometry. ISO: Geneva, Switzerland, 2000. (In Romanian)
29. *SR EN ISO 6878:2005*; Water Quality—Determination of Phosphor Content. ISO: Geneva, Switzerland, 2005. (In Romanian)
30. *SR ISO 6332:1996*; Water Quality—Determination of Iron Content—Spectrometric Method with 1,10—Phenanthroline. ISO: Geneva, Switzerland, 1996. (In Romanian)
31. *SR ISO 8662/2:1997*; Determination of Manganese Content—The Atomic Absorption Spectrometric Method. ISO: Geneva, Switzerland, 1997. (In Romanian)
32. *SR EN ISO 15586:2004*; Water Quality—Determination of Trace Elements by Atomic Absorption Spectrometry with a Graphite Furnace. ISO: Geneva, Switzerland, 2004. (In Romanian)
33. *SR ISO 8288:2001*; Water Quality—Determination of Zinc Content—The Atomic Absorption Spectrometric Method. ISO: Geneva, Switzerland, 2001. (In Romanian)
34. Das, K.R.; Bhoominathan, S.D.; Kanagaraj, S.; Govindaraju, M. Development of a water quality index (WQI) for the Loktak Lake in India. *Appl. Water Sci.* **2017**, *7*, 2907–2918.
35. Uugwanga, M.N.; Kgabi, N.A. Heavy metal pollution index of surface and groundwater from around an abandoned mine site, Klein Aub. *Phys. Chem. Earth Parts A/B/C* **2021**, *124*, 103067. [CrossRef]
36. Eid, M.H.; Eissa, M.; Mohamed, E.A.; Hatem, S.R.; Tamas, M.; Kovács, A.; Szűcs, P. New approach into human health risk assessment associated with heavy metals in surface water and groundwater using Monte Carlo Method. *Sci. Rep.* **2024**, *14*, 1008. [CrossRef]
37. Guney, M.; Zagury, G.J.; Dogan, N.; Onay, T.T. Exposure Assessment and Risk Characterization from Trace Elements Following Soil Ingestion by Children Exposed to Playgrounds, Parks and Picnic Areas. *J. Hazard. Mater.* **2010**, *182*, 656–664. [CrossRef] [PubMed]
38. Taddeo, S.; Dronova, I.; Depsky, N. Spectral vegetation indices of wetland greenness: Responses to vegetation structure, composition, and spatial distribution. *Remote Sens. Environ.* **2019**, *234*, 111467. [CrossRef]
39. Chatanga, P.; Nituli, V.; Mugomeri, E.; Keketsi, T.; Chikow, T.V.N. Situational analysis of phisico chemical and microbiological quality of water along Mohokare River, Lesorho. *Egyp. J. Aquat. Res.* **2019**, *45*, 45–51. [CrossRef]
40. Mohamed, I.; Al-Khalaf, H.; Alwan, H.; Naje, S.A. Envinoramental assessment of Karbala water treatment plant using water quality index. *Mater. Today Proc.* **2021**, *60*, 1154–1560.
41. Dippong, T.; Resz, M.A. Chemical Assessment of Drinking Water Quality and Associated Human Health Risk of Heavy Metals in Gutai Mountains, Romania. *Toxics* **2024**, *12*, 168. [CrossRef]
42. WHO, World Health Organization. *Guidelines for Drinking-Water Quality*, 4th ed.; Incorporating the 1st Addendum; WHO: Geneva, Switzerland, 2017.
43. Romanian Government. *Related to the Water Quality Destined for Human Consumption*; Regulation No. 7 from 25 January 2023; Official Gazette. No. 63/25.01.2023; Romanian Government: Bucharest, Romania, 2023.
44. Loucif, K.; Neffar, S.; Menasriae, T.; Cherif Maazi, M.; Houhamdi, M.; Chenchounid, H. Physico-chemical and bacteriological quality assessment of surface water at Lake Tonga in Algeria. *Environ. Nanotechnol. Monit. Manag.* **2020**, *13*, 100284. [CrossRef]
45. Zdorovennova, G.; Palshin, N.; Golosov, S.; Efremova, T.; Belashev, B.; Bogdanov, S.; Fedorova, I.; Zverev, I.; Zdorovennov, R.; Terzhevik, A. Dissolved Oxygen in a Shallow Ice-Covered Lake in Winter: Effect of Changes in Light, Thermal and Ice Re-gimes. *Water* **2021**, *13*, 2435. [CrossRef]
46. Constantin, S.; Doxaran, D.; Constantinescu, Ș. Estimation of water turbidity and analysis of its spatio-temporal variability in the Danube River plume (Black Sea) using MODIS satellite data. *Cont. Shelf Res.* **2016**, *112*, 14–30. [CrossRef]
47. Ammann, B.; Birks, H.J.B.; Brooks, S.J.; Eicher, U.; von Grafenstein, U.; Hofmann, W.; Wick, L. Quantification of biotic responses to rapid climatic changes around the Younger Dryas—A synthesis. *Palaeogeogr. Palaeoclimatol. Palaeoecol.* **2000**, *159*, 313–347. [CrossRef]
48. Begum, W.; Rai, S.; Banerjee, S.; Bhattacharjee, S.; Mondal, M.H.; Bhattarai, A.; Saha, B. A comprehensive review on the sources, essentiality and toxicological profile of nickel. *RCS Adv.* **2022**, *12*, 9139. [CrossRef]
49. Zhou, W.; Lan, T.; Shang, G.; Li, J.; Geng, J. Adsorption performance of phosphate in water by mixed precursor base geopolymers. *J. Contam. Hydrol.* **2023**, *255*, 104166. [CrossRef] [PubMed]
50. Algül, F.; Beyhan, M. Concentrations and sources of heavy metals in shallow sediments in Lake Bafa, Turkey. *Sci. Rep.* **2020**, *10*, 11782. [CrossRef] [PubMed]

51. Aliu, S.; Aliu, A.; Mustafi, M.; Selmani, L.; Ibrahimi, B. Physico-chemical paramethers in rivers and in lakeshore of Lake Ohrid-law, economic and social aspects. *Proc. Soc. Behav. Sci.* **2011**, *19*, 499–503. [CrossRef]
52. Moya, S.M.; Botella, N.B. Review of Techniques to Reduce and Prevent Carbonate Scale. Prospecting in water treatment by magnetism and electromagnetism. *Water* **2021**, *13*, 2365. [CrossRef]
53. Okonkwo, O.J.; Mothiba, M. Physico-chemical characteristics and pollution levels of heavy metals in the rivers in Thohoyandou, South Africa. *Glob. Ecol. Conserv.* **2023**, *308*, 122–127. [CrossRef]
54. Ramachandran, A.; Sivakumar, K.; Shanmugasundharam, A.; Sangunathan, U.; Krishnamurthy, R. Evaluation of potable groundwater zones identification basedon WQI and GIS techniques in Adyar river basin, Chennai, Tamilnadu, India. *Acta Ecol. Sin.* **2020**, *41*, 285–295. [CrossRef]
55. Udhayakumar, R.; Manivannan, P.; Raghu, K.; Vaideki, S. Assessment of physico-chemical characteristics of water in Tamilnadu. *Ecotoxicol. Environ. Saf.* **2016**, *134*, 474–477. [CrossRef]
56. Zedler, J.B. Wetlands at your service: Reducing impacts of agriculture at the watershed scale. *Front. Ecol. Environ.* **2023**, *1*, 65–72. [CrossRef]
57. Nguyen, P.M.; Mulligan, G.N. Study on the influence of water composition on iron nail corrosion and arsenic removal performace of the Kanchan arsenic filter (KAF). *J. Hazard. Mater. Adv.* **2023**, *10*, 100285.
58. World Health Organization. *Guidelines for Drinking-Water Quality*, 4th ed.; WHO: Geneva, Switzerland, 2011; pp. 1–541, ISBN 978-92-4-154815-1.
59. Nemerow, N.L.; Agardy, F.J.; Sullivan, P.; Salvato, J.A. *Enviromental Engineering, Water, Wasterwater, Soil and Grpundwater Treatment and Remediation*, 6th ed.; John Wiley & Sons, Inc.: Hoboken, NJ, USA, 2009; pp. 300–380.
60. Egbueri, J.C. A multi-model study for understanding the contamination mechanisms, toxicity and health risks of hardness, sulfate, and nitrate in natural water resources. *Environ. Sci. Pollut. Res.* **2023**, *30*, 61626–61658. [CrossRef]
61. Eslami, H.; Esmaeili, A.; Razaeian, M.; Salari, M.; Hosseini, A.N.; Mobini, M.; Barani, A. Potentially toxic metal concentration, spatial distribution, and health risk assessment in drinking groundwater resources of southeast Iran. *Geosci. Front.* **2022**, *13*, 101276. [CrossRef]
62. Krok, B.; Mohammadian, S.; Noll, H.M.; Surau, C.; Markwort, S.; Fritzsche, A.; Nachev, M.; Sures, B.; Meckenstock, R.U. Remediation of zinc-contaminated groundwater by iron oxide in situ adsorption barriers–From lab to the field. *Sci. Total Environ.* **2022**, *897*, 151066. [CrossRef] [PubMed]
63. Abbaszade, G.; Tserendorj, D.; Salazar-Yanez, N.; Zachary, D.; Volgyesi, P.; Toth, E. Lead and stable lead isotopes as tracers of soil pollution and human health risk assessment in former industrail citisen of Hungary. *Appl. Geochem.* **2022**, *145*, 105397. [CrossRef]
64. Muneer, J.; AlObaid, A.; Ullah, R.; Rehman, K.U.; Erinle, K.O. Appraisal of toxic metals in water bottom sediments and fish of fresh water lake. *J. King Saud Univ. Sci.* **2022**, *34*, 101685. [CrossRef]
65. Khan, M.; Din, I.; Aziz, F.; Qureshi, I.U.; Zahid, M.; Mustafa, G.; Sher, A.; Hakim, S. Chromium adsorption from water using mesoporous magnetic iron oxide-aluminum silicate absorbent: An investigation of adsorbtion isotherms and kinetics. *Curr. Res. Green Sustain. Chem.* **2023**, *7*, 100368. [CrossRef]
66. Gibbs, J.R. Mechanisms Controlling World Water Chemistry. *Science* **1970**, *80*, 1088–1090. [CrossRef]
67. He, X.; Wei, H. Biodiversity conservation and ecological value of protected areas: A review of current situation and future prospects. *Front. Ecol. Evol.* **2023**, *11*, 1261265. [CrossRef]
68. Zakir, H.M.; Sharmin, S.; Akter, A.; Rahman, M.S. Assessment of health risk of heavy metals and water quality indices for irrigation and drinking suitability of waters: A case study of Jamalpur Sadar area, Bangladesh. *Environ. Adv.* **2020**, *2*, 100005. [CrossRef]
69. Abdo, M.H.; Ahmed, H.B.; Helal, M.; Fekry, M.; Abdelhamid, A.E. Water Quality Index and Environmental Assessment of Rosetta Branch Aquatic System, Nile River, Egypt. *Egypt. J. Chem.* **2022**, *65*, 321–331. [CrossRef]
70. Sadee, B.A.; Zebari, S.M.S.; Galali, Y.; Saleem, M.F. A review on arsenic contamination in drinking water: Sources, health impacts, and remediation approaches. *RSC Adv.* **2025**, *15*, 2684–2703. [CrossRef]
71. Hörnberg, G.; Zackrisson, O.; Segerström, U.; Svensson, B.W.; Ohlson, M.; Bradshaw, R.H. Boreal swamp forests: Biodiversity "hotspots" in an impoverished forest landscape. *BioScience* **1998**, *48*, 795–802.
72. Flinn, M.B.; Adams, S.R.; Whiles, M.R.; Garvey, J.E. Biological responses to contrasting hydrology in backwaters of Upper Mississippi River Navigation Pool 25. *Environ. Manag.* **2008**, *41*, 468–486. [CrossRef] [PubMed]
73. Mitsch, W.J.; Gosselink, J.G. *Wetlands*, 6th ed.; John Wiley & Sons: Hoboken, NJ, USA, 2023; ISBN 978-1-119-82695-8.
74. Tiner, R.W. Geographically isolated wetlands of the United States. *Wetlands* **2003**, *23*, 494–516. [CrossRef]
75. Vose, J.M.; Miniat, C.F.; Luce, C.H.; Asbjornsen, H.; Caldwell, P.V.; Campbell, J.L.; Grant, G.E.; Isaak, D.J.; Loheide, S.P., II; Sun, G. Ecohydrological implications of drought for forests in the United States. *Forest Ecol. Manag.* **2016**, *380*, 335–345. [CrossRef]
76. Bonanno, G.; Vymazal, J.; Cirelli, G.L. Translocation, accumulation and bioindication of trace elements in wetland plants. *Sci. Total Environ.* **2018**, *631*, 252–261. [CrossRef]

77. Yan, X.; An, J.; Yin, Y.; Gao, C.; Wang, B.; Wei, S. Heavy metals uptake and translocation of typical wetland plants and their ecological effects on the coastal soil of a contaminated bay in Northeast China. *Sci. Total Environ.* **2022**, *803*, 149871. [CrossRef]
78. Greger, M.; Landberg, T. Use of willow in phytoextraction. *Int. J. Phytoremediat.* **1998**, *1*, 115–123. [CrossRef]
79. Guittonny-Philippe, A.; Petit, M.E.; Masotti, V.; Monnier, Y.; Malleret, L.; Coulomb, B.; Combroux, J.; Baumnerger, T.; Viglione, J.; Laffont-Schwob, I. Selection of wild macrophytes for use in constructed wetlands for phytoremediation of contaminant mixtures. *J. Environ. Manag.* **2015**, *147*, 108–123. [CrossRef]
80. Vymazal, J.; Březinová, T. Accumulation of heavy metals in Phragmites australis for wastewater treatment. *Chem. Eng. J.* **2016**, *290*, 232–242. [CrossRef]
81. Bothe, H.; Słomka, A. Divergent biology of facultative heavy metal plants. *J. Plant Physiol.* **2017**, *219*, 45–61. [CrossRef]
82. Ricotta, C.; Carboni, M.; Acosta, A.T. Let the concept of indicator species be functional! *J. Veg. Sci.* **2015**, *26*, 839–847. [CrossRef]
83. Cakaj, A.; Drzewiecka, K.; Hanć, A.; Lisiak-Zielińska, M.; Ciszewska, L.; Drapikowska, M. Plants as effective bioindicators for heavy metal pollution monitoring. *Environ. Res.* **2024**, *256*, 119222. [CrossRef] [PubMed]
84. Szyczewski, P.; Siepak, J.; Niedzielski, P.; Sobczyński, T. Research on heavy metals in Poland. *Pol. J. Environ. Stud.* **2009**, *18*, 755.
85. Charlesworth, S.; De Miguel, E.; Ordóñez, A. A review of the distribution of particulate trace elements in urban terrestrial environments and its application to considerations of risk. *Environ. Geochem. Health* **2011**, *33*, 103–123. [CrossRef] [PubMed]
86. Bonanno, G.; Orlando-Bonaca, M. Trace elements in Mediterranean seagrasses and macroalgae. A review. *Sci. Total Environ.* **2018**, *618*, 1152–1159. [CrossRef]
87. Kabata-Pendias, A.; Mukherjee, A.B. Trace elements of group 12 (Previously group IIb). In *Trace Elements from Soil to Human*; Springer: Berlin/Heidelberg, Germany, 2017; pp. 283–319.
88. Dhote, S.; Dixit, S. Water quality improvement through macrophytes—A review. *Environ. Monit. Assess.* **2009**, *152*, 149–153. [CrossRef]
89. Bonanno, G.; Pavone, P. Leaves of Phragmites australis as potential atmospheric biomonitors of Platinum Group Elements. *Ecotox. Environ. Saf.* **2015**, *114*, 31–37. [CrossRef]

Disclaimer/Publisher's Note: The statements, opinions and data contained in all publications are solely those of the individual author(s) and contributor(s) and not of MDPI and/or the editor(s). MDPI and/or the editor(s) disclaim responsibility for any injury to people or property resulting from any ideas, methods, instructions or products referred to in the content.

water

MDPI

Article

Evaluation of Water Quality in the Tamiš River in Serbia Using the Water Pollution Index: Key Pollutants and Their Sources

Dragana Milijašević Joksimović, Dejana Jakovljević * and Tamara Jojić Glavonjić

Geographical Institute Jovan Cvijić, Serbian Academy of Sciences and Arts, 11000 Belgrade, Serbia; d.milijasevic@gi.sanu.ac.rs (D.M.J.); t.jojic@gi.sanu.ac.rs (T.J.G.)
* Correspondence: d.jakovljevic@gi.sanu.ac.rs

Abstract: This study evaluates the water quality of the Tamiš River, a tributary of the Danube, at the Jaša Tomić and Pančevo hydrological stations from 2011 to 2016 and from 2018 to 2022, using the Water Pollution Index (WPI) and seasonal analysis. The analysis revealed elevated nitrite and orthophosphate concentrations at both stations, with Pančevo exhibiting extreme microbial contamination in 2015, attributed to urban runoff, agricultural activities, and inadequate wastewater treatment. Seasonal results indicate that while spring conditions align with Class I water standards, summer presents critical risks, especially at Pančevo, where the highest WPI value (26.47 in 2015) was recorded. Autumn shows stabilization, though sporadic WPI peaks reflect the impact of nutrient runoff. Winter conditions are marked by stability, with favorable dissolved oxygen levels but occasional exceedances in heavy metals, particularly at Jaša Tomić. Increased concentrations of suspended solids and heavy metals at Jaša Tomić emphasize diverse pollution sources, including industrial discharges and soil erosion. These findings underscore the necessity of integrated water management strategies, such as wastewater treatment upgrades and sustainable agricultural practices, to mitigate pollution. Protecting the Tamiš River is crucial for supporting biodiversity, safeguarding public health, and ensuring sustainable use of this vital Danube tributary.

Keywords: Tamiš River; water pollution index; anthropogenic pollution; surface water quality assessment; physico-chemical parameters

1. Introduction

Water quality is a critical factor in assessing the sustainability of aquatic ecosystems and is essential not only for drinking water supply but also for tourism, transportation, and various other uses. Surface water quality is highly sensitive to both natural processes and human activities. Natural factors such as geological formations, soil types, climatic conditions, and hydrological characteristics inherently influence water quality parameters. However, anthropogenic activities—including urbanization, industrialization, mining, agriculture, and increased water consumption—significantly exacerbate the degradation of surface waters.

Numerous studies have documented the impact of the petrochemical industry on the quality and pollution of surface and groundwater, highlighting the release of nitrates, chlorides, sulphates and petroleum products into aquatic systems [1–4]. Agricultural practices, particularly intensive farming, contribute substantial amounts of nitrogen and phosphorus to water bodies through runoff, making agricultural nutrients a significant component of diffuse water pollution [5–7]. These factors often act as continuous pollution sources, severely degrading aquatic ecosystems and diminishing the usability of water for various activities.

Monitoring water quality and identifying sanitary risks are essential for protecting populations from waterborne diseases and developing appropriate preventive measures. Global water contamination poses a severe threat to public health, with the impact and seriousness of health issues depending on factors such as the chemical composition of contaminants, exposure duration, and pollutant concentration levels [8–10].

Water quality indices serve as valuable tools for assessing and communicating water quality status. Their primary purpose is to condense numerous physical, chemical, and biological measurements into a single value that reflects the ecological condition of a specific water body. These indices facilitate the comparison of water quality across different regions and time periods, forming the basis for extensive scientific research [11–29].

One such index, the Water Pollution Index (WPI), is employed to measure the extent of water pollution caused by various contaminants. The WPI helps in understanding the pollution load, identifying primary pollution sources, and assessing the effectiveness of pollution control measures. This index is applied in water pollution assessment worldwide: rivers in China [30]; Lake Ulansuhai in China [31]; Grenjeng River in Indonesia [32]; Nambul River in Manipur and groundwater in West Bengal in India [33,34]; Mekong River in Vietnam [35]. The WPI is also widely used in various studies in Serbia, including the Danube River in the Djerdap National Park and downstream until the Serbian–Romanian border [19,36]; the Sava River [37]; the Timok River [38]; the Porečka River [39]; the Hydrosystem Danube–Tisa–Danube [40]; and Vlasina Lake [41].

The Tamiš River is a significant watercourse originating in the western part of Romania and flowing through the Banat region before joining the Danube near Pančevo in northern Serbia. The river holds substantial cultural and economic importance, supporting activities such as transportation, recreation, and tourism. However, its water quality faces challenges due to various pollution sources. In Romania, the Tamiš is polluted starting from Caransebeș due to wastewater discharges containing chlorides, fats, detergents, and petroleum products, as well as industrial activities and agricultural runoff [42]. In Serbia, previous assessments have indicated that pollutants enter the river through agricultural runoff and industrial and municipal sewage systems located along its course [43]. Besides this pollution, Lujić et al. [44] addressed inadequate maintenance of irrigation channels as well as effluents from fish and livestock farms as a main pollution sources.

In 2023, the establishment of a protected area within the Tamiš basin (Potamišje Region of Exceptional Characteristics) emphasized the ecological significance of the region. The designation of this area underscores the need for continuous monitoring and effective management of the river's water quality to preserve its ecological integrity [45].

The Serbian section of the Tamiš River has not been subject to water quality research since 2009 [43]. This study aims to provide new insights into the water quality status of the Tamiš River in Serbia by calculating the Water Pollution Index (WPI) at two hydrological stations, namely Jaša Tomić and Pančevo, for the periods 2011–2015 and 2018–2022. Due to the cessation of operations at the Pančevo hydrological station in 2015, the WPI for this station was calculated only for the first period. By evaluating key physico-chemical parameters, this research aims to provide valuable insights into pollution levels, identify the main sources of contamination, and contribute to a comprehensive understanding of the river's ecological status. The findings can contribute to future water management strategies and support efforts to protect and restore the Tamiš River basin.

2. Materials and Methods

2.1. Study Area

The Tamiš River, a tributary of the Danube in Serbia, originates in Romania under the highest peak of the Semenic Mountain. It flows for 340 km, of which 118 km is in Serbia, before joining the Danube near Pančevo. The river's catchment area covers approximately 7319 km^2, with 1529 km^2 within Serbia [46].

Due to frequent meandering caused by the flat terrain and gentle river gradient, the Tamiš has a slow water flow, which historically led to significant flooding. To mitigate this, regulatory works began in the first half of the 18th century, cutting through numerous meanders to control flooding and improve navigation. Significant changes to the Tamiš water regime occurred in 1967 with the diversion of water from the Brzava River basin into the Danube and the construction of three sluice gates on the Tamiš River in Serbia [46].

The river exhibits its highest average monthly flows in March and April, while the lowest flows occur in September and October, which is characteristic of rivers with Central Europe's rain–snow regime. The average annual flow of the Tamiš at Jaša Tomić is 39.1 m^3/s [47].

A systematic analysis of river water quality was conducted at two hydrological stations (Figure 1): the Jaša Tomić hydrological station (UTM coordinates 5,031,950 m N and 7,489,150 m E, zone 34 N), and the Pančevo hydrological station (UTM coordinates 4,969,525 m N and 7,471,325 m E, zone 34 N) [48]. The Jaša Tomić Hydrological Station, located in northern Serbia near the border with Romania, primarily reflects pollution originating from upstream Romanian sources. Key contributors include industrial facilities, agricultural runoff, and urban wastewater. In contrast, the Pančevo Hydrological Station, situated near the confluence of the Tamiš with the Danube, integrates both upstream pollution from Romania and additional contributions from urban, industrial, and agricultural sources within Serbia (Table 1).

Table 1. Anthropogenic pollution sources in Serbia.

Company	Facility	Location
Stary Tamiš D.O.O.	Nadel Farm	Pančevo, Serbia
HIP Petrohemija d.o.o. Pančevo	Petrochemical industry	Pančevo, Serbia
Timkok Proizvodno d.o.o.	Poultry Farm	Pančevo, Serbia
Higijena Javno komunalno preduzeće Pančevo	New landfill	Pančevo, Serbia
HIP Azotara DOO	Nitrogen fertilizer production	Pančevo, Serbia
Naftna Industrija Srbije a.d. Novi Sad	Oil refinery	Pančevo, Serbia
MASSAGRAR	Pig Farm	Sečanj, Serbia

Note: Source: SEPA [49].

This dual influence highlights the complexity of managing pollution in transboundary rivers like the Tamiš river.

The geographic distribution of these stations underscores the necessity for collaborative transboundary water management between Romania and Serbia. Effective measures require not only addressing local pollution sources but also implementing upstream interventions to mitigate the overall impact on the river's ecological health.

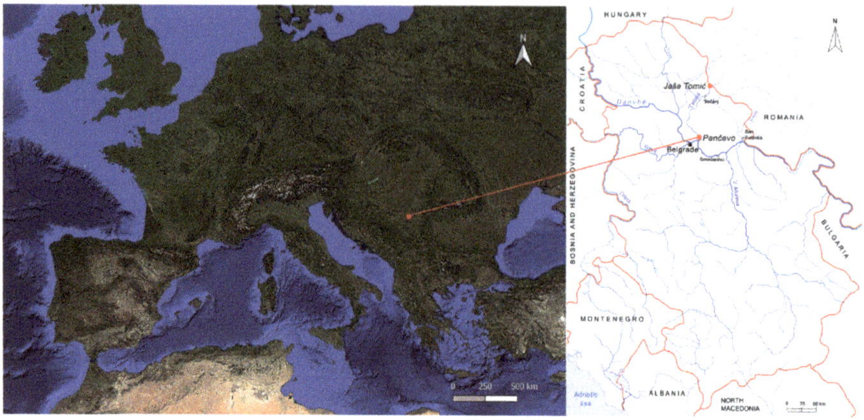

Figure 1. Map of the study area. Source of data: [50].

An Integrated Polluters Cadastre of the Tamiš River was created as part of the "Eco-status of the Tamiš River" project, funded by the EU IPA funds under the Neighbourhood Programme Romania–Serbia. This project identified pollutants—including both natural and artificial substances—that disrupt the natural composition and properties of the environment when released into water, air, and soil. The concentrated pollution sources for the Tamiš River include industrial facilities, farms, public utility landfills, sewage discharges, and fishponds. Additionally, diffuse pollution sources encompass the use of artificial fertilizers and agrochemicals in agriculture, transportation (as diffuse sources), septic tanks, and forest resource utilization [51].

Water quality monitoring of the Tamiš River is of great importance, particularly due to the establishment of the Potamišje Region of Exceptional Characteristics. Covering 22,633.18 ha, this first-category natural asset of international, national, and exceptional significance boasts a unique mosaic of ecosystems, including alluvial forests, wet meadows and pastures, ancient river sections, marshes, and both mesophilic and saline meadows. It is celebrated for its rich biodiversity, being a hotspot with 606 taxa of higher plants, 42 species of fish, 11 species of amphibians, 9 species of reptiles, 49 species of invertebrates of national and international importance, 257 species of birds, and 58 species of mammals. Additionally, it encompasses two internationally significant bird areas (Important Bird Areas—IBA) and an internationally significant plant area (Important Plant Area—IPA). The created values of this area are no less important. There are 17 villages evenly distributed around the perimeter of the protected area. All are known for their traditional use of resources, indigenous culture, and preserved rural architecture [52]. The protected area holds immense value not only for the local community and, also, for the broader scientific community, offering a unique opportunity to study ecological processes and interactions within preserved natural habitats.

2.2. Data and Methods

This study systematically evaluated river water quality at the Jaša Tomić and Pančevo hydrological stations. The river water quality data were analyzed for two separate periods, 2011–2015 and 2018–2022, from the Serbian Environmental Protection Agency [49]. The dataset for Pančevo was available only for the period 2011–2015 due to the cessation of station operation in 2015.

The water quality evaluation included key physico-chemical parameters, namely dissolved oxygen (DO), oxygen saturation (OS), pH, suspended solids (SS), biochemical oxygen demand (BOD), chemical oxygen demand (COD_{Mn}), nitrites (NO_2^-), ammonium (NH_4^+), orthophosphates (PO_4^{3-}), sulphates (SO_4^{2-}), dissolved metals (Fe, Mn, Ni, Hg Cu, Pb, Cd), and total coliforms (TC).

Based on water sampling, the Serbian Environmental Protection Agency (SEPA), under the Ministry of Environmental Protection of the Republic of Serbia, applied the following methods for these analyses: SRPS ISO 5813:1994 (for dissolved oxygen); SRPS H.Z1.160:1987 (for oxygen saturation); SRPS H.Z1.111:1987 (for pH); SRPS H.Z1.160:1987 (for suspended solids); BOD JUS ISO 5815 (for biochemical oxygen demand). determination using potassium permanganate: Kubel–Tiemann method (for chemical oxygen demand); HACH Method 8507—EPA 353.2 (for nitrites); HACH Method 8155 (for ammonium); HACH Method 8048—EPA 365.1 (for orthophosphates); HACH Method 8051 (for sulphates); EPA 6020 A:2014 (for Fe, Mn, Ni, Cu, and Cd); EPA 6020 A:2007 (for Pb); EPA Method 245.7 (for Hg); SRPS EN ISO 9308-1:2010 (for total coliforms) [49].

2.3. Water Pollution Index (WPI)

The water pollution index (WPI) was calculated to quantify water quality status based on measured parameters. WPI is defined as the sum of ratios between observed parameter values and corresponding standard limits:

$$\mathrm{WPI} = \sum_{i=1}^{n} \frac{Ci}{SFQS} \cdot \frac{1}{n}$$

In this formula, *Ci* represents the average annual concentration of the analyzed parameters based on river water quality data [53]. *SFQS* denotes the standard values (Table 2) for the first class of water quality in Serbia, while *n* presents the total number of analyzed parameters. In Serbia, the *SFQS* values correspond to Class I of surface water according to national legislation [54–57]. Water quality Class I could be used for drinking (after appropriate filtration and disinfection), recreation, irrigation, and various industrial uses.

Table 2. Parameter standard values for Class I.

Parameter	Unit	Class I
Dissolved Oxygen (DO)	mg/L	8.5
Oxygen Saturation (OS)	%	90
pH		6.5–8.5
Suspended solids (SS)	mg/L	25
Biochemical Oxygen Demand (BOD)	mg/L	2
Chemical Oxygen Demand (COD_{Mn})	mg/L	10
Nitrites (NO_2^-),	mg/L	0.01
Ammonium (NH_4^+)	mg/L	0.1
Orthophosphates (PO_4^{3-})	mg/L	0.02
Sulphates (SO_4^{2-})	mg/L	50
Iron (Fe)	mg/L	0.3
Manganese (Mn)	mg/L	0.05
Nickel (Ni)	μg/L	20
Mercury (Hg)	μg/L	1
Copper (Cu)	μg/L	2000
Lead (Pb)	μg/L	50
Cadmium (Cd)	μg/L	5
Total Coliforms (TC)	CFU/100 mL	500

Note: Source: [54–57].

Classification of water quality status based on WPI values is presented in Table 3, following Lyulko et al. [53].

Table 3. Classification of surface water using the WPI.

Class	Characteristics	WPI
I	Very pure	$\leq$0.3
II	Pure	0.3–1
III	Moderately polluted	1–2
IV	Polluted	2–4
V	Impure	4–6
VI	Heavily impure	>6

Note: Source: [53].

One of the advantages of the WPI is its flexibility to incorporate various parameters. It is not restricted by the quantity or types of parameters utilized.

Understanding the deviations from class I water quality standards contributes to the identification of pollution levels and sources affecting the river ecosystem. By analyzing the ratio of mean annual concentrations, the parameters that exceed the acceptable limits are addressed, providing a comprehensive assessment of the river's water quality over the studied periods.

3. Results and Discussion

3.1. Parameters Deviation from Class I

Table 4 presents the deviations from Class I Water Quality Standards at the Pančevo hydrological station for the period 2011–2015 and at the Jaša Tomić hydrological station for the periods 2011–2016 and 2018–2022. It should be noted that, due to the cessation of data collection at the Pančevo station in 2015, the results for this location cover only the first period. Additionally, dissolved metal measurements were unavailable or incomplete for certain years, thus excluded from parts of the analysis.

Table 4. Ratio of mean annual concentration and standards of the Class I of water quality for analyzed parameters.

Pančevo																		
Year	DO	OS	pH	SS	BOD	COD_{Mn}	NO_2^-	NH_4^+	PO_4^{3-}	SO_4^{2-}	Fe	Mn	Ni	Hg	Cu	Pb	Cd	TC
2011	1.138	0.943	0.929	0.732	0.895	0.48	1.9	1.33	5.3	1.326	0.347	0.6	0.168	0.1	0.002	0.01	0.005	21.067
2012	1.175	1.022	0.941	0.648	0.885	0.488	1.7	0.8	4.1	0.72	0.17	0.28	0.511	0.1	0.002	0.01	0.004	6.3
2013	1.15	0.959	0.929	0.604	0.82	0.53	1.8	1.18	4.9	0.748								1.34
2014	1.013	0.891	0.929	1.128	0.845	0.605	1.6	1.25	4.35	0.75	0.353	0.56	0.511	0.1	0.005	0.014	0.006	39.8
2015	1.125	0.942	0.929	1.3	1.055	0.513	1.8	1.14	3.8	0.948	0.29	0.48	0.171	0.1	0.003	0.014	0.006	149.5
Jaša Tomić																		
Year	DO	OS	pH	SS	BOD	COD_{Mn}	NO_2^-	NH_4^+	PO_4^{3-}	SO_4^{2-}	Fe	Mn	Ni	Hg	Cu	Pb	Cd	TC
2011	2.088	1.022	0.929	1.456	0.922	0.41	1.3	0.84	1.95	1.062	0.247	1.68	0.59	0.046	0.002	0.01	0.005	2.096
2012	1.175	1.03	0.918	1.064	0.841	0.36	1.6	2.8	1.15	1.442	0.177	0.82	0.26	0.114	0.001	0.01	0.004	0.25
2013	1.254	1.071	0.926	1.953	0.795	0.456	1.7	1.1	2.35	0.757								7.55
2014	1.203	1.011	0.935	2.107	0.763	0.575	1.4	0.6	2.5	0.715	0.41	0.64	0.495	0.1	0.004	0.013	0.005	13.333
2015	1.188	1.036	0.941	1.272	0.704	0.415	1.5	0.69	1.75	0.884	0.28	0.42	0.206	0.1	0.002	0.014	0.005	3.69
2018	1.2	1.009	0.918	1.852	0.625	0.498	1.6	0.89	1.7	0.845	2	1	0.26	0.07	0.007	0.015	0.005	1.823
2019	1.200	0.912	0.906	0.828	0.692	0.449	1.3	1.25	1.85	1.035	0.697	1.14	1.745	0.07	0.01	0.016	0.005	1.488
2020	1.225	1.028	0.918	0.808	0.823	0.328	1.9	1.9	1	0.855	0.613	0.82	0.494	0.07	0.003	0.014	0.005	1.84
2021	1.249	1.061	0.908	1.252	0.775	0.435	1	0.4	1	0.953	0.183	0.48	0.246	0.07	0.003	0.01	0.015	3.895
2022	1.231	1.025	0.902	1.340	0.84	0.498	2	2	1	0.8	0.367	0.8	0.34	0.07	0.005	0.013	0.022	1.584

The results at the Pančevo hydrological station for the period 2011–2015 show that the dissolved oxygen (DO) ratios ranged from 1.013 to 1.175. The dissolved oxygen ratios at the Jaša Tomić hydrological station ranged from 1.175 to 1.355, relative to the Class I standard. These increased levels indicate supersaturation in the water, which is generally not harmful to the water quality or to aquatic organisms.

The oxygen saturation (OS) ratios ranged from 0.891 to 1.022 at the Pančevo hydrological station and from 0.912 to 1.071 at the Jaša Tomić hydrological station, indicating that oxygen levels fluctuated near optimal conditions. Although slightly lower oxygen saturation values were observed occasionally, they remained close to the optimal range.

The pH ratios were consistently between 0.929 and 0.941 at the Pančevo hydrological station and between 0.902 and 0.941 at the Jaša Tomić hydrological station throughout the examined years, indicating that the water acidity levels remained stable and well within the Class I water quality standards. This stability suggests a balanced aquatic environment without significant acidic or alkaline fluctuations, which is essential for the health of diverse aquatic species.

Suspended solids (SS) ratios at the Pančevo hydrological station varied between 0.604 and 1.300, with exceedances in 2014 (1.128) and 2015 (1.300). The increased levels may have reduced water clarity, affecting photosynthesis by limiting light penetration and potentially degrading habitat quality for aquatic organisms by smothering benthic habitats [58]. Similarly, SS ratios at the Jaša Tomić hydrological station varied between 0.808 and 2.107, with notable deviations in 2013 (1.953), 2014 (2.107), and 2018 (1.852), which likely resulted from increased soil erosion, runoff, or anthropogenic activities upstream.

The ratios for biological oxygen demand (BOD) ranged from 0.820 to 1.055 at the Pančevo hydrological station. A slight increase was observed in 2015, when the ratio peaked at 1.055, slightly surpassing the Class I standard. This increase indicates that the amount of biodegradable organic matter in the water was occasionally high enough to cause minor oxygen depletion. BOD ratios at the Jaša Tomić hydrological station ranged from 0.625 to 0.922, remaining within the Class I water quality standards throughout the studied periods, indicating no significant risk of oxygen depletion.

Chemical oxygen demand (COD_{Mn}) values ranged from 0.800 to 1.300 at the Pančevo hydrological station, with some years surpassing the Class I thresholds. Elevated COD_{Mn} levels indicate the presence of oxidizable organic and inorganic substances, indicating potential sources such as industrial effluents or urban runoff. The COD_{Mn} ratios were between 0.328 and 0.575 at the Jaša Tomić hydrological station, remaining within the Class I standards. This indicates that the levels of oxidizable compounds were low and chemical pollutants were not a significant concern during the analyzed period.

Nitrite (NO_2^-) ratios were consistently high, ranging from 1.600 to 1.900 at the Pančevo hydrological station and from 1.000 to 2.000 at the Jaša Tomić hydrological station, significantly exceeding the Class I standards throughout the studied period. These elevated nitrite levels signal substantial nitrogen pollution, raising concerns since nitrites can be toxic to aquatic life even at low concentrations. The likely sources include agricultural runoff rich in fertilizers and inadequate sewage treatment [59], emphasizing the need for improved nutrient management in the region. According to the Ministry of Agriculture, Forestry and Water Management of the Republic of Serbia [60], there is currently no organized collection of data on pesticide use, nor an appropriate database that records the quantities of chemical agents used on different crops. Water bodies at significant potential risk of pesticide pollution are predominantly located in Vojvodina, northern Serbia. Therefore, further research and improvements in monitoring and analysis regarding the chemical use are necessary to accurately assess the actual impact from hazardous substances from agriculture.

Ammonium (NH_4^+) ratios fluctuated between 0.890 and 1.330 at the Pančevo hydrological station. Notably, in 2011 (1.330), 2013 (1.180), 2014 (1.250), and 2015 (1.140), the ratios exceeded the Class I standard. At the Jaša Tomić hydrological station, ammonium ratios varied between 0.400 and 2.800, with significant deviations recorded in 2012 (2.800) and 2022 (2.000). Increased ammonium concentrations can be detrimental, as ammonium is toxic to fish and invertebrates at higher levels, affecting their growth, reproduction, and survival. These findings highlight potential issues with wastewater discharge and agricultural practices leading to ammonia pollution [61]. Decomposition of organic matter could also cause increased ammonium concentrations.

Orthophosphate (PO_4^{3-}) ratios were significantly above the Class I standards during the entire period, ranging from 3.800 to 5.300 at the Pančevo hydrological station and from 1.000 to 2.500 at the Jaša Tomić hydrological station. Such high concentrations indicate considerable phosphorus pollution, likely from agricultural fertilizers, detergents, and sewage effluent. According to Biliani et al. [62] increased orthophosphates promote excessive algal growth, leading to eutrophication, which degrades water quality by causing oxygen depletion and harming aquatic life.

Sulphate (SO_4^{2-}) ratios were mostly within standards, except in 2011 at the Pančevo hydrological station, when the ratio peaked at 1.326, surpassing the Class I limit. This peak suggests a possible influx of sulphate-rich effluents during that year, necessitating investigation into industrial discharges, agricultural runoff, or natural sources. Sulphate ratios ranged from 0.715 to 1.442 at the Jaša Tomić hydrological station, with a deviation in 2012 (1.442) exceeding the standard.

The analysis of metal concentrations at the Pančevo hydrological station from 2011 to 2015 reveals that metal pollution was not a significant concern during this period, with all measured metals remaining within Class I water quality standards. However, the absence of data for 2013 underscores the need for continuous monitoring. Slight increases in certain metals, such as nickel, lead, and cadmium, highlight the importance of ongoing attention to maintain the river's health and prevent future contamination. At the Jaša Tomić hydrological station, notable deviations from Class I standards were recorded for dissolved metals. Iron (Fe) ratios ranged from 0.177 to 2.000, with a notable deviation in 2018 (2.000) exceeding the Class I standard. Possible sources include natural geological leaching, industrial discharges, or mining activities upstream. Manganese (Mn) ratios were between 0.420 and 1.680, with deviations in 2011 (1.680) exceeding the standards. Nickel (Ni) ratios were notably elevated in 2019 (1.745), exceeding the class I standard. The elevated levels indicate possible industrial influence or contamination from urban runoff.

Total coliform (TC) ratios at the Pančevo hydrological station ranged from 1.340 to an alarming 149,500 in 2015. The mean annual concentration of total coliforms in 2015 was 747,500 CFU/100 mL vastly exceeding Class I standards. This substantial increase indicates severe fecal contamination, posing significant health risks for human contact and signaling potential outbreaks of pathogenic organisms. Contamination can stem from untreated sewage discharge, septic system failures, or runoff from livestock areas. The extraordinarily high level of coliforms necessitates urgent attention and remediation efforts. Total coliform ratios at the Jaša Tomić hydrological station varied from 0.250 to 13.333, with elevated levels in 2014 (13.333) and 2021 (3.895) exceeding the Class I standards.

The data from the Pančevo Hydrological Station between 2011 and 2015 highlight several concerning deviations from Class I water quality standards. Parameters such as nitrites, ammonium, orthophosphates, suspended solids, BOD, sulphates, and total coliforms exceeded acceptable levels in various years, suggesting significant pollution from agricultural runoff, inadequate wastewater treatment, and possibly industrial activities.

Based on the comparative analysis of the results from both hydrological stations, we can conclude that both stations exhibited high levels of nitrites and orthophosphates, with Pančevo showing higher concentrations of orthophosphates. Pančevo also displayed significantly higher levels of total coliforms, particularly in 2015. Regarding heavy metals, Jaša Tomić exhibited significant deviations in iron, manganese, and nickel. Analyses have shown that Pančevo had fewer deviations concerning heavy metals. Suspended solids were higher at the Jaša Tomić station, indicating upstream erosion or construction activities.

The data reveal that both hydrological stations deviate from Class I water quality standards, albeit with different predominant pollutants. Pančevo exhibited elevated nutrients and microbial pollution with likely sources including urban runoff, inadequate wastewater treatment, and agricultural activities. Agriculture land use includes poultry and pig farms, cultivation of cereals, legumes, and oilseeds. Besides agriculture, industrial activities (petrochemical industry, oil refinery, nitrogen fertilizer production) also contribute to water quality impairment. Regional landfill also affects water quality [49]. Jaša Tomić had elevated levels of suspended solids, heavy metals, and nutrients. Probable sources are soil erosion, industrial discharges, and agricultural practices.

3.2. *Seasonal Analysis*

By comparing seasonal data to Class I water quality standards, the analysis highlights temporal fluctuations influenced by natural and anthropogenic factors. Spring was analyzed based on the mean values of parameters measured in March, April, and May; summer as the mean of June, July, and August; autumn as the mean of September, October, and November; and winter as the mean of December, January, and February. These seasonal analyses provide insights into temporal variations across spring, summer, autumn, and winter, offering a detailed evaluation of how water quality fluctuates throughout the year in relation to regulatory standards. By analyzing the ratio of mean annual and seasonal concentrations, the parameters that exceed the acceptable limits are addressed, providing a comprehensive assessment of the river's water quality over the studied periods. Understanding seasonal dynamics is crucial for water resource management, as it reveals patterns that inform mitigation strategies and environmental protection measures.

During the spring, water quality at both Jaša Tomić and Pančevo stations generally aligns with Class I standards, with minor seasonal fluctuations. Dissolved oxygen levels at the Jaša Tomić hydrological station consistently exceed thresholds (greater than 1.10), driven by increased photosynthesis and algal activity. The pH remains neutral, reflecting balanced aquatic systems, while suspended solids exhibit moderate variability linked to spring runoff and sediment transport. Nutrient concentrations, such as nitrites, ammonium, and orthophosphates, show slight increases due to agricultural runoff but remain within the acceptable limits. Overall, spring highlights balanced conditions at both stations, with minimal anthropogenic impacts (Table 5).

Summer introduces notable changes at both stations, with elevated temperatures and intensified biological activity. Dissolved oxygen levels decline at both Jaša Tomić and Pančevo hydrological stations, with ratios slightly below Class I thresholds (for example, ranging from 0.67 to 0.92 at Pančevo). Nutrient levels, especially nitrates and phosphates, increase, reflecting eutrophication risks. At Pančevo, orthophosphates reach consistently high ratios, from 4.95 to 6.05. Manganese concentrations also rise, exceeding regulatory thresholds at Pančevo (up to 1.96 in 2015). Despite increased pressures, key parameters like suspended solids and organic pollutants remain relatively stable at both stations. Summer conditions emphasize the need for careful monitoring of nutrients and metals (Table 6).

In autumn, water quality stabilizes at both stations as temperatures drop. Dissolved oxygen ratios improve at the Jaša Tomić (approximately 1.30) and Pančevo hydrological stations (ranging from 1.17 to 1.44), reflecting cooler conditions and reduced biological activity. Nutrient levels, particularly ammonium and orthophosphates, show increases due to organic matter degradation and runoff, with Pančevo exhibiting peaks (ammonium reaching 2.10 and orthophosphates reaching 4.3). Despite localized risks, autumn represents a transitional phase with overall improvements compared to summer (Table 7).

Table 5. Ratio of seasonal mean concentration and standards of the Class I of water quality for analyzed parameters (Spring).

Year	DO	OS	pH	SS	BOD	COD_{Mn}	NO_2^-	NH_4^+	PO_4^{3-}	SO_4^{2-}	Fe	Mn	Ni	Hg	Cu	Pb	Cd	TC
Pančevo																		
2011	0.963	0.926	0.933	0.906	0.734	0.463	2.5	0.77	6.15	1.313	0.200	1.00	0.255	0.100	0.001	0.010	0.005	7.600
2012	0.996	0.944	0.918	1.346	0.784	0.577	2.3	1.00	3.35	0.467	0.253	0.28	0.062	0.100	0.001	0.010	0.004	
2013	0.958	0.796	0.909	0.613	1.015	0.727	1.9	0.87	5.50	0.653								1.000
2014	1.208	1.100	0.953	0.800	1.217	0.427	1.7	1.73	3.50	0.827	0.143	0.58	0.110	0.100	0.003	0.013	0.005	20.000
2015	1.171	0.967	0.941	0.960	0.950	0.460	2.0	0.90	3.50	1.040	0.267	0.40	0.209	0.100	0.002	0.017	0.006	24.000
Jaša Tomić																		
Year	DO	OS	pH	SS	BOD	COD_{Mn}	NO_2^-	NH_4^+	PO_4^{3-}	SO_4^{2-}	Fe	Mn	Ni	Hg	Cu	Pb	Cd	TC
2011	3.450	1.041	0.929	1.373	1.067	0.420	1.2	0.53	3.50	0.780	0.167	1.00	1.500	0.055	0.002	0.010	0.005	1.000
2012	1.213	1.041	0.906	1.933	0.700	0.470	1.6	0.80	1.60	0.640	0.190	0.44	0.075	0.100	0.001	0.010	0.004	
2013	1.204	1.081	0.902	5.240	0.917	0.717	1.4	0.93	2.90	0.673								8.200
2014	1.263	1.081	0.953	1.386	0.665	0.480	0.8	0.53	1.50	0.880	0.370	0.94	0.152	0.100	0.002	0.016	0.004	2.000
2015	1.125	0.967	0.954	0.573	0.634	0.350	1.2	0.30	2.10	0.827	0.220	0.32	0.194	0.100	0.002	0.017	0.006	0.700
2018	1.171	1.015	0.918	2.813	0.634	0.593	1.2	0.40	1.95	0.367	4.110	0.66	0.150	0.070	0.010	0.021	0.006	4.200
2019	1.263	1.037	0.912	0.560	0.434	0.373	1.5	0.53	0.95	0.427	0.403	0.60	1.102	0.070	0.013	0.010	0.005	1.600
2020	1.138	1.026	0.904	0.653	0.517	0.303	2.2	0.33	1.00	0.627	0.660	0.60	0.592	0.070	0.003	0.018	0.006	
2021	1.329	1.059	0.908	2.253	0.965	0.600	1.0	0.50	1.00	0.647	0.200	0.40	0.103	0.070	0.002	0.010	0.004	0.460
2022	1.329	1.048	0.898	1.573	0.667	0.343	1.4	0.33	0.90	0.700	0.583	0.58	0.001	0.070	0.009	0.017	0.040	2.400

Table 6. Ratio of seasonal mean concentration and standards of the Class I of water quality for analyzed parameters (Summer).

Year	DO	OS	pH	SS	BOD	COD_{Mn}	NO_2^-	NH_4^+	PO_4^{3-}	SO_4^{2-}	Fe	Mn	Ni	Hg	Cu	Pb	Cd	TC
Pančevo																		
2011	0.891	0.907	0.926	0.800	0.700	0.437	1.5	0.73	4.95	0.774	0.300	0.40	0.098	0.100	0.004	0.010	0.005	27.8
2012	0.900	1.011	0.927	0.546	0.700	0.370	1.2	0.53	5.45	0.693	0.194	0.29	0.065	0.100	0.003	0.010	0.004	
2013	0.921	0.985	0.906	0.506	0.734	0.510	1.4	0.47	5.65	0.913								0.58
2014	0.671	0.693	0.914	1.853	0.734	0.927	1.9	0.83	6.05	0.607	0.573	0.44	1.064	0.100	0.007	0.017	0.006	9.6
2015	0.766	0.874	0.931	1.227	1.415	0.563	1.0	0.60	5.00	1.007	0.683	1.96	0.295	0.100	0.004	0.010	0.014	460
Jaša Tomić																		
Year	DO	OS	pH	SS	BOD	COD_{Mn}	NO_2^-	NH_4^+	PO_4^{3-}	SO_4^{2-}	Fe	Mn	Ni	Hg	Cu	Pb	Cd	TC
2011	1.254	0.967	0.945	1.213	0.900	0.473	1.5	0.60	0.90	1.313	0.167	1.00	0.215	0.010	0.002	0.010	0.006	0.44
2012	1.000	1.107	0.929	0.826	0.984	0.370	0.4	0.40	1.30	1.333	0.150	0.70	0.608	0.150	0.002	0.010	0.004	0.4
2013	1.017	1.133	0.968	0.200	0.780	0.320	1.3	0.27	3.85	1.093								2.4
2014	0.925	0.952	0.933	2.733	0.867	0.623	1.6	0.70	3.25	0.667	0.583	0.42	0.475	0.100	0.004	0.011	0.005	4
2015	0.966	1.081	0.962	0.586	0.615	0.353	1.0	0.33	1.70	0.847	0.227	0.72	0.282	0.100	0.003	0.013	0.005	11
2018	0.888	0.951	0.907	1.600	0.584	0.493	1.7	0.67	2.50	0.760	2.443	1.10	0.360	0.070	0.005	0.016	0.005	4.
2019	0.863	0.896	0.842	1.200	0.934	0.730	1.1	1.13	3.65	0.600	0.407	2.36	1.600	0.070	0.008	0.011	0.006	4.4
2020	0.950	1.000	0.921	1.480	1.134	0.397	1.5	0.37	1.05	1.047	0.493	0.82	0.977	0.070	0.003	0.014	0.005	2.6
2021	1.029	1.122	0.906	0.360	0.817	0.337	1.8	0.23	1.45	0.847	0.153	0.62	0.189	0.070	0.001	0.010	0.004	0.56
2022	0.929	1.011	0.898	1.147	0.867	0.513	1.6	3.37	1.55	0.660	0.407	0.72	0.448	0.070	0.002	0.012	0.013	1.373

Winter brings stability to water quality at both stations. Dissolved oxygen ratios remain favorable, with both the Jaša Tomić (ranging from 1.42 to 1.66) and Pančevo hydrological stations (ranging from 1.38 to 1.64) showing sufficient aeration under low temperatures. Ammonium concentrations increase significantly at Pančevo (peaks of up to 3.30 in 2011), reflecting slower degradation rates. Heavy metals, including manganese, occasionally approach or exceed standards. The pH values and suspended solids remain stable, supporting balanced aquatic ecosystems. It is noteworthy that total coliform bacteria were not analyzed in winter (Table 8).

Table 7. Ratio of seasonal mean concentration and standards of the Class I of water quality for analyzed parameters (Autumn).

Pančevo

Year	DO	OS	pH	SS	BOD	COD_{Mn}	NO_2^-	NH_4^+	PO_4^{3-}	SO_4^{2-}	Fe	Mn	Ni	Hg	Cu	Pb	Cd	TC
2011	1.438	1.050	0.959	0.360	1.100	0.460	1.8	1.10	4.30	1.410	0.333	0.40	0.110	0.10	0.002	0.010	0.004	
2012	1.338	1.044	0.955	0.293	0.800	0.440	1.5	0.83	4.00	0.720	0.033	0.30	0.083	0.10	0.002	0.010	0.004	0.6
2013	1.171	0.952	0.949	0.747	0.617	0.513	1.8	1.77	4.25	0.680								3.2
2014	0.916	0.804	0.902	0.586	0.700	0.557	1.3	1.50	4.25	0.727	0.323	0.64	0.354	0.10	0.005	0.013	0.005	120
2015	1.063	0.926	0.939	0.640	0.850	0.410	3.0	2.10	4.00	0.847	0.100	0.40	0.129	0.10	0.002	0.010	0.004	57

Jaša Tomić

Year	DO	OS	pH	SS	BOD	COD_{Mn}	NO_2^-	NH_4^+	PO_4^{3-}	SO_4^{2-}	Fe	Mn	Ni	Hg	Cu	Pb	Cd	TC
2011	1.538	1.111	0.929	0.480	0.800	0.270	2.0	0.70	0.40	2.280	0.133	5.00	0.485	0.010	0.001	0.010	0.004	0.44
2012	1.091	0.981	0.948	0.772	0.665	0.290	2.9	1.20	0.80	2.346	0.047	0.82	0.143	0.100	0.001	0.010	0.004	0.1
2013	1.191	1.026	0.901	0.760	0.800	0.353	1.9	1.07	0.85	0.787								8.333
2014	1.138	0.978	0.911	3.680	0.784	0.860	1.6	0.50	3.25	0.553	0.433	0.46	0.919	0.100	0.007	0.012	0.008	29
2015	1.241	1.059	0.954	0.293	0.650	0.323	2.0	0.87	0.95	1.040	0.183	0.34	0.195	0.100	0.001	0.010	0.004	0.26
2018	1.200	1.041	0.927	0.480	0.717	0.297	1.9	0.53	1.15	1.373	0.187	0.88	0.292	0.070	0.005	0.010	0.004	0.58
2019	1.154	0.993	0.938	0.266	0.634	0.293	1.0	0.33	0.70	2.047	0.163	0.76	3.838	0.070	0.010	0.031	0.004	0.32
2020	1.295	1.011	0.906	0.413	0.517	0.247	1.7	1.80	1.35	0.793	0.590	0.68	0.108	0.070	0.002	0.010	0.004	0.62
2021	1.179	1.030	0.919	0.373	0.500	0.263	0.6	0.30	0.55	1.327	0.053	0.32	0.354	0.070	0.005	0.010	0.042	
2022	1.196	1.022	0.926	1.386	0.800	0.413	1.6	1.00	1.10	0.893	0.240	0.50	0.083	0.070	0.003	0.010	0.004	1.4

Table 8. Ratio of seasonal mean concentration and standards of the Class I of water quality for analyzed parameters (Winter).

Pančevo

Year	DO	OS	pH	SS	BOD	COD_{Mn}	NO_2^-	NH_4^+	PO_4^{3-}	SO_4^{2-}	Fe	Mn	Ni	Hg	Cu	Pb	Cd	TC
2011	1.450	0.917	0.918	0.780	1.225	0.590	1.9	3.30	5.55	2.090	0.60	0.80	0.280	0.100	0.002	0.010	0.005	
2012	1.638	1.122	0.947	0.280	1.450	0.605	2.0	1.35	3.40	1.140								
2013	1.566	1.100	0.948	0.520	0.915	0.376	2.0	1.60	4.25	0.746								
2014	1.381	1.006	0.927	1.560	0.675	0.465	1.7	0.80	3.28	0.890	0.39	0.55	0.318	0.100	0.005	0.015	0.006	
2015	1.483	1.000	0.905	2.373	1.000	0.620	1.1	0.90	2.65	0.900	0.43	0.20	0.118	0.100	0.003	0.020	0.006	

Jaša Tomić

Year	DO	OS	pH	SS	BOD	COD_{Mn}	NO_2^-	NH_4^+	PO_4^{3-}	SO_4^{2-}	Fe	Mn	Ni	Hg	Cu	Pb	Cd	TC
2011	1.581	1.044	0.918	3.400	0.800	0.330	0.9	1.75	1.95	0.500	0.600	0.40	0.160	0.100		0.012	0.005	7.6
2012	1.538	0.972	0.865	0.560	1.100	0.285	1.7	11.85	0.60	1.450	0.453	1.86	0.170	0.100	0.001	0.010	0.004	
2013	1.604	1.044	0.929	1.613	0.684	0.433	2.0	2.13	1.75	0.473								11.2
2014	1.484	1.033	0.944	0.626	0.734	0.337	1.8	0.80	1.10	0.760	0.263	0.68	0.383	0.100	0.003	0.010	0.004	12
2015	1.421	1.033	0.914	3.640	0.915	0.633	1.9	1.27	2.20	0.820	0.487	0.32	0.154	0.100	0.001	0.017	0.005	2.8
2018	1.538	1.030	0.900	2.508	0.565	0.610	1.5	1.97	1.15	0.880	0.880	1.60	0.228	0.070	0.007	0.012	0.006	1
2019	1.500	1.000	0.899	1.280	0.765	0.400	1.7	3.00	2.10	1.067	1.800	0.84	0.438	0.070	0.008	0.013	0.005	0.8
2020	1.656	1.094	0.920	0.620	1.275	0.385	2.2	6.70	0.65	1.000	0.763	1.37	0.200	0.070	0.003	0.012	0.004	3.4
2021	1.563	1.017	0.894	2.400	0.825	0.595	1.0	0.70	2.25	1.010	0.403	0.49	0.228	0.070	0.005	0.010	0.004	14
2022	1.470	0.973	0.888	1.252	1.015	0.723	1.6	3.33	1.10	0.947	0.290	1.14	0.393	0.070	0.004	0.013	0.029	

3.3. Water Pollution Index (WPI) Analysis

Table 9 shows the WPI values for the analyzed hydrological stations. According to these values, water quality at the Jaša Tomić hydrological station shows slight variations in WPI classes. Most years were rated as Class II, indicating moderate pollution, while a few years shifted to Class III, indicating an increase in pollution. Overall, pollution remained moderate throughout the analyzed period.

Notably, at the Pančevo hydrological station, the highest WPI class (Class VI) was recorded in 2015, indicating extremely high pollution. During the other years (2011–2014), the WPI classes varied between Class III and Class IV, indicating changes in pollution levels. The extremely high pollution level in 2015 highlights the urgent need for targeted remediation efforts.

Table 9. Mean annual WPI values and water quality classifications at the analyzed hydrological stations.

Hydrological Station	2011	2012	2013	2014	2015	2018	2019	2020	2021	2022
Pančevo	2.1 IV	1.1 III	1.4 III	3 IV	9.1 VI	-	-	-	-	-
Jaša Tomić	0.93 II	0.78 II	1.81 III	1.49 III	0.83 II	0.91 II	0.87 II	0.81 II	0.77 II	0.82 II

Note: "-" indicates data not available due to the cessation of monitoring.

The data clearly indicate higher pollution levels at the Pančevo station, notably due to extremely elevated total coliforms, nutrients, and organic pollutants. However, analyzing long-term pollution trends at this station is complicated because of the cessation of monitoring after 2015, which emphasizes the necessity for renewed monitoring.

The stabilization of water quality from 2015 to 2022 at the Jaša Tomić hydrological station indicates that pollution control measures may have been effective. This improvement could be attributed to better agricultural practices, improved wastewater management, or decreased industrial discharges upstream.

Previous studies of nearby rivers have obtained similar WPI results. The WPI values for the Danube River showed similar trends: the Danube River had moderate to high pollution (Class II–III), with a mean WPI of 0.84 (Class II) in 2014 at the Pančevo area, while the Karaš River showed a relatively lower pollution level (Class II, WPI = 0.75) in 2019 [63,64]. These comparisons highlight that pollution in the Tamiš River aligns with broader regional trends, underscoring the importance of integrated water resource management across the region.

The Water Pollution Index (WPI) was also calculated for each season, providing a comprehensive overview of ecological conditions. The seasonal analysis of the Water Pollution Index (WPI) at the Jaša Tomić hydrological station reveals distinct patterns. During spring, WPI values generally fall into Class II (pure) or Class III (moderately polluted), reflecting balanced ecological conditions with occasional spikes in pollution. In summer, WPI values often reach the upper range of Class III or even the lower limits of Class IV (polluted), driven by increased biological activity and nutrient enrichment due to elevated temperatures. Autumn exhibits stabilization, with WPI values predominantly within Class III, though sporadic peaks at the threshold of Class IV highlight nutrient runoff from seasonal rains. Winter, however, records elevated WPI values, frequently near or within Class IV, attributed to reduced water flow and sediment interactions, signaling a need for careful observation during this period. Detailed seasonal WPI values for Jaša Tomić hydrological station are presented in Table 10.

The seasonal analysis of the Water Pollution Index (WPI) at the Pančevo hydrological station shows clear seasonal fluctuations (Table 11). In spring, WPI values predominantly fall within Class III (moderately polluted), indicating moderate ecological balance with occasional stress. Notably, 2015 recorded a Class IV (polluted) level, reflecting increased seasonal runoff and sediment transport. In summer, values display significant variability, ranging from Class III to as high as Class VI (severely polluted) in 2015, driven by intensified biological activity, nutrient enrichment, and elevated temperatures. The extreme WPI value of 26.47 recorded in 2015, due to high total coliform concentrations (measured in n/L), is also summarized in Table 11. In autumn, a stabilization is observed, with most WPI values in Class III; however, sporadic peaks, such as the Class VI levels in 2014 and 2015, indicate nutrient runoff exacerbated by seasonal rainfall. Winter highlights relatively stable conditions, with WPI values consistently in Class III across the years analyzed, reflecting reduced biological activity and favorable oxygen conditions despite sediment dynamics. The complete seasonal WPI dataset for the Pančevo station is also included in Table 11.

Table 10. Seasonal WPI Values and Water Classifications at Jaša Tomić hydrological station.

Jaša Tomić	2011	2012	2013	2014	2015	2018	2019	2020	2021	2022
Spring	1.00	0.69	2.20	0.73	0.59	1.13	0.65	0.63	0.64	0.72
Class	II	II	IV	II	II	III	II	II	II	II
Summer	0.66	0.59	1.21	1.05	1.16	1.06	1.16	0.82	0.58	0.87
Class	II	II	III	III	III	III	III	II	II	II
Autumn	0.92	0.73	1.63	2.51	0.58	0.65	0.75	0.67	0.46	0.7
Class	II	II	III	IV	II	II	II	II	II	II
Winter	1.30	1.38	2.17	1.28	1.03	0.91	0.98	1.24	1.53	0.90
Class	II	III	IV	III	III	II	II	III	III	II

Table 11. Seasonal WPI Values and Water Classifications at the Pančevo hydrological station.

Pančevo	2011	2012	2013	2014	2015
Spring	1.38	0.79	1.36	1.91	2.1
Class	III	II	III	III	IV
Summer	2.3	0.76	1.23	1.5	26.47
Class	IV	II	III	III	VI
Autumn	0.88	0.73	1.51	7.43	4.03
Class	II	II	III	VI	VI
Winter	1.21	1.39	1.4	0.82	0.81
Class	III	III	III	II	II

This seasonal variation emphasizes the need for targeted monitoring during critical periods, particularly in summer and autumn, to mitigate pollution risks and ensure sustainable water quality management.

4. Conclusions

The consistent monitoring and analysis of water quality in the Tamiš River have highlighted significant challenges related to anthropogenic pressures, particularly urban runoff, agricultural practices, and inadequate wastewater treatment. The key findings include elevated levels of nitrites and orthophosphates at both hydrological stations, with the Pančevo hydrological station showing extreme microbial contamination in 2015. In contrast, Jaša Tomić hydrological station exhibited higher levels of heavy metals and suspended sediments, emphasizing diverse pollution sources along the river's course.

Seasonal analysis of water quality parameters and the Water Pollution Index (WPI) further underscores temporal variability and critical periods of heightened ecological vulnerability. During spring, water quality at both stations generally aligns with Class I standards, highlighting balanced ecological conditions. However, summer presents significant stress, particularly at the Pančevo hydrological station, where the WPI values peaked at an alarming 26.47 in 2015, reflecting severe microbial contamination. Autumn represents a period of stabilization, although the sporadic peaks in WPI values, especially at the Pančevo hydrological station, emphasize the impact of nutrient runoff and rainfall. Winter exhibits stability in water quality, though occasional exceedances in heavy metals and elevated WPI values at the Jaša Tomić hydrological station indicate sediment interactions and reduced flow dynamics. These findings underscore the seasonal complexity of pollution dynamics in the Tamiš River system.

This study underscores the critical importance of implementing integrated water management strategies to address these issues. Upgrading wastewater treatment facilities, enforcing stricter industrial discharge regulations, and promoting sustainable agricultural practices are essential steps to mitigate pollution. Furthermore, restoring monitoring at the Pančevo station is imperative for comprehensive long-term assessments. Taking into account that the Tamiš River is an international river and the location of Jaša Tomić hydrological station is near the Serbian–Romanian border, joint measures of Serbia and Romania should be implemented to improve water quality.

Additionally, it is essential to engage the local community in efforts to protect water quality by implementing educational initiatives, encouraging sustainable practices, and increasing awareness of how human activities affect the river. Providing incentives for farmers and industrial sectors that adopt eco-friendly technologies could help minimize pollution. Additionally, authorities should establish comprehensive long-term water management strategies, ensuring consistent river monitoring and prompt action in response to detected issues.

Protecting the Tamiš River is vital for maintaining its ecological integrity, supporting local biodiversity, and safeguarding public health. Future research should focus on identifying the impacts of specific pollution sources and evaluating the effectiveness of implemented mitigation measures. By adopting a multi-faceted approach, it is possible to ensure the sustainable use of this crucial water resource and contribute to broader ecological stability.

Author Contributions: Conceptualization, D.M.J., D.J. and T.J.G.; methodology, D.M.J.; software, D.M.J.; validation, D.M.J., D.J. and T.J.G.; formal analysis, D.M.J.; investigation, D.M.J., D.J. and T.J.G.; resources, D.M.J., D.J. and T.J.G.; data curation, D.M.J.; writing—original draft preparation, D.M.J.; writing—review and editing, D.M.J., D.J. and T.J.G.; visualization, D.M.J., D.J. and T.J.G.; supervision, D.M.J.; project administration, D.J.; funding acquisition, D.M.J., D.J. and T.J.G. All authors have read and agreed to the published version of the manuscript.

Funding: This research was funded by the Ministry of Science, Technological Development and Innovation of the Republic of Serbia (grant number 451-03-136/2025-03/200172), and the APC was funded by the Geographical Institute "Jovan Cvijić" SASA.

Data Availability Statement: The data are contained within the article.

Acknowledgments: We would like to thank our colleague Dejan Doljak, from Geographical Institute "Jovan Cvijić" SASA who helped us in map quality improvement.

Conflicts of Interest: The authors declare no conflicts of interest.

References

1. Sun, D.; Shao, S.; Zhang, Y.; Yang, Q.; Hou, H.; Quan, X. Integrated analysis of the water–energy–environmental pollutant nexus in the petrochemical industry. *Environ. Sci. Technol.* **2020**, *54*, 14830–14842. [CrossRef] [PubMed]
2. Valeev, T.K.; Rakhmanin, Y.A.; Suleimanov, R.A.; Malysheva, A.G.; Bakirov, A.B.; Rakhmatullin, N.R.; Rakhmatullina, L.R.; Daukaev, R.A.; Baktybaeva, Z.B. Experience on the environmental and hygienic assessment of water pollution in the territories referred to oil refining and petrochemical complexes. *Hyg. Sanit.* **2020**, *99*, 886–893. [CrossRef]
3. Radelyuk, I.; Tussupova, K.; Klemeš, J.J.; Persson, K.M. Oil refinery and water pollution in the context of sustainable development: Developing and developed countries. *J. Clean. Prod.* **2021**, *302*, 126987. [CrossRef]
4. Alzboon, K.; Abu Salem, Z.; Alshboul, Z.; Al-Tabbal, J.; Al Tarazi, E.; Alrawashdeh, K. Impacts of the petrochemical industries on groundwater quality. *J. Ecol. Eng.* **2024**, *25*, 286–299. [CrossRef]

5. Kostadinov, S.; Zlatic, M.; Dragicevic, S.; Novkovic, I.; Kosanin, O.; Borisavljevic, A.; Lakicevic, M.; Mladjan, D. Anthropogenic influence on erosion intensity changes in the Rasina river watershed-Central Serbia. *Fresenius Environ. Bull.* **2014**, *23*, 254–263.
6. Yu, G.; Wang, J.; Liu, L.; Li, Y.; Zhang, Y.; Wang, S. The analysis of groundwater nitrate pollution and health risk assessment in rural areas of Yantai, China. *BMC Public Health* **2020**, *20*, 437. [CrossRef]
7. Rey-Romero, D.C.; Domínguez, I.; Oviedo-Ocaña, E.R. Effect of agricultural activities on surface water quality from páramo ecosystems. Environ. *Sci. Pollut. Res.* **2022**, *29*, 83169–83190. [CrossRef]
8. Mukate, S.; Wagh, V.; Panaskar, D.; Jacobs, J.A.; Sawant, A. Development of new integrated water quality index (IWQI) model to evaluate the drinking suitability of water. *Ecol. Ind.* **2019**, *101*, 348–354. [CrossRef]
9. Njuguna, S.M.; Onyango, J.A.; Githaiga, K.B.; Gituru, R.W.; Yan, X. Application of multivariate statistical analysis and water quality index in health risk assessment by domestic use of river water. Case study of Tana River in Kenya. Process. *Saf. Environ. Prot.* **2020**, *133*, 149–158. [CrossRef]
10. Babuji, P.; Thirumalaisamy, S.; Duraisamy, K.; Periyasamy, G. Human health risks due to exposure to water pollution: A review. *Water* **2023**, *15*, 2532. [CrossRef]
11. Jakovljević, D. Serbian and Canadian water quality index of Danube river in Serbia in 2010. *J. Geogr. Inst. Jovan Cvijić SASA* **2012**, *62*, 1–18. [CrossRef]
12. Li, R.; Zou, Z.; Yan, A. Water quality assessment in Qu River based on fuzzy water pollution index method. *Acta Sci. Circumst.* **2016**, *50*, 87–92. [CrossRef]
13. Milijašević Joksimović, D.; Gavrilović, B.; Obradović Lović, S. Application of the water quality index in the Timok River basin (Serbia). *J. Geogr. Inst. Jovan Cvijić SASA* **2018**, *68*, 333–344. [CrossRef]
14. Mladenović-Ranisavljević, I.I.; Žerajić, S.A. Comparison of different models of water quality index in the assessment of surface water quality. *Int. J. Environ. Sci. Technol.* **2018**, *15*, 665–674. [CrossRef]
15. Ismail, A.H.; Robescu, D. Assessment of water quality of the Danube River using water quality indices technique. *Environ. Eng. Manag. J.* **2019**, *18*, 1727–1737. [CrossRef]
16. Kachroud, M.; Trolard, F.; Kefi, M.; Jebari, S.; Bourrié, G. Water quality indices: Challenges and application limits in the literature. *Water* **2019**, *11*, 361. [CrossRef]
17. Banda, T.D.; Kumarasamy, V.M. Development of water quality indices (WQIs): A review. *Pol. J. Environ. Stud.* **2020**, *29*, 2011–2021. [CrossRef]
18. Calmuc, M.; Calmuc, V.; Arseni, M.; Topa, C.; Timofti, M.; Georgescu, L.P.; Iticescu, C. A comparative approach to a series of physico-chemical quality indices used in assessing water quality in the Lower Danube. *Water* **2020**, *12*, 3239. [CrossRef]
19. Milanović Pešić, A.; Brankov, J.; Milijašević Joksimović, D. Water quality assessment and populations' perceptions in the National park Djerdap (Serbia): Key factors affecting the environment. *Environ. Dev. Sustain.* **2020**, *22*, 2365–2383. [CrossRef]
20. Khanday, S.A.; Bhat, S.U.; Islam, S.T.; Sabha, I. Identifying lithogenic and anthropogenic factors responsible for spatio-seasonal patterns and quality evaluation of snow melt waters of the River Jhelum Basin in Kashmir Himalaya. *Catena* **2021**, *196*, 104853. [CrossRef]
21. Bouimouass, H.; Fakir, Y.; Tweed, S.; Sahraoui, H.; Leblanc, M.; Chehbouni, A. Traditional irrigation practices sustain groundwater quality in a semiarid piedmont. *Catena* **2022**, *210*, 105923. [CrossRef]
22. Qi, J.; Yang, L.; Liu, E. A holistic framework of water quality evaluation using water quality index (wqi) in the Yihe river (China). *Environ. Sci. Pollut.* **2022**, *29*, 80937–80951. [CrossRef]
23. Khan, M.H.R.B.; Ahsan, A.; Imteaz, M.; Shafiquzzaman, M.; Al-Ansari, N. Evaluation of the surface water quality using global water quality index (WQI) models: Perspective of river water pollution. *Sci. Rep.* **2023**, *13*, 20454. [CrossRef]
24. Lukhabi, D.K.; Mensah, P.K.; Asare, N.K.; Pulumuka-Kamanga, T.; Ouma, K.O. Adapted water quality indices: Limitations and potential for water quality monitoring in Africa. *Water* **2023**, *15*, 1736. [CrossRef]
25. Patel, P.S.; Pandya, D.M.; Shah, M. A systematic and comparative study of Water Quality Index (WQI) for groundwater quality analysis and assessment. Environ. *Sci. Pollut. Res.* **2023**, *30*, 54303–54323. [CrossRef]
26. Dippong, T.; Mihali, C.; Avram, A. Evaluating Groundwater Metal and Arsenic Content in Piatra, North-West of Romania. *Water* **2024**, *16*, 539. [CrossRef]
27. Khazheeva, Z.I.; Plyusnin, A.M.; Dampilova, B.V. Changes in water quality in the Modonkul River assessed by combinatory pollution index. *Water Resour.* **2024**, *51*, 513–524. [CrossRef]
28. Dass, B.; Rao, M.S.; Sen, S. Hydrogeochemical characterization and water quality assessment of mountain springs: Insights for strategizing water management in the lesser Indian Himalayas. *J. Hydrol. Reg. Stud.* **2025**, *57*, 102126. [CrossRef]
29. Xiao, W.; Ren, B.; Wu, B.; Deng, X. Spatiotemporal dynamics and optimization of water quality assessment in the Nantong section of the Yangtze River Basin: A WQImin approach. *J. Hydrol. Reg. Stud.* **2025**, *57*, 102106. [CrossRef]
30. Li, J.; Meng, X.; Zhang, Y.; Li, J.; Xia, L.; Zheng, H. Analysis of the temporal and spatial distribution of water quality in China's major river basins, and trends between 2005 and 2010. *Front. Earth Sci.* **2015**, *9*, 463–472. [CrossRef]

31. Zhang, Q.; Yu, R.; Jin, Y.; Zhang, Z.; Liu, X.; Xue, H.; Hao, Y.; Wang, L. Temporal and Spatial Variation Trends in Water Quality Based on the WPI Index in the Shallow Lake of an Arid Area: A Case Study of Lake Ulansuhai, China. *Water* **2019**, *11*, 1410. [CrossRef]
32. Widodo, T.; Budiastuti, M.T.S.; Komariah, K. Water Quality and Pollution Index in the Grenjeng River, Boyolali Regency, Indonesia. *Caraka Tani J. Sustain. Agric.* **2019**, *34*, 150–161. [CrossRef]
33. Hossain, M.; Patra, P.K. Water pollution index—A new integrated approach to rank water quality. *Ecol. Indic.* **2020**, *117*, 106668. [CrossRef]
34. Singh, K.K.K.M.B.; Singh, L.C.; Singh, N.D. Analysis of Seasonal Variation in Water Pollution Index of Nambul River, Manipur, North-East, India. *Int. J. Lakes Rivers* **2023**, *16*, 141–158. [CrossRef]
35. Tran, T.T.; Liem, N.D.; Hieu, H.H.; Tam, H.T.; Mong, N.V.; Yen, N.T.M.; Yen, T.T.H.; Quang, N.X.; Luu, P.T. Assessment of long-term surface water quality in Mekong River estuaries using a comprehensive water pollution index. *Environ. Nat. Resour. J.* **2023**, *21*, 524–533. [CrossRef]
36. Jakovljević, D.; Milanović Pešić, A.; Miljanović, D. Human impacts on water resources in the Lower Danube River Basin in Serbia. In *The Lower Danube River: Hydro-Environmental Issues and Sustainability*; Negm, A., Zaharia, L., Ioana-Toroimac, G., Eds.; Springer: Cham, Switzerland, 2022; pp. 195–251. [CrossRef]
37. Popović, N.; Đuknić, J.; Čanak Atlagić, J.; Raković, M.; Marinković, N.; Tubić, B.; Paunović, M. Application of the Water Pollution Index in the Assessment of the Ecological Status of Rivers: A Case Study of the Sava River, Serbia. *Acta Zool. Bulg.* **2016**, *68*, 97–102.
38. Brankov, J.; Milijašević, D.; Milanović, A. The assessment of the surface water quality using the water pollution index: A case study of the Timok river (The Danube River basin), Serbia. *Arch. Environ. Prot.* **2012**, *38*, 49–61. [CrossRef]
39. Milijašević Joksimović, D.; Vuletić, M. Assessing surface water quality of the Porečka River. In Proceedings of the Zbornik Radova—VI Kongres Geografa Srbije sa Medunarodnim Ucešcem, Zlatibor, Serbia, 29–31 August 2024; 2024. [CrossRef]
40. Milanović, A.; Milijašević, D.; Brankov, J. Assessment of polluting effects and surface water quality using water pollution index: A case study of hydro-system Danube-Tisa-Danube, Serbia. *Carpath. J. Earth Environ. Sci.* **2011**, *6*, 269–277.
41. Milanović Pešić, A.; Jojić Glavonjić, T.; Denda, S.; Jakovljević, D. Sustainable Tourism Development and Ramsar Sites in Serbia: Exploring Residents' Attitudes and Water Quality Assessment in the Vlasina Protected Area. *Sustainability* **2023**, *15*, 15391. [CrossRef]
42. Filimon, M.N.; Gotia, S.R.; Gotia, S.L.; Borozan, A.; Popescu, R.; Gherman, V.D. Bacteriological studies on river Timis with a role in evaluating pollution. *Ann. Rom. Soc. Cell Biol.* **2009**, *14*, 284–288.
43. Babović, N.; Marković, D.; Dimitrijević, V.; Marković, D. Some indicators of water quality of the Tamiš river. *Chem. Ind. Chem. Eng. Q.* **2011**, *17*, 107–115. [CrossRef]
44. Lujić, J.; Matavulj, M.; Poleksić, V.; Rašković, B.; Marinović, Z.; Kostić, D.; Miljanović, B. Gill reaction to pollutants from the Tamiš River in three freshwater fish species, *Esox lucius* L. 1758, *Sander lucioperca* (L. 1758) and *Silurus glanis* L. 1758: A comparative study. *Anat. Histol. Embryol.* **2015**, *44*, 128–137. [CrossRef] [PubMed]
45. Institute for Nature Conservation of Serbia. Pregled Zaštićenih Područja. 2024. (In Serbian). Available online: https://zzps.rs/wp-content/uploads/2024/10/11-12-2023-Pregled-zasticenih-podrucja-RS.docx (accessed on 3 December 2024).
46. Gavrilović, L.; Dukić, D. *Reke Srbije*; Zavod za udžbenike: Belgrade, Serbia, 2002. (In Serbian)
47. Republic Hydrometeorological Service of Serbia. Hidrološki Godišnjaci—1. Površinske Vode, 2003–2022. (In Serbian). Available online: https://www.hidmet.gov.rs/latin/hidrologija/povrsinske_godisnjaci.php (accessed on 3 December 2024).
48. Serbian Environmental Protection Agency. Rezultati Ispitivanja Kvaliteta Površinskih i Podzemnih Voda, 2011–2015, 2018–2022. (In Serbian). Available online: https://sepa.gov.rs/publikacije/ (accessed on 3 December 2024).
49. Serbian Environmental Protection Agency. Available online: https://sepa.gov.rs/wp-content/uploads/2024/10/Veliki-zagadjivaci-2023.htm (accessed on 12 March 2025).
50. Google. Google Maps Satellite Imagery. Available online: https://www.google.com/maps (accessed on 16 March 2025).
51. Marković, D.; Dražić, M.; Mitić, M.; Babović, N.; Jakovljev, Z.; Marković, D.; Dimitrijević, V.; Aranđelović, M.; Đorđević, S.; Aleksić, D.; et al. *Integralni Katastar Zagađivača Reke Tamiš*; Faculty of Applied Ecology Futura, Singidunum University: Belgrade, Serbia, 2010.
52. Official Gazette of the Republic of Serbia, No. 93/2023. Uredba o Proglašenju Predela Izuzetnih Odlika "Potamišje". (In Serbian). Available online: https://pravno-informacioni-sistem.rs/eli/rep/sgrs/vlada/uredba/2023/93/5 (accessed on 3 December 2024).
53. Lyulko, I.; Ambalova, T.; Vasiljeva, T. To integrated water quality assessment in Latvia. In *MTM (Monitoring Tailor-Made) III, Proceedings of International Workshop on Information for Sustainable Water Management*; Institute for Inland Water Management and Waste Water Treatment (RIZA): Lelystad, The Netherlands, 2000.
54. Official Gazette of the RS, No 24/2014. Uredba o Graničnim Vrednostima Prioritetnih i Prioritetnih Hazardnih Supstanci Koje Zagađuju Površinske Vode i Rokovima za Njihovo Dostizanje. (In Serbian). Available online: https://pravno-informacioni-sistem.rs/eli/rep/sgrs/vlada/uredba/2014/24/3/reg (accessed on 3 December 2024).

55. Official Gazette of the RS, No 50/2012. Uredba o Graničnim Vrednostima Zagađujućih Materija u Površinskim i Podzemnim Vodama i Sedimentu i Rokovima za Njihovo Dostizanje. (In Serbian). Available online: https://pravno-informacioni-sistem.rs/eli/rep/sgrs/vlada/uredba/2012/50/1/reg (accessed on 3 December 2024).
56. Official Gazette of the RS, No. 74/2011 Pravilnik o Parametrima Ekološkog i Hemijskog Statusa Površinskih Voda i Parametrima Hemijskog i Kvantitativnog Statusa Podzemnih Voda. (In Serbian). Available online: https://pravno-informacioni-sistem.rs/eli/rep/sgrs/ministarstva/pravilnik/2011/74/4/reg (accessed on 3 December 2024).
57. Official Gazete of the SRS, No 31/1982. Pravilnik o Opasnim Materijama u Vodama. (In Serbian). Available online: https://pravno-informacioni-sistem.rs/eli/rep/sgsrs/ministarstva/pravilnik/1982/31/1 (accessed on 3 December 2024).
58. Bilotta, G.S.; Brazier, R.E. Understanding the influence of suspended solids on water quality and aquatic biota. *Water Res.* **2008**, *42*, 2849–2861. [CrossRef] [PubMed]
59. Wang, J.; Tang, H.; Zhang, X.; Xue, X.; Zhu, X.; Chen, Y.; Yang, Z. Mitigation of nitrite toxicity by increased salinity is associated with multiple physiological responses: A case study using an economically important model species, the juvenile obscure puffer (*Takifugu obscurus*). *Environ. Pollut.* **2018**, *232*, 137–145. [CrossRef]
60. Ministry of Agriculture, Forestry and Water Management of the Republic of Serbia. Plan Vodoprivrede na Teritoriji Republike Srbije za Period 2021. do 2027. Godine, 2021. (In Serbian). Available online: http://www.minpolj.gov.rs/download/Plan_upravljanja_-vodama_do_2027-FINAL.pdf?script=lat (accessed on 3 December 2024).
61. Ding, T.T.; Du, S.L.; Huang, Z.Y.; Wang, Z.J.; Zhang, J.; Zhang, Y.H.; Liu, S.S.; He, L.S. Water quality criteria and ecological risk assessment for ammonia in the Shaying River Basin, China. *Ecotox. Environ Safe.* **2021**, *215*, 112141. [CrossRef]
62. Biliani, I.; Tsavatopoulou, V.; Zacharias, I. Comparative Study of Ammonium and Orthophosphate Removal Efficiency with Natural and Modified Clay-Based Materials, for Sustainable Management of Eutrophic Water Bodies. *Sustainability* **2024**, *16*, 10214. [CrossRef]
63. Takić, L.; Mladenović-Ranisavljević, I.; Vasović, D.; Đorđević, L. The Assessment of the Danube River Water Pollution in Serbia. *Water Air Soil Pollut.* **2017**, *228*, 380. [CrossRef]
64. Popović, N.; Raković, M.; Marinković, N.; Đuknić, J.; Tubić, B.; Čanak Atlagić, J.; Paunović, M. Assessment of Surface Water Quality Using the Water Pollution Index (WPI)—Possibilities and Advantages. In Proceedings of the Conference Proceedings of the 51st Annual Conference of the Serbian Water Pollution Control Society Water 2022, Vrnjačka Banja, Serbia, 26–28 October 2022.

Disclaimer/Publisher's Note: The statements, opinions and data contained in all publications are solely those of the individual author(s) and contributor(s) and not of MDPI and/or the editor(s). MDPI and/or the editor(s) disclaim responsibility for any injury to people or property resulting from any ideas, methods, instructions or products referred to in the content.

water

MDPI

Article

Microstructural Comparison of the Mineralization Within Borsec and Tusnad Public Springs

Simona Elena Avram [1], Lucian Barbu Tudoran [2,3], Gheorghe Borodi [3] and Ioan Petean [4,*]

[1] Faculty of Materials and Environmental Engineering, Technical University of Cluj-Napoca, 103-105 Muncii Bd., 400641 Cluj-Napoca, Romania; simona.avram@imadd.utcluj.ro
[2] Faculty of Biology and Geology, Babes-Bolyai University, 44 Gheorghe Bilaşcu Street, 400015 Cluj-Napoca, Romania; lucian.barbu@ubbcluj.ro
[3] National Institute for Research and Development of Isotopic and Molecular Technologies, 65-103 Donath Street, 400293 Cluj-Napoca, Romania; borodi@itim-cj.ro
[4] Faculty of Chemistry and Chemical Engineering, Babes-Bolyai University, 11 Arany Janos Street, 400028 Cluj-Napoca, Romania
* Correspondence: ioan.petean@ubbcluj.ro

Abstract

Mineral water content strongly depends on the geologic layer characteristics. Therefore, the aim of the present study is to make a comparison between two renowned mineral water sources in Romania, Borsec and Tusnad. Two public springs were selected from each location: Boldizsar (about 6600 L/day) and Lazar (about 500 L/day) from Borsec and Mikes (about 5000 L/day) and Young's spring (about 600 L/day) from Tusnad. All investigated springs are naturally carbonated. Water properties were measured in situ and in laboratory for the collected samples; the results found that Borsec mineral water has a pH of about 7.5, while Tusnad mineral water is slightly acid (pH = 6.5). TDS strongly depends on the spring's flow (for instance, Boldizsar has a TDS of about 900 mg/L, while Lazar has a TDS of about 1529 mg/L due to its high mineralization, while Young's spring has a TDS of 165 mg/L due to its low mineralization, although it has low flow). Borsec mineral water has a lower salinity of about 1.22 PSU, while Tusnad water has a salinity of about 2 PSU, caused by a high amount of Na and Fe ions. Mineral waters dissolve ions from the geological layers, which react with carbonic acid during drying, generating specific crystallized compounds. The crystallized matter was investigated using XRD coupled with mineralogical optical microscopy (MOM); their microstructural features were observed using SEM coupled with elemental spectroscopy. Borsec water generates mainly Ca, Mg, and Na minerals like calcite, aragonite, pseudo-dolomite, natron, and traces of halite. Tusnad mineral waters have significant amounts of Ca, but also have Fe and much more Cl, since calcite and aragonite are mixed up with large amounts of halite and iron compounds. It looks like the presence of iron ions in the Tusnad mineral water collected from Mikes and Young's spring explains the acidic pH. All these aspects are useful for further investigation regarding specific therapeutic purposes like chronic colitis and biliary lithiasis symptom amelioration (Boldizsar), chronic colitis, and enterocolitis symptoms (Lazar). Tusnad waters, like the water from Mikes spring, are recommended for anemia and neurasthenia, while Young's spring is recommended for renal lithiasis amelioration.

Keywords: Borsec; Tusnad; mineral water; crystallization

1. Introduction

The deeper water deposit interaction with geological layers causes the dissolving of hydro-soluble minerals like carbonates or evaporites, generating mineral waters which also might contain rare earth elements [1,2]. The excessive presence of evaporitic deposits interacting with water generates brines with different concentrations depending on the local geologic conditions [3,4]. Mineral waters are marketed as bottled water, representing a wide share of the worldwide bottled water consumption. For instance, Parag et al. report that 443 billion liters of bottled water were consumed worldwide in 2020 according to Statista [5]. Following this path, we found that Statista reported the consumption of about 39.44 billion liters in China and 38.34 billion liters in United States of America, followed by Mexico with 31.15 billion liters of consumed bottled water [6]. The increased worldwide demand for mineral water and the huge amount bottled yearly raise a great interest regarding their mineralization and its complex implication in drinking behavior and possible therapeutic effects.

Novel approaches have appeared regarding the mineral content within mineral water and geothermal water that imply the extraction of a powder to be used further for specific purposes. For instance, the famous Onsen geothermal waters from Japan [7] are now transformed into a mineral powder by Le Furo Company and used as a powder for mineral baths [8]. One of our previous studies found a similar possibility for collecting mineral powder from Borsec (Romania) public springs through water evaporation, finding that it could be used in a similar manner for personalized baths [9].

Natural carbogaseous mineral waters are formed by the interaction of volcanic gaseous emissions, which interact with the mineral water formed in contact with the geological layers [10,11]. The rock fissures generated by tectonic movements facilitate the movement of deeper mineralized water and volcanic gases to rich upper layers, merging with other aquifers and increasing their mineral content and carbogaseous characteristics [10]. The literature data reveal that CO_2, N_2, and CH_4 are the gases dissolved into the natural carbogaseous waters [11].

Romania has several locations with geothermal and/or mineral water, due to the geological conditions of the Carpathian Mountains and their sub mountain areas. Two locations, Borsec and Tusnad, are renowned for their natural carbogaseous waters for drinking and spa treatments. Both are situated in the Eastern Carpathians (Romania), with a distance of about 100 km. Borsec is situated on the eastern side of Giurgeului Mountain and Tusnad is situated on the western side of the Ciomadu Mare Mountain (a former volcano).

A geological survey found two main aquifers in the Borsec depression: the first occurs at 65 m depth and is formed on a compact dark dolomitic base, with the upper layers containing gray clay, shale, and sand layers; the second aquifer occurs at 60 m depth and is formed on a dolomitic limestone substrate, with the upper layers containing a green shale (chlorite) intercalation with sand and dolomitic limestone [12].

The literature data indicate that the Ciomadu Mare Mountain and Tusnad Bath geologies were investigated by Lazar and Arghiz in 1964 and by Screiber in 1972 [13,14]. The local volcanic activity in the Pliocene era had three or four massive eruptions which raised the local mountain peaks, especially Ciomadu Mare Mountain, causing large deposits of volcanic ash and andesitic lava. The later eruptions involved a more viscous andesitic lava, which intercalated within the sedimentary formations [14]. Unfortunately, there are no available public data regarding the aquifer situation, but several carbogaseous mineral water springs were captured by specific drillings and are publicly available at the Tusnad Bath touristic station. The diffuse emission of CO_2 within the Ciomadu Mare Mountain cone's respective Sf. Ana crater has a variable flux ranging from 2 to 90 g/m^2 [15]. The literature indicates carbon dioxide sources from the endogeneous degradation of carbonaceous rocks [16] or from the magma gases trespassing through the tectonic regions and their subsequent fractures [17]. However, the gaseous emissions related to the Sf. Ana crater contain 97.69% CO_2, followed by small quantities of nitrogen, methane, oxygen, argon, and helium [18]. These aspects are related to the carbogaseous behavior of the mineral waters for the public springs at the Tusnad Bath touristic station.

Carbonated rocks within sedimentary layers in both Borsec and Tusnad contain calcite and dolomite, having a water solubility of 0.013 g/L [19,20] and being able to release Ca^{2+} and Mg^{2+} ions. On the other hand, some investigations regarding dispersed particulate matter in the Sf. Ana volcanic lake revealed a significant amount of biotite and hornblende mixed with feldspar and muscovite [21]. Such clay minerals might be able to release iron ions when in contact with water through the reduction of Fe (III) within their structure via a complex mechanism [22]. In consequence, it is expected that the mineral water formed in the Tusnad Bath has an increased iron content compared to the Borsec mineral waters.

Considering the progress achieved by Le Furo Company with Onsen water, there remains a lack of information regarding the crystallization behavior of Borsec and Tusnad mineral waters. Therefore, the aim of the current research is to compare the crystallization processes of Borsec and Tusnad mineral waters.

Therefore, a comparison between some relevant springs from Borsec and Tusnad is of great interest to reveal their mineralization as a consequence of water evaporation. Since the spring flow influences the mineral concentration, it should be considered as the main parameter for selecting the springs. Therefore, the study should compare a high-flow spring from Borsec with a similar one from Tusnad, with the same comparison for a low-flow spring. Our study indicates that dissolved CO_2 interacts directly with dissolved ions, facilitating the precipitation of crystallization germs, which further form crystalline powders [9]. Therefore, the physicochemical properties of the collected water samples must be measured, and their values must be correlated with the crystallization process. Mineralogical optical microscopy (MOM) effectuated in cross-polarized light allows for a proper view of each crystalline particle and ensures a good fit with the X-ray diffraction (XRD) results. The morphological aspects are optimally observed by Scanning Electron Microscopy (SEM) coupled with Energy Dispersive Spectroscopy (EDS), which allows the measuring of the elemental composition of certain microscopic sites and also allows the viewing of an elemental distribution map. Finally, the obtained microstructural data are useful for the identification and potential application of the obtained powder for further treatments.

2. Materials and Methods

2.1. Water Sample Collection and Storage

The selected springs, having a flow of about 5000–6000 L/day, are Boldizsar in Borsec and Mikes in Tusnad. The selected springs having a low flow (e.g., about 500 L/day) are Lazar in Borsec and Young's in Tusnad. Their position is marked on the maps displayed in Figure 1.

Each of the selected springs has displayed at its entrance the chemical composition determined by the National Mineral Water Company of Romania [23]; these values are centralized in Table 1. The specific values for Young's spring were not made public, and therefore the specific values cannot be displayed in Table 1.

The sampling campaign was held on 21–22 April 2025 starting in Borsec and finishing in Tusnad. The water samples were collected directly from the spring by filling glass vials of 1 L equipped with a metallic cap, being properly labeled with the sample's name. Three individual samples were collected for each spring.

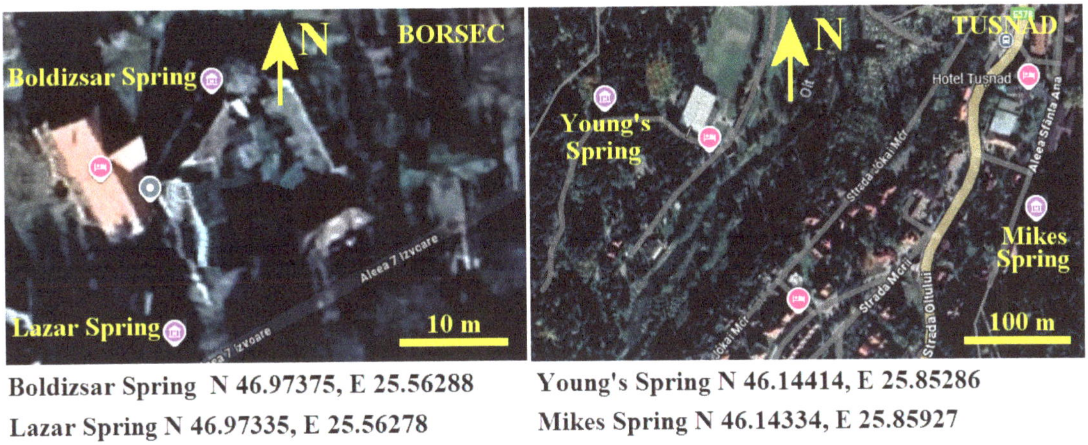

Figure 1. Investigated spring positioning on Borsec and Tusnad maps [Google Maps 2025]; the precise GPS position of each spring is displayed below maps.

Table 1. Spring mineral composition according to the public declaration displayed nearby spring.

Parameter	Measuring Unit	Boldizsar	Lazar	Mikes
pH	pH	6.25	6.43	6.0
CO_3^{2-}	mg/L	1470	2430	378.2
CO_2 free	mg/L	1310	1450	1397.7
Ca^{2+}	mg/L	250	465	91.4
Fe^{2+}	mg/L	2.84	0.194	25.7
Mg^{2+}	mg/L	78.4	163	35.5
K^{+}	mg/L	17.3	29	74.0
Na^{+}	mg/L	64.4	111	300.9
Si^{4+}	mg/L	18.8	28.5	103.6
Cl^{-}	mg/L	bdl	bdl	509.9
SO_4^{2-}	mg/L	bdl	bdl	6.7
NH_4	mg/L	bdl	bdl	3.0
Mn^{2+}	mg/L	bdl	bdl	bdl
BO_2^{-}	mg/L	bdl	bdl	89.1

bdl—below detection limit.

They were stored in a refrigerating container at 4 °C immediately after collection and transported to the laboratory where they were tested. The vials were gently brought to room temperature before measuring the water properties.

Glass slides for microscopy were prepared by deposing about three drops on the fresh surface of the lamella, followed by a primary inspection after 6 h to reveal the incipient stage of crystallization, followed by the final stage after 24 h when the water is completely evaporated and the crystallized matter is completely formed.

Larger amounts of water samples were placed in Petri dishes for longer-time crystallization in order to obtain thicker crystallized deposits, which were inspected microscopically before their erasing to obtain powder for XRD and SEM–EDX investigations.

2.2. Investigation Methods

Water physicochemical properties like pH, electrical conductivity, total dissolved solids (TDS), and salinity were measured with a HANNA HI 9829 multi parameter measuring device (Hanna Instruments Co., Bedfordshire, UK). Turbidity was measured with a nephelometric turbidimeter Hanna 93703 (Hanna Instruments Co., Bedfordshire, UK) with an accuracy of 0–1000 FTU (NTU) and a resolution of ±5% F.S. Turbidity was measured immediately after opening the sample's vial; since the reading value was influenced by the CO_2 bubbles, the measurement was repeated after complete degassing of the sample. Total chloride was measured using photo-colorimeter detection performed on an HI 96711 (Hanna Instruments Co., Bedfordshire, UK) device having a resolution of 0.01 mg/L. Each parameter was measured in triplicate (for each collected sample), and the mean values were calculated along with the standard deviation. The values were statistically analyzed by Anova, followed by a Tukey post hoc test, at a significance level a = 0.05 using Origin 6.0 software, which was used for the graphical representations (Microcal Co., Amherst, MA, USA).

Mineralogical optical microscopy (MOM) was effectuated under cross-polarized light inspection on a Laboval 2 microscope (Carl Zeiss, Oberkochen, Germany) having a digital image capture system with resolution of 10 MPx (Samsung Co., Hangul, Republic of Korea).

X-ray diffraction (XRD) was performed on a Bruker D8 Advance, produced by Bruker Company (Karlsruhe, Germany), diffractometer having a Bragg–Brentano system. It was operated using Cu kα monochromatic radiation (1.540562 Å) in the range of 10° to 80° at a speed of 1°/min. The XRD peak identification was effectuated with Match 1.0 software produced by the Crystal Impact Company (Bonn, Germany) using the PDF 2.0 database.

Scanning Electron Microscopy (SEM) was performed with a Hitachi SU8230 Scanning Electron Microscope (SEM) (Hitachi Co., Tokyo, Japan) operated in high vacuum mode at an acceleration voltage of 30 kV. The Energy Dispersive Spectroscopy (EDS) was performed with an X-Max 1160 detector (Oxford Instruments, Oxford, UK) installed on the SEM microscope. The sample surfaces were coated with a thin layer of Au for ensuring a proper electrical conduction; this component was subtracted from the elemental analysis.

3. Results

3.1. Water Sample Properties

The key role parameters of the water sample mineralization are the spring's water flow and its dissolved mineral amounts. It was observed that the higher flux of fresh water infiltrating into the aquifer dilutes the dissolved mineral concentration, affecting the water properties. Thus, a spring with a high flow is supported by larger amounts of fresh-water infiltrations, and subsequently has a milder mineralization compared to the springs with a lower flow, such as our previous observation in Borsec regarding the Boldizsar and Lazar springs [9].

This supposition is assumed as a basic hypothesis for the current investigation. Therefore, electrical conductivity of the water samples is affected by their total dissolved solids (TDS) and salinity. Moreover, the total chloride amount is a quantitative measure of the dissolved halite (sodium chloride crystallized in cubic system). The mineral types and their dissolution influence the water samples' pH values, indicating certain characteristics of the source area.

The variation in pH reveals two distinct statistical groups, as shown in Figure 2a. The first group is formed by the slightly alkaline waters collected from the Boldizsar and Lazar springs from Borsec. The second relevant statistical group is formed by the waters collected from Mikes and Young's spring in Tusnad, having a slightly acidic pH. Although their values are very close to the neutral pH of 7, the observed differences are statistically representative. It should be further noted that Borsec and Tusnad waters have a different chemistry, induced by the specific dissolved minerals.

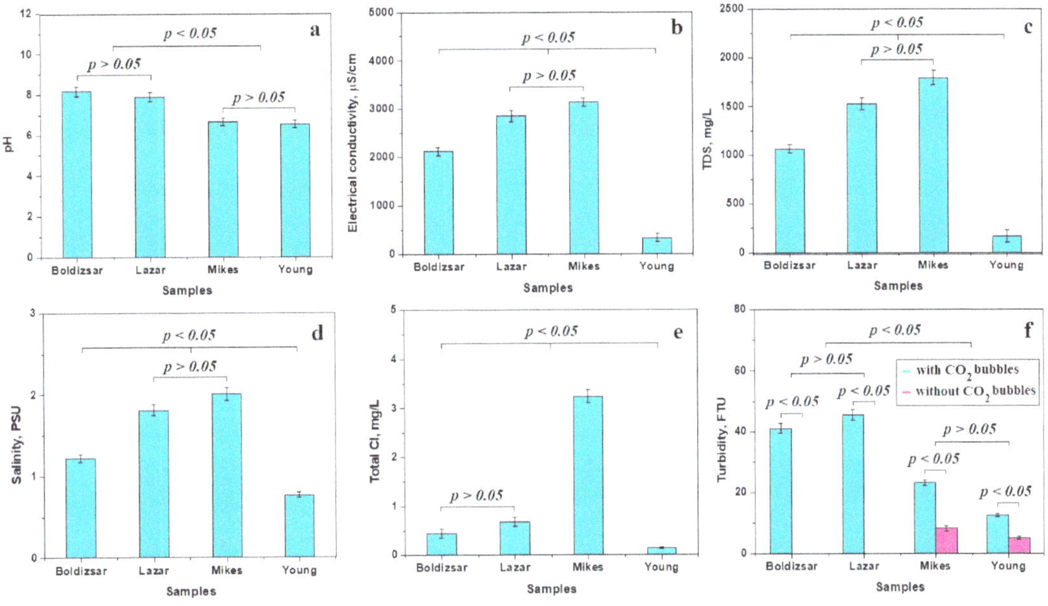

Figure 2. Variation in the water properties mean values: (**a**) pH, (**b**) electrical conductivity, (**c**) total dissolved solids, (**d**) salinity, (**e**) total Cl, and (**f**) turbidity.

Electrical conductivity, Figure 2b, reveals very interesting values that form three different statistical groups. The first group is formed by the Boldizsar sample, having a mean electrical conductivity of 2121 µS/cm; this is an increased value compared to the regular drinking water, but it is still moderate compared to the second statistical group including the Lazar and Mikes samples. These water samples have a higher electrical conductivity of about 3000 µS/cm. This is a puzzling finding, since the Lazar spring has a low flow and the Mikes spring has a high flow. Finally, the third relevant statistical group formed by Young's spring, with a lower electrical conductivity of about 329 µS/cm, is a more puzzling result. At first sight, it looks like the Tusnad waters have an opposite behavior to the Borsec waters regarding the mineralization behavior dependence on the spring's flow. This mysterious reverted behavior could be explained by the other specific physicochemical property variations.

The total dissolved solids (TDS) is a quantitative value summing all dissolved minerals like carbonate rocks and evaporites. It cannot show the difference between dissolved matter, but clearly indicates the mineralization level of the investigated waters. Figure 2c reveals a moderate mineralization within the Boldizsar spring due to its high flow, making it a statistical relevant group. Lazar and Mikes springs have higher TDS values, indicating an increased mineralization in the second statistically relevant group. These high values are due to the low flow of the Lazar spring, but indicate a large amount of hydro-soluble salt dissolved in the Mikes spring regardless of its high flow. Young's spring has the lowest TDS, indicating a low mineralization despite its reduced flow.

Salinity values, shown in Figure 2d, are complementary to TDS, revealing the presence of dissolved salts, including carbonates and halite, without an ability to make a difference. Still, three relevant statistical groups are evidenced for water salinity: the first one, having moderate salinity, consists of Boldizsar spring; the group having a high salinity contains Lazar and Mikes springs; and the third group, with a low salinity, contains Young's spring.

The HCO_3^- anion influences the samples' salinity by its interaction with other species, facilitating carbonate precipitation, but the lack of a proper measuring kit during this stage of research left us relying on the public declaration in Table 1.

The total chloride is strictly related to dissolved chlorides such as NaCl, since it is natural water which was not treated by chlorination. Thus, the explanation starts to be revealed in Figure 2e. The first statistical group contains water samples collected from Boldizsar and Lazar springs, having a moderate amount of Cl. The second relevant statistical group contains only the water collected from Mikes spring, which has a very high amount of chloride. On the other hand, the Young's spring water has a very low chloride amount, forming the third relevant statistical group. Significant differences were observed between these three statistical groups, $p < 0.05$. This clearly indicates an uneven availability of halite deposits within Mikes and Young's springs that might indicate their sourcing in different geological layers.

Gas bubbles within the carbogaseous waters scatter the incident light beam of the nephelometric sensor for measuring turbidity, increasing its value [24,25]. Therefore, the water tested from Borsec has an increased turbidity of about 40 FTU in its initial state and a turbidity below the detection limit after complete degassing, revealing a significant statistical difference between the initial and final state, $p < 0.05$, as shown in Figure 2f. Both the Boldizsar and Lazar springs feature similar values and behavior, and form a relevant statistical group. The Tusnad samples have a lower turbidity of about 15–20 FTU in their initial state, which decreases significantly after the complete removal of CO_2 bubbles to about 5.15 to 8.24 FTU. The statistical difference between the initial and final state of turbidity is significant, $p < 0.05$, and also indicates that the remnant turbidity is generated by solid micro-dispersoids within both Mikes and Young's waters. Both springs investigated from Tusnad have the same turbidity behavior, representing a distinct statistical group. The turbidity difference between Borsec and Tusnad is evidenced by the statistical analysis value $p < 0.05$. Finally, the solid dispersoids within Tusnad water might be responsible for the slightly acidic pH.

3.2. *Mineral Crystallization Assessement*

The crystallized powders obtained by natural drying of the water samples were subjected to an XRD investigation. The Borsec samples, shown in Figure 3a, reveal well-developed diffraction peaks with narrow aspects and intensities depending on the amount of each mineral corresponding to a high crystalline characteristic. The mineralization within both Borsec springs is dominated by calcite, pseudo-dolomite, and aragonite, while natron and halite have low intensities and appear as traces.

The Tusnad samples, shown in Figure 3b, also reveal well-developed peaks with narrow aspects and strong intensities corresponding to the crystallized minerals, but also have some nanocrystaline components, most likely related to the pre-existing dispersoids detected by the remnant turbidity. Thus, the Tusnad springs have a divergent dominance of minerals: the Mikes spring is dominated by halite, followed by sylvite, calcite, and pseudo-dolomite as accompanying minerals. On the other hand, Young's spring is dominated by calcite as a main mineral, followed by pseudo-dolomite, and having only small traces of halite. A small amount of amorphous or nanocrystaline components influence the baseline of the Mikes XRD pattern in Figure 3b and significantly increased that of Young's spring by forming of an amorphous bump ranging from about 18° to 22°. These nanocrystaline features are convergent with the observations for the orange sediment precipitated out of the Ciomadu Mare Mountain slope beside the Mikes spring location. Its XRD pattern shows broadened peaks of goethite, indicating the presence of a nanostructural iron hydroxide which corresponds to the amorphous–nanocrystaline features observed in the Mikes and

Young's springs. The iron hydroxide nanoparticles are about 50 nm according to the Scherrer formula, which was applied to the relevant XRD peaks.

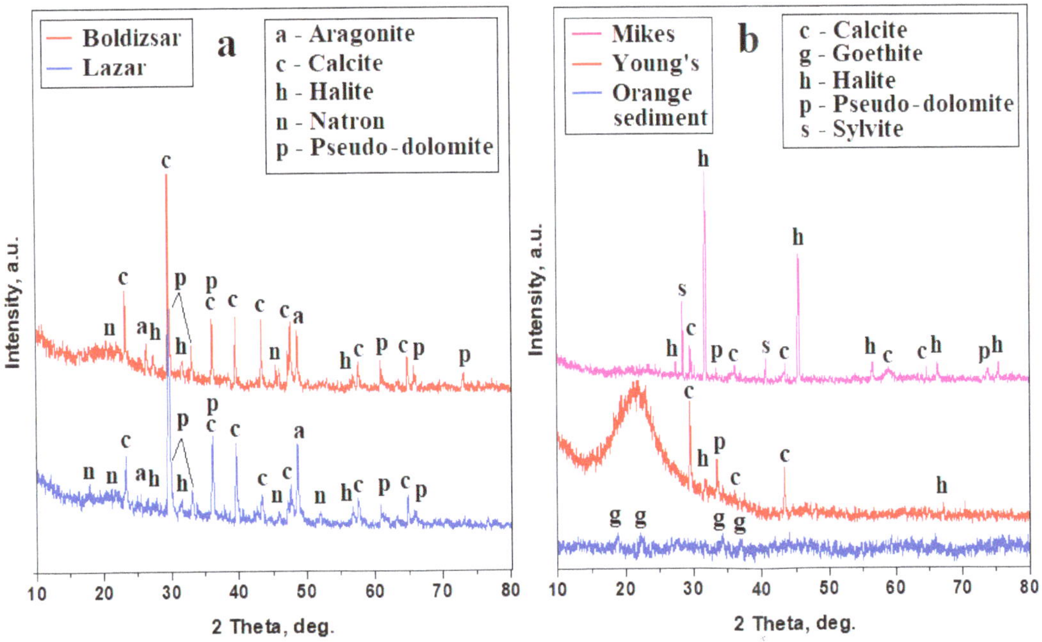

Figure 3. XRD patterns for the powders obtained from mineral water samples: (**a**) Borsec and (**b**) Tusnad.

The precise amount of each mineral formed by the water evaporation can be calculated with the Relative Intensity Ratio (RIR) method, which correlates the corundum factor of each identified mineral with the relative intensities of its specific diffraction peaks [26]. The RIR method was applied to the XRD patterns in Figure 3, and the obtained values are centralized in Table 2.

Table 2. XRD data correlation with MOM observations.

Characteristics		Mineral Composition					
		Aragonite	**Calcite**	**Pseudo-Dolomite**	**Halite**	**Sylvite**	**Natron**
Chemical formula		$CaCO_3$	$CaCO_3$	$CaMg(CO_3)_2$	NaCl	KCl	$Na_2CO_3{\cdot}10H_2O$
Crystal system		Orthorhombic	Trigonal	Trigonal	Cubic	Cubic	Monoclinic
Habit		rhombic granular	pseudo-hexagonal	columnar granular	rectangular plates	rectangular plates	prismatic columnar
MOM colors		Violet–blue	Yellow white –brown	Yellow white –brown	Transparent pale blue	Transparent pale blue	Yellow–yellow gray
Boldizsar	Amount, wt.%	12	58	15	9	-	6
	Size range, μm	2.5–5	5–10	5–12	*	-	5–10
Lazar	Amount, wt.%	23	37	20	12	-	8
	Size range, μm	3–10	5–30	5–22	*	-	5–15
Mikes	Amount, wt.%	-	13	10	45	32	-
	Size range, μm	-	5–15	15–25	25–250	25–250	-
Young's	Amount, wt.%	-	53	32	15	-	-
	Size range, μm	-	5–20	10–25	*	-	-

* Halite crystallizes as crusts beside the other minerals, and their size range cannot be measured.

The crystallization process was monitored by mineralogical optical microscopy (MOM) under cross-polarized light, revealing the specific colors of each mineral. It was observed on the water drops spread onto glass slides. The initial crystallization state becomes visible after 12 h, revealing well-formed crystals which continue growing until the final stage at about 24 h. On the other hand, the minerals formed on the glass slide are not enough for the XRD investigation; therefore, the long-term crystallization of about 30 mL of water was conducted in Petri dishes for about 7 days. The obtained MOM images are presented in Figure 4. The size range of each mineral's crystals was measured in the final stage of crystallization on the glass slide, and the values are centralized in Table 2.

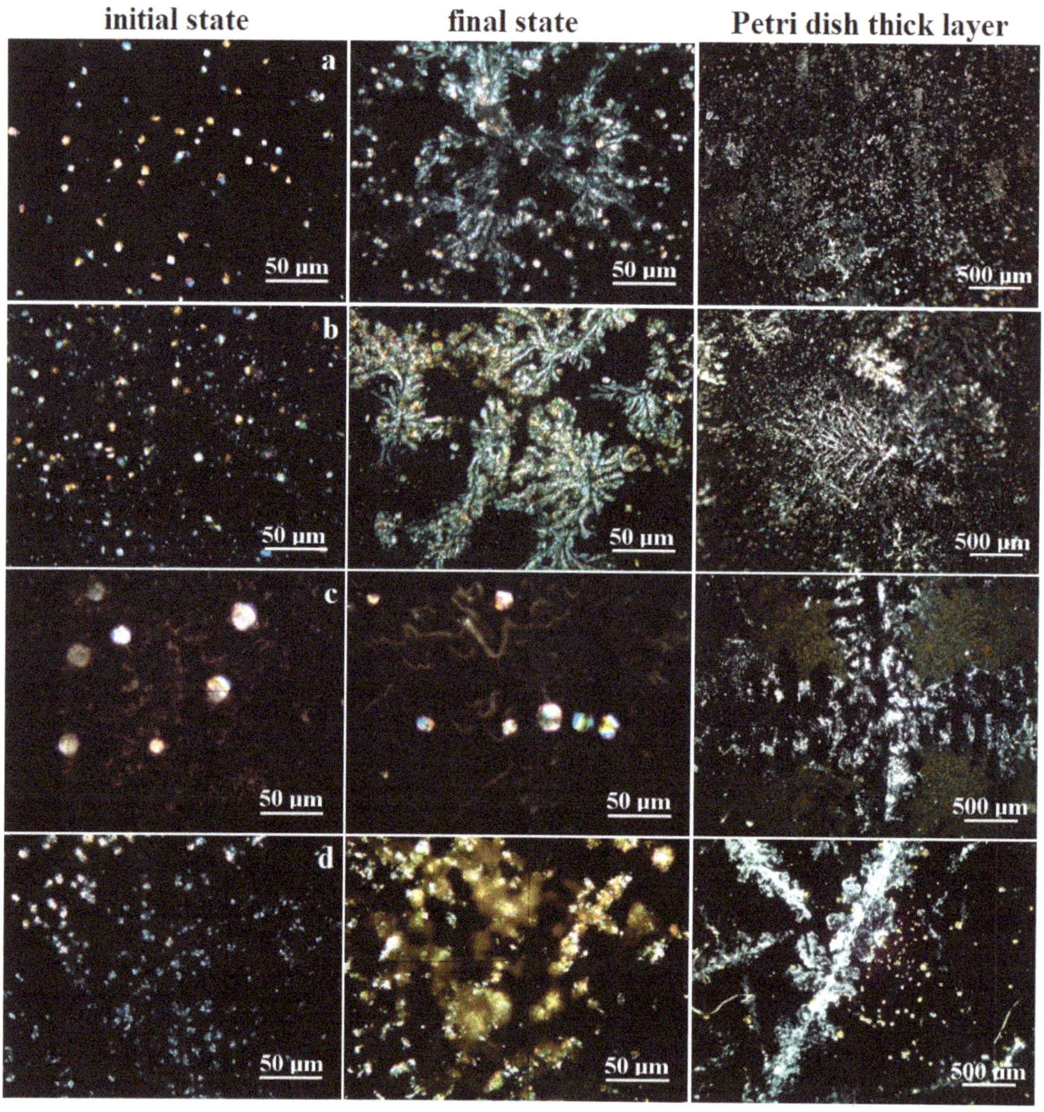

Figure 4. Crystallization process evidenced by mineralogical optical microscopy for the investigated water samples: (**a**) Boldizsar, (**b**) Lazar, (**c**) Mikes, and (**d**) Young's spring.

Boldizsar water has a turbidity below the detection limit; thus, the crystallization process occurs via homogeneous germination. CO_2 reacts with dissolved ions within the water samples, causing the precipitation of carbonates, which grow as well-individualized particles of calcite (predominantly white, having pseudo-hexagonal shapes), aragonite (rhombic particles colored in blue–magenta depending on the face position of the optical axis of microscope), and pseudo-dolomite (predominantly yellow with granular aspect, having elongation tendency to columnar bodies), as shown in Figure 4a.

The limited solubility of carbonate into water facilitates its crystallization during the initial state when sodium chloride remains dissolved. The carbonate mineral crystallization ends when the concentration reaches its supersaturated level. The formed calcite, pseudo-dolomite, and aragonite crystals self-assemble into dendritic structures sustained by the remnant saline solution. Several prismatic yellow crystals of natron are observed as a consequence of the direct reaction between Na ions and CO_2. The dissolved sodium chloride crystallizes at the end, due to its high hydro-solubility, and forms thin crusts that develop beside the carbonate dendrites, giving a blue hue over their sides as observed in the final stage in Figure 4a. The long-time crystallization in Petri dishes allows for the formation of thicker carbonate dendrite deposits embedded in the remnant halite.

The Lazar spring also has a turbidity below the detection limit, facilitating a homogeneous crystallization following the path described for the Boldizsar spring. Its early stage of crystallization is characterized by a denser precipitation of crystallization germs, leading to a larger number of well-individualized crystals of calcite, pseudo-dolomite, and aragonite, followed by traces of natron, shown in Figure 4b. The dissolved sodium chloride amount is too small for initializing crystals at this stage. The carbonate dendrites developed over the final stage are better-defined in the case of the Lazar spring because of its higher mineral content. Their sides also have a blue hue caused by the formation of halite crusts crystallized from the remnant liquid during the end stage. The better-defined shape of the carbonate dendrites superimpose the long-time crystallization pattern in the Petri dishes, being clearly visible and defined.

The Mikes spring from Tusnad Bath has a TDS and salinity similar to the Lazar spring in Borsec, forming a relevant statistical group in the matter of values. The difference is given by the increased level of chloride. The fact is sustained by the crystallization process, as shown in Figure 4c. The initial stage causes a rapid crystallization of calcite and pseudo-dolomite because of their limited hydro-solubility, which occurs on the heterogeneous germination grains favored by the remnant turbidity of the sample given by the solid dispersoids. The calcite crystals have a pseudo-hexagonal shape and small sizes, while the pseudo-dolomite crystals are sparse, with bigger sizes and a granular shape. These different crystallizations are caused by the higher salinity of the sample, which promotes carbonate clustering instead of dendritic disposal. This fact is further evidenced by the coalescence of calcite and pseudo-dolomite crystals in micro crystalline clusters, surrounded by the halite and sylvite crystals with well-individualized shapes as squares, rectangles, and less octahedral forms. The long-term crystallization promotes an even harder segregation of the halite/sylvite dendrites, embedding at their borders calcite and pseudo-dolomite crystals. Several brown clusters of about 50–100 μm are observed beside the other discussed crystals. These belong to the iron hydroxide nanoparticles.

Young's spring, Figure 4d, has a prominent carbonated characteristic based on the white calcite and slightly white–yellow pseudo-dolomite crystals formed during the early stage. These are most likely promoted by heterogeneous germs consisting in micro-dispersoids related to the remnant turbidity. The carbonate minerals end their crystallization, since the supersaturated limit is achieved by the water evaporation, while sodium chloride remains dissolved. This influences the coalescence of carbonate crystals, which do not have enough mobility to form well-structured dendrites, facilitating the appearance of rounded micro-clusters of about 25–40 μm. The halite crystallizes as thin crusts at the end of the final stage, consisting of a pale blue hue beside the fine carbonate formations. The long-term crystallization in the Petri dish allows carbonate crystals such as calcite and pseudo-dolomite to re-organize into dendritic shapes, surrounded by the halite crusts. Some bright yellow spots are observed as distinct segregations. Their color is insensitive to their position in relation to optical axis of the microscope, proving their nanocrystaline nature. Their color indicates most likely a mixture of iron hydroxide with sulfur traces. Such an assumption requires further investigations. First of all, we found a fine orange sediment carried out from the Ciomadu Mare slope on the back of Mikes spring through a minor water flow (below 5 L/day). Its XRD pattern reveals merely an amorphous state, with some broadened peaks for goethite, shown in Figure 3b. The dried powder was inspected by MOM, as shown in Figure 5.

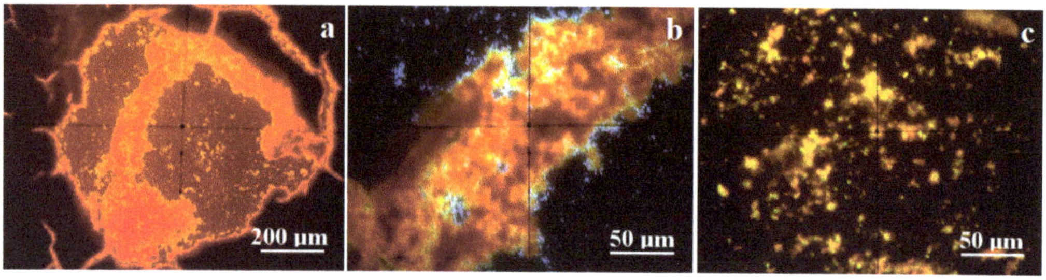

Figure 5. Mineralogical optical microscopy aspect of the iron orange sediment: (**a**) overall aspect, (**b**) microstructural detail, and (**c**) individualized clusters of goethite nanoparticles.

The drying process led to a compact scale crust, shown in Figure 5a, having a uniform brown nuance specific to iron oxides and hydroxides. The excoriation of central scales allows for the observation of a thin layer of constituent particles still attached onto the glass slide. They have a fine granular matter colored brown–orange, indicating the massive presence of iron hydroxide (goethite). The higher-magnification image in Figure 5b clearly shows that in fact they are clusters of even more fine particles. Therefore, a re-dispersion in ultra-pure water, followed by a magnetic agitation for 30 min at 4000 rpm and adsorption onto a fresh glass slide, allows for a better dispersion of these clusters, as shown in Figure 5c. They feature a uniform orange nuance insensitive to their relative position to the optical axis of the microscope. This clearly indicates that the observed particles are fine clusters of goethite nanoparticles, which ensure the uniform color without maximums and extinctions. This fine mineral matter is insoluble in water [27], causing the observed remnant turbidity in Mikes and Young's spring water samples and corresponding to the MOM observations in Figure 4c,d after long-time crystallization.

3.3. Microstructural Observation and Elemental Spectroscopy

The SEM investigation of the powder samples, which were previously subjected to XRD, allows for the correlation of the morphology with the element distribution. Thus, the Mikes powder is predominantly granular and derived from the broken pseudo-hexagonal crystals of calcite, rhombic aragonite, and granular–elongated crystals of pseudo-dolomite, as shown in Figure 6a. The elemental map clearly shows the calcite and aragonite crystals in a pale green color (due to the green label of Ca, turquoise label of O, and yellow label of C). Pseudo-dolomite has a violet nuance because the Mg label and acicular natron slivers appear in pink nuances because of the Na label color. Minor traces of sulfur are observed as a segregation at the carbonate crystals borders. A halite crust was not detected in the observed micro-site for the Mikes sample; therefore, Cl is missing from the elemental composition.

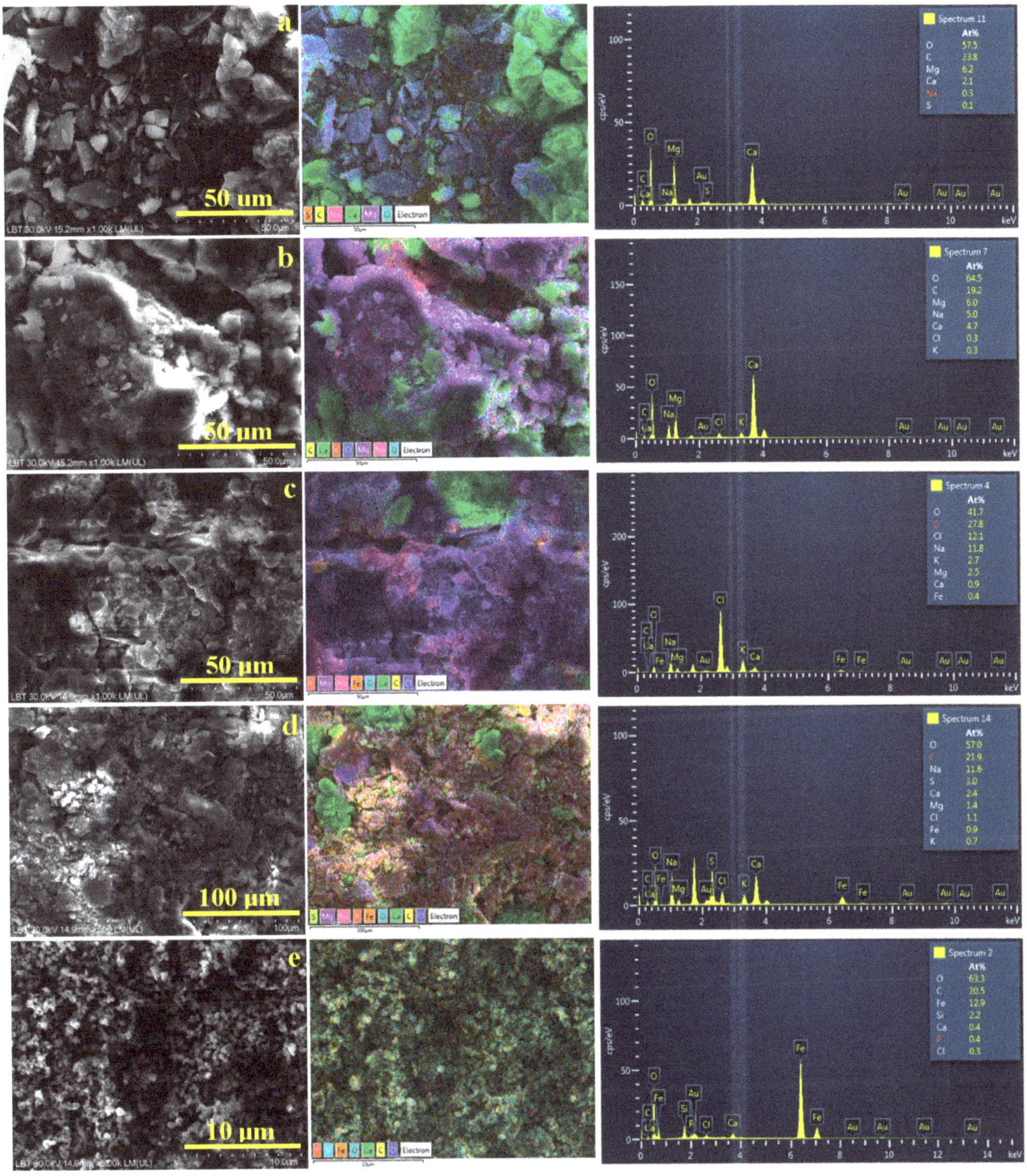

Figure 6. SEM investigation of the powder crystallized from the water samples: (**a**) Boldizsar, (**b**) Lazar, (**c**) Mikes, (**d**) Young's spring and (**e**) iron orange sediment. EDS elemental map and elemental spectrum are displayed on the right side of each SEM image.

The Lazar powder has a more crusty shape because of the slightly increased amount of halite which embeds well in the carbonate crystals having various shapes derived from the pseudo-hexagonal and rhombic habits of the crystals prior to fracturing. Thus, calcite and aragonite particles (green nuance) and pseudo-dolomite particles (violet nuance) are embedded throughout the halite crust, appearing in a dark blue nuance specific to the Cl label, shown in Figure 6b. Some local concentrations of natron particles are observed beside the calcite and pseudo-dolomite. The mineral disposal in the elemental map confirms the MOM observation made for the long-term crystallization deposits within the Lazar sample. Overall, the SEM investigation proves a congruent crystallization within the Borsec springs (e.g., Boldizsar and Lazar), respecting the initial assumption of the higher mineralization due to the lower flow of the Lazar spring.

XRD and MOM observations reveal a higher content of dissolved chlorides (e.g., halite and sylvite) in the Tusnad waters compared to the Borsec ones. Thus, the mineralization changes affect the sample's microstructure. Figure 6c shows fragments of compact crusts formed by the synergistic crystallization of halite and sylvite (deep-blue alternating with reddish areas in the elemental map as consequence of Na and K labels). The formed crusts confirm the presence of evaporitic dendrites observed in Figure 4c after long-term crystallization. The carbonate crystals have a predominantly clustered shape induced by the high concentration of brine during water evaporation. Both the calcium carbonate crystals (green color) and pseudo-dolomite (violet color) are predominantly rounded, ranging from about 10 to 25 μm, in good agreement with the MOM observations. Fe deposits are slightly difficult to observe, partly because of their action as crystallization germs for carbonate minerals and the local clustering process of the remnant turbid particles. For instance, such clusters are visible on the upper left side the elemental map in Figure 6c, having an orange nuance induced by the Fe label.

Young's spring has a lower flow, and thus it is expected to have a greater amount of dissolved minerals; but the results looks different. The SEM image in Figure 6d reveals a dense mixture of broken crystals and less evidence of embedding crusts. The larger formations (15–20 μm) have broken borders intersected at specific angles of pseudo-hexagonal crystallization, representing calcite particles (green nuance in elemental map), while the pseudo-dolomite particles are predominantly broken into fine fractions of about 10–15 μm, only a few being about 25 μm. The iron and sulfur combined labels give a light orange nuance in the elemental map, which is well dispersed among the broken pseudo-dolomite particles, indicating their action as heterogeneous germination seeds. Still, some small clusters of iron and sulfur are observed within the finest particles, having a light-orange nuance and sizes of about 5–10 μm, in good agreement with the MOM observations.

The elemental map in Figure 6d reveals some irregular deep-blue lines at the crystals' borders corresponding to the halite crust traces formed during the final stage of crystallization. The elemental compositions of the investigated micro-sites in Figure 6 are displayed in Table 3. They are expressed in atomic percents for a better correlation with the minerals' stoichiometric chemical formulas.

All samples are dominated by oxygen as the main element because of their oxide nature, followed by carbon due to the high amount of carbonate minerals. The chloride and potassium amounts are strongly related to evaporitic minerals like halite and sylvite. The investigated springs' elemental composition comparisons reveal the dominance of carbonate minerals in the Boldizsar and Lazar springs, proving to be the main characteristic of the Borsec area. Tusnad waters are characterized by an increased amount of evaporitic minerals like halite and sylvite diminishing carbonate mineral amounts. Another specific footprint of the Tusnad waters proves to be iron dispersoids, which are completely lacking in the Borsec waters. Another elemental difference is observed between the Mikes and Young's springs, mainly regarding the Cal and Cl amount, which suggests a mineralization source from different aquifers affected by different geological layers.

Table 3. The elemental composition of the crystalline powders.

Element, at.%	Boldizsar	Lazar	Mikes	Young's
O	57.5	64.5	41.7	57.0
C	33.8	19.2	27.8	21.9
Ca	2.1	6.0	0.9	2.4
Mg	6.2	5.0	2.5	1.4
Na	0.3	4.7	11.8	11.6
Cl	-	0.3	12.1	1.1
K	-	0.3	2.7	0.7
S	0.1	-	-	3.0
Fe	-	-	0.4	0.7

Finally, the iron dispersoids observed in the orange sediment were analyzed using SEM, as shown in Figure 6d. They are formed by very small clusters with a rounded shape and sizes ranging from 125 to 300 nm, surrounding fine micro particles of about 1–3 μm. The elemental map reveals that these micro particles are SiO_2 (having a pale blue nuance induced by the specific labels of Si and O) and calcite particles with a pale green nuance. These are too few (perhaps below 1% of the total weight of the sample), becoming undetectable by the XRD investigation. The nano-clusters are made of iron, tightly surrounding the other mineral particles within the sediment. The traces of Cl and P most likely occur from some intercalation of volcanic rock particles, along with the detected SiO_2 exuded from the mountain slope by the weak water flow bringing them to the surface. These sample characteristics are important because both Mikes and Young's spring feature such dispersoids in their waters, being a characteristic of the local mineralization.

4. Discussion

The initial assumption that a spring's flow is directly responsible for the water mineralization is partly rejected based on the obtained results. At first sight, a slower flow of the spring would keep an increased mineralization, while a high flow supported by larger infiltrations of top water would decrease the mineralization grade, a fact already observed in our previous study regarding the Borsec public springs [9]. Thus, the Boldizsar and Lazar spring behavior sustains this assumption, but it is contradicted by the Tusnad springs having the opposite behavior. Young's spring, having a slower flow, has a lower mineralization than the Mikes spring, which has a high flow. Moreover, this contradiction is highlighted by the Lazar and Mikes springs forming statistically relevant groups for electrical conductivity, TDS, and salinity in Figure 2b–d.

The first clue resolving the contradiction is given by the total Cl amount, which forms two distinct relevant groups for the Tusnad springs: Mikes represents the highest level and Young's spring the lowest level, while the Borsec springs form another statistically relevant group with a mild Cl amount. This indicates the different mineralization within Tusnad compared to Borsec springs.

XRD coupled with MOM gives a complete overview regarding all samples' mineralization, observations confirmed by SEM coupled with EDS elemental analysis. It was found that there is a complex behavior of mineral crystallization as a consequence of water evaporation. First of all, CO_2 dissolved in water samples generates carbonic acid, which reacts with the dissolved mineral ions, especially with Ca, Mg, and Na, causing the precipitation of $CaCO_3$ crystallized as calcite and aragonite, along with magnesian calcite $CaMg(CaCO_3)_2$, while sodium carbonate precipitates and crystallizes as natron [28,29]. Magnesian calcite has the same chemical formula as dolomite, but the crystal structure cannot be replicated by a crystallization from an aqueous solution because of the incapacity of ensuring successively ordered crystal planes of Mg and Ca within the crystal lattice [30,31]. Both Ca and Mg atoms take random places within the magnesian calcite lattice, named suggestive pseudo-dolomite [31,32]. Tsuchiya et al. report the successful crystallization of dolomite from concentrated brine by a slow injection of fine CO_2 bubbles [33]. It looks like simply degassing through CO_2 bubbles within our samples is not enough to ensure the proper dolomitization of Ca and Mg ions. We found that carbonates precipitate and grow in the first stage of crystallization due to their limited aqueous solubility.

Evaporitic minerals like halite and sylvite have a high aqueous solubility, and therefore the supersaturated solution forms only at the final stage of crystallization [34,35]. We found that low amounts of halite have a limited influence on the carbonate crystallization, being limited to crust formation beside the calcite–pseudo-dolomite dendrites during the final stage of crystallization (Boldizsar, Lazar, and Young's spring). Higher amounts of evaporites (e.g., halite and sylvite) have a greater influence on the carbonate crystallization, causing their clusterization perhaps due to the increased superficial tension caused by the considerably greater density of the solution. Thus, the Tusnad water samples are subjected to different crystallization conditions compared to the Borsec samples.

Another major difference is given by the remnant turbidity observed in the Mikes and Young's springs, which proves to be the fine dispersion of nanostructural iron hydroxide (e.g., goethite), partly accompanied with sulfur traces. These solid nanoparticles act as heterogeneous crystallization germs, promoting the growth of carbonate minerals. The literature data show that calcium carbonate crusts are involved with goethite in the corrosion of mild steel pipes at pH = 8.8, a fact favored by the increased temperature of geothermal water [36]. Some of our previous observations reveal nano iron hydroxide recrystallization in a ferrous conglomerate, with calcite occurring in a carwash slurry [37]. Other research reports the sequential precipitation of goethite–calcite nano-composites mediated by $Ca(OH)_2$ [38]; the solution pH depends on the calcium hydroxide concentration. Overall, the literature data sustain the goethite affinity for calcite being able to act as heterogeneous germination centers.

Papp and Nitoi reveal that the Tusnad mineralization of the geothermal and mineral water within the area is caused by an intense underground circulation of water through the Neogene fault system, occurring in magmatites [39]. The observation is in good agreement with the other literature data [13,14]. These magmatites are represented by andesite deposits in close contact with the underground water. Andesite is known for its rich content in plagioclase, pyroxenes, and hornblende [40,41], explaining the rich iron content of water from Tusnad. Since hornblende is a complex mineral with the conventional formula $(Ca,Na)_{2-3}(Mg,Fe,Al)_5(Al,Si)_8O_{22}(OH,F)_2$, this explains the presence of almost all identified dissolved elements within the Mikes and Young's springs. Their reaction with carbonic acid induced by CO_2 explains the crystallization of the carbonate minerals without pre-existing sedimentary rocks like dolomite substrate, as in Borsec. The higher amount of chloride found in the Mikes spring and significant level in Young's spring would require water contact with some evaporitic sedimentary deposit interlocked with magmatic bursts. Unfortunately, we assume that this is most likely possible, but we did not find any reference to confirm this assumption.

Overall, the results showed that the crystallization process within the Borsec water is dominated by the carbonate mineral formation started in the first stage; thus, it is classified as carbonated mineral water. The saline characteristics of the Tusnad water strongly influence the carbonate minerals' crystallization during the first stage and dominate the end stage of crystallization. Nanostructured iron hydroxide particles are another special characteristic of the Tusnad mineralization. Thus, the Mikes spring can be categorized as salted mineral water, while Young's spring can be categorized as iron-carbonated mineral water. Of course, each of the spring's source water can be used for its specific balneotherapy. For instance, carbonated mineral waters are recommended for adjusting stomach hyper acidity [42]. These are also efficient for balneology, relieving rheumatic pains by a proper adjustment of the temperature and adapting the cure parameters to the patient's needs [43,44].

Saline water can be useful for ingestion cures prescribed by therapists for adjusting symptoms related to colitis and acute enterocolitis affection because of the antibacterial effects of saline solutions [45,46]. On the other hand, these waters have a lot of balneologic benefits, alleviating the symptoms of various skin diseases such as fungal infection [47], atopic dermatitis, and psoriasis [48].

Drinking water with a significant iron content is useful for the amelioration of anemia, especially for infants [49]. This fact is confirmed by data from the other literature [50,51]. Thus, Tusnad mineral water proves to contain enough nanostructural iron to be effective in anemia cures under strict medical observation.

The investigated springs have a continuous flow, which is publicly available to anyone without charge. Anyone can enter into the spring's room and fill a bottle with mineral water. The water which is not taken by the visitors and the bath's patients is discharged into the adjacent rivers. The increased mineral content of the spilled water affects the mineral equilibrium of the surface water, which might increase its hardness. What if this mineral water had been collected in a buffer tank and subjected to assisted evaporation? This would result in mineral powders like the ones obtained in this research:

- Boldizsar spring: calcite, pseudo-dolomite, aragonite, halite, and natron;
- Lazar spring: calcite, aragonite, pseudo-dolomite, halite, and natron;
- Mikes spring: halite, sylvite, calcite, pseudo-dolomite, and iron nano-dispersoids;
- Young's spring: calcite, pseudo–dolomite, halite, and iron nano-dispersoids.

Following the example of the Le Furo application of the Onsen mineral waters, our obtained mineral powders might be further used for home cures or balneology treatments under direct medical prescription, having therapeutic benefits and local ecological enhancements by avoiding mineral amounts being spilled in the surface waters. This is a challenge for further research.

The specific parameters depend on each spring's flow and the degree of mineralization, which could depend on the seasonality and precipitation water infiltration. This is a limitation of the current investigation that will be addressed in the next investigations, which will be focused on each season—spring, summer, autumn, and winter—revealing their influence on crystallization behavior. Also, the iron content influence on the pH value is an implicit function revealed by the measured values, which will be further investigated in regard to the seasonality chemistry of the considered springs.

5. Conclusions

The crystallization process is driven by the formation of carbonate minerals (calcite, aragonite, and pseudo-dolomite) in the early stage, followed by their growth and arrangement into dendritic structures until the supersaturated solution is obtained and the crystallization of evaporitic minerals (e.g., halite and sylvite).

The Boldizsar and Lazar springs in Borsec are predominantly carbonated because of their source on a dolomitic substrate.

The Mikes spring in Tusnad has a high content of evaporitic minerals like halite and sylvite, imposing a salted characteristic. Young's spring in Tusnad is rich in its mineralization from other geological layers, having less halite, causing a more carbonated characteristic. Both of the investigated Tusnad springs contain iron hydroxide nano-dispersoids, imposing a ferric characteristic.

The Borsec waters form Boldizsar and Lazar springs are predominantly carbonated and are useful for alleviating gastric hyper acidity, while also having balneologic benefits, alleviating rheumatic symptoms.

Tusnad waters could be effective in anemia treatments because of their amount of iron nano-dispersoids. Young's spring might be a good candidate for this purpose because of its low halite content and moderate carbonates. The high content of halite and sylvite within the Mikes spring make it a good candidate for balenologic purposes for alleviating skin affection symptoms.

The obtained mineral powders are ready to be re-dissolved at a patient's home for personalized cures under strict medical prescription, or could be used as dietary supplements for adjusting electrolytic balances on hot days.

Author Contributions: Conceptualization, S.E.A. and I.P.; methodology, S.E.A. and I.P.; software, G.B.; validation, S.E.A. and I.P.; formal analysis, I.P.; investigation, S.E.A., L.B.T., G.B., and I.P.; resources, S.E.A.; data curation, I.P.; writing—original draft preparation, S.E.A. and I.P.; writing—review and editing, I.P.; visualization, L.B.T., and I.P.; supervision, S.E.A. and I.P.; project administration, S.E.A. and I.P. All authors have read and agreed to the published version of the manuscript.

Funding: This research received no external funding.

Data Availability Statement: The original contributions presented in the study are included in the article.

Acknowledgments: The authors gratefully acknowledge the XRD diffractometer maintenance which was supported by the Ministry of Research, Innovation, and Digitization through Program 1-Development of the National Research and Development System, Subprogram 1.2-Institutional Performance-Funding Projects for Excellence in RDI, Contract No. 37PFE/30.12.2021.

Conflicts of Interest: The authors declare no conflicts of interest.

Abbreviations

The following abbreviations are used in this manuscript:

EDS	Energy Dispersive Spectroscopy
MOM	Mineralogical Optical Microscopy
RIR	Relative Intensity Ratio
SEM	Scanning Electron Microscopy
TDS	Total Dissolved Solids
XRD	X-ray Diffraction

References

1. Li, C.; Shan, X.; Li, C.; Hao, S.; Cheng, B.; Lu, C.; Zhao, J.; Wang, X.; Su, Z. Analysis of the Occurrence Conditions and Formation Mechanism of Mineral Water in the Southern Region of Yaoquan Mountain, Wudalianchi. *Water* **2024**, *16*, 3130. [CrossRef]
2. Xu, Y.; Li, Y.; Wang, P.; Feng, C.; Cao, S.; Li, J. Unraveling the geochemical behaviors of rare earth elements (REEs) in Chinese drinking natural mineral waters: Environmental and health perspectives. *J. Hazard. Mater.* **2025**, *494*, 138731. [CrossRef] [PubMed]
3. Stiller, M.; Yechieli, Y.; Gavrieli, I. Rates of halite dissolution in natural brines: Dead Sea solutions as a case study. *Chem. Geol.* **2016**, *447*, 161–172. [CrossRef]
4. Bai, H.; Pan, T.; Han, G.; Fan, Q.; Miao, Q.; Bu, H. Hydrochemical Characteristics and Genesis of Sand–Gravel Brine Deposits in the Mahai Basin of the Northern Qinghai–Tibetan Plateau. *Water* **2024**, *16*, 3562. [CrossRef]
5. Parag, Y.; Elimelech, E.; Opher, T. Bottled Water: An Evidence-Based Overview of Economic Viability, Environmental Impact, and Social Equity. *Sustainability* **2023**, *15*, 9760. [CrossRef]
6. Ridder, Consumption Volume of Bottled Water Worldwide 2020, by Leading Countries, Statista, 26 June 2025. Available online: https://www.statista.com/statistics/1307883/consumption-of-bottled-water-worldwide-in-2009/?srsltid=AfmBOorZx9WuHBSGSNXNhDqHqQL2CzCVuNyE742p2ZS1R2klahWaJPW8 (accessed on 14 August 2025).

7. Serbulea, M.; Payyappallimana, U. Onsen (hot springs) in Japan—Transforming terrain into healing landscapes. *Health Place* **2012**, *18*, 1366–1373. [CrossRef]
8. Le Furo, Mineral Mist Bath. Available online: https://aquaignis.jp/en/spa_lefuro.php (accessed on 15 August 2025).
9. Avram, S.E.; Platon, D.V.; Tudoran, L.B.; Borodi, G.; Petean, I. Mineral Powder Extraction by the Natural Drying of Water from the Public Springs in Borsec. *Appl. Sci.* **2024**, *14*, 10806. [CrossRef]
10. Dewandel, B.; Alazard, M.; Lachassagne, P.; Bailly-Comte, V.; Couëffé, R.; Grataloup, S.; Ladouche, B.; Lanini, S.; Maréchal, J.C.; Wyns, R. Respective roles of the weathering profile and the tectonic fractures in the structure and functioning of crystalline thermo-mineral carbo-gaseous aquifers. *J. Hydrol.* **2017**, *547*, 690–707. [CrossRef]
11. Dupuy, M.; Garel, E.; Chatton, E.; Labasque, T.; Mattei, A.; Santoni, S.; Vergnaud, V.; Aquilina, L.; Huneau, F. Using natural gas content of groundwater to improve the understanding of complex thermo-mineral spring systems. *J. Hydrol.* **2024**, *634*, 130956. [CrossRef]
12. Pricajan, A. *Mineral and Thermal Water from Romania (in Romanian: Apele Minerale Si Termale Din Romania)*; Technical Publishing House: Bucharest, Romania, 1972; 296p.
13. Gal, E.E. Tusnad Baths Touristic Station and its Surroundings—Touristic Potential and Development Perspectives. Bachelor's Thesis, Faculty of Economic, Law and Administrative Sciences, Petru Maior University, Targu Mures, Romania, 2008. (In Romanian)
14. Schreiber, W.E. Geographical classification and genesis of massive Ciomadu. *Studia Univ. Babes-Bolyai Geogr.* **1972**, *1*, 15–22.
15. Frunzeti, N.; Baciu, C. Diffuse CO_2 Emission at Sfânta Ana Lake-Filled Crater (Eastern Carpathians, Romania). *Procedia Environ. Sci.* **2012**, *14*, 188–194. [CrossRef]
16. Keller, C.K.; Bacon, D.H. Soil respiration and georespiration distinguished by transport analyses of vadose CO_2, $^{13}CO_2$ and $^{14}CO_2$, *Glob. Biogeochem. Cycles* **1998**, *12*, 361–372. [CrossRef]
17. Giggenbach, W.F.; Sano, Y.; Wakita, H. Isotopic composition of helium and CO_2 and CH_4 contents in gases produced along the New Zealand part of a convergent plate boundary. *Geochim. Cosmochim. Acta* **1993**, *57*, 3427–3455. [CrossRef]
18. Vaselli, O.; Minissale, A.; Tassi, F.; Magro, G.; Seghedi, I.; Ioane, D.; Szakacs, A. A geochemical traverse across the Eastern Carpathians (Romania): Constraints on the origin and evolution of the mineral water and gas discharges. *Chem. Geol.* **2002**, *182*, 637–654. [CrossRef]
19. Yang, M.; Tan, L.; Batchelor-McAuley, C.; Compton, R.G. The solubility product controls the rate of calcite dissolution in pure water and seawater. *Chem. Sci.* **2024**, *15*, 2464–2472. [CrossRef]
20. Liao, Y.; Liu, Y.; Wang, G.; Li, T.; Liu, F.; Wei, S.; Yan, X.; Gan, H.; Zhang, W. Genesis Mechanisms of Geothermal Resources in Mangkang Geothermal Field, Tibet, China: Evidence from Hydrochemical Characteristics of Geothermal Water. *Water* **2022**, *14*, 4041. [CrossRef]
21. Campean, R.F.; Petean, I.; Baraian, M.; Hosu-Prack, A.G.; Ristoiu, D.; Arghir, G. Mineral particulate matter from the St. ana lake sand related to the water suspensions. *Carpathian J. Earth Environ. Sci.* **2012**, *7*, 57–66.
22. Liu, Y.; Ma, T.; Chen, J.; Peng, Z.; Xiao, C.; Qiu, W.; Liu, R.; Du, Y. Impact of clayey sediment compaction on pore water evolution and the release of iron. *Appl. Geochem.* **2023**, *152*, 105635. [CrossRef]
23. Borsec Public Springs Composition Declaration Displayed on the Entrance Front, National Mineral Water Company of Romania. Available online: https://www.snam.ro/ (accessed on 21 February 2024).
24. Hsieh, Z.-H.; Fan, C.-H.; Tu, H.-H.; Hu, C.-V.; Li, M.L.; Yeh, C.-K. Acoustic vortex induced bubble cage: An on-demand light guide in turbid media. *Ultrasonics* **2025**, *156*, 107761. [CrossRef] [PubMed]
25. Choe, U.H.; Choe, J.H.; Kim, Y.J. Removal of deposits and improvement of shelf life in CO_2 rich mineral water by ozone microbubbles. *Qimica Nova* **2025**, *48*, e-20250045. [CrossRef]
26. Avram, S.E.; Birle, B.V.; Tudoran, L.B.; Borodi, G.; Petean, I. Investigation of Used Water Sediments from Ceramic Tile Fabrication. *Water* **2024**, *16*, 1027. [CrossRef]
27. Villalobos, M.; Cruz-Valladares, A.X. Towards Building a Unified Adsorption Model for Goethite Based on Variable Crystal Face Contributions: III Carbonate Adsorption. *Colloids Interfaces* **2025**, *9*, 51. [CrossRef]
28. Nie, X.; Wang, Z.; Wan, J.; Wang, G.; Li, Y.; Ouyang, S. Competition between homogeneous and heterogeneous crystallization of CaCO3 during water softening. *Water Res.* **2024**, *250*, 121061. [CrossRef] [PubMed]
29. Ali, A.; Mendes, C.E.; de Melo, L.G.T.C.; Wang, J.; Santos, R.M. Production of Sodium Bicarbonate with Saline Brine and CO_2 Co-Utilization: Comparing Modified Solvay Approaches. *Crystals* **2023**, *13*, 470. [CrossRef]
30. Gregg, J.M.; Bish, D.L.; Kaczmarek, S.E.; Machel, H.G. Mineralogy, nucleation and growth of dolomite in the laboratory and sedimentary environment: A review. *Sedimentology* **2015**, *62*, 1749–1769. [CrossRef]
31. Land, L.S. Failure to precipitate dolomite at 25 °C from dilute solution despite 1000-fold oversaturation after 32 years. *Aquat. Geochem.* **1998**, *4*, 361–368. [CrossRef]
32. Pimentel, C.; Pina, C.M. Reaction pathways towards the formation of dolomite-analogues at ambient conditions. *Geochim. Cosmochim. Acta* **2016**, *178*, 259–267. [CrossRef]

33. Tsuchiya, Y.; Wada, Y.; Hiaki, T.; Onoe, K.; Matsumoto, M. Effects of CO_2 fine bubble injection on reactive crystallization of dolomite from concentrated brine. *J. Cryst. Growth* **2017**, *469*, 36–41. [CrossRef]
34. Chruszcz-Lipska, K.; Zelek-Pogudz, S.; Solecka, U.; Solecki, M.L.; Szostak, E.; Zborowski, K.K.; Zając, M. Use of the Far Infrared Spectroscopy for NaCl and KCl Minerals Characterization—A Case Study of Halides from Kłodawa in Poland. *Minerals* **2022**, *12*, 1561. [CrossRef]
35. Petean, I.; Arghir, G.; Câmpean, R.F.; Bărăian, M.; Hosu Prack, A.G. Crystalographyc relations applied to homogeneous cristalization of badenian salt. *Acta Tech. Napoc. Ser. Appl. Mech. Math.* **2011**, *54*, 193–200.
36. Maior, I.; Badea, G.E.; Stănășel, O.D.; Sebeșan, M.; Cojocaru, A.; Petrehele, A.I.G.; Creț, P.; Blidar, C.F. Chemical Composition and Corrosion—Contributions to a Sustainable Use of Geothermal Water. *Energies* **2025**, *18*, 3634. [CrossRef]
37. Avram, S.E.; Filip, M.R.; Barbu Tudoran, L.; Borodi, G.; Petean, I. Investigation of ferrous conglomerate particles found in carwash slurry and their environmental implications. *Stud. UBB Chem.* **2023**, *68*, 57–70. [CrossRef]
38. Montes-Hernandez, G.; Renard, F.; Chiriac, R.; Findling, N.; Ghanbaja, J.; Toche, F. Sequential precipitation of a new goethite–calcite nanocomposite and its possible application in the removal of toxic ions from polluted water. *Chem. Eng. J.* **2013**, *214*, 139–148. [CrossRef]
39. Papp, D.C.; Nițoi, E. Isotopic composition and origin of mineral and geothermal waters from Tuşnad Băi Spa, Harghita Mountains, Romania. *J. Geochem. Explor.* **2006**, *89*, 314–317. [CrossRef]
40. Kao, N.-C.; Wang, M.-K.; Chiang, P.-N.; Chang, S.-S. Characterization of wheat-rice-stone developed from porphyritic hornblende andesite. *Appl. Clay Sci.* **2003**, *23*, 337–346. [CrossRef]
41. Topaz, A.; Golan, T.; Boneh, Y. Natural fabrics of a hornblende-rich amphibolite: Implications for hornblende crystallographic preferred orientation and seismic anisotropy of the lower crust. *Tectonophysics* **2023**, *865*, 230036. [CrossRef]
42. Linc, R.; Pantea, E.; Șerban, E.; Ciurba, A.-P.; Serban, G. Hydrochemical and Microbiological Investigations and the Therapeutic Potential of Some Mineral Waters from Bihor County, Romania. *Sustainability* **2023**, *15*, 15640. [CrossRef]
43. Romay-Barrero, H.; Herrero-López, J.; Llorente-González, J.A.; Melgar Del Corral, G.; Palomo-Carrión, R.; Martínez-Galán, I. Balneotherapy and health-related quality of life in individuals with Rheumatoid arthritis: An observational study under real clinical practice conditions. *Balneo PRM Res. J.* **2022**, *13*, 527. [CrossRef]
44. Verhagen, A.P.; Bierma-Zeinstra, S.M.; Boers, M.; Cardoso, J.R.; Lambeck, J.; Bie, R.; Vet, H.C. Balneotherapy (or spa therapy) for rheumatoid arthritis. *Cochrane Database Syst. Rev.* **2015**, *11*, CD000518. [CrossRef]
45. Dickson-Gomez, J.; Nyabigambo, A.; Rudd, A.; Ssentongo, J.; Kiconco, A.; Mayega, R.W. Water, Sanitation, and Hygiene Challenges in Informal Settlements in Kampala, Uganda: A Qualitative Study. *Int. J. Environ. Res. Public Health* **2023**, *20*, 6181. [CrossRef]
46. Wangchuk, P.; Yeshi, K.; Ugyen, K.; Dorji, J.; Wangdi, K.; Samten; Tshering, P.; Nugraha, A.S. Water-Based Therapies of Bhutan: Current Practices and the Recorded Clinical Evidence of Balneotherapy. *Water* **2021**, *13*, 9. [CrossRef]
47. Huang, A.; Seité, S.; Adar, T. The use of balneotherapy in dermatology, *Clin. Dermatol.* **2018**, *36*, 363–368. [CrossRef]
48. Milanković, V.; Djuriš, J.; Tubić, A.; Agbaba, J.; Forkapić, S.; Lukić, M. Assessing the safety of thermal mineral water for cosmetic applications: An integrated approach using physicochemical, cheminformatics, and bioinformatics techniques. *RSC Adv.* **2025**, *15*, 17755–17775. [CrossRef]
49. Choudhury, N.; Siddiqua, T.J.; Ahmed, S.M.T.; Haque, M.A.; Ali, M.; Dil Farzana, F.; Naz, F.; Rahman, S.S.; Faruque, A.S.G.; Rahman, S.; et al. Iron content of drinking water is associated with anaemia status among children in high groundwater iron areas in Bangladesh. *Trop. Med. Int. Health* **2022**, *27*, 149–157. [CrossRef] [PubMed]
50. Noubactep, C.; Kenmogne-Tchidjo, J.F.; Vollmer, S. Iron-fortified water: A new approach for reducing iron deficiency anemia in resource-constrained settings. *Sci. Rep.* **2023**, *13*, 13565. [CrossRef]
51. Dutra-de-Oliveira, J.E.; Lamounier, J.A.; de Almeida, C.N.A.; Marchini, J.S. Fortification of Drinking Water to Control Iron-Deficiency Anemia in Preschool Children. *Food Nutr. Bull.* **2007**, *28*, 173–180. [CrossRef] [PubMed]

Disclaimer/Publisher's Note: The statements, opinions and data contained in all publications are solely those of the individual author(s) and contributor(s) and not of MDPI and/or the editor(s). MDPI and/or the editor(s) disclaim responsibility for any injury to people or property resulting from any ideas, methods, instructions or products referred to in the content.

water

MDPI

Article

Concentration Characteristics, Source Apportionment, and Health Risk Assessment of Groundwater Contaminants in a Karst Coal Mining Area (Guizhou Province, China)

Chendong Yang [1,2], Xiujuan Feng [2,3,*], Liangfu Jiang [1,2], Haotong Guo [2,3], Benqing He [1,2], Ziqi Wang [1,2], Zhiyuan Tang [1,2], Chunfang Chen [1,2], Xinyu Luo [1,2] and Lindani Ncube [4]

1 Guizhou Coal Mine Geological Engineering Consulting and Environmental Monitoring Center, Guiyang 550081, China
2 Guizhou Academician Workstation for Rich Ore Mining and Environmental Protection, Guiyang 550081, China
3 State Key Laboratory of Safe Mining of Deep Coal and Environmental Protection, Huainan 232000, China
4 Department of Environmental Sciences, University of South Africa, Roodepoort 1714, South Africa
* Correspondence: fengxj.2011@tsinghua.org.cn

Abstract

The contamination of groundwater in karst coal mining areas presents a unique environmental challenge due to the interplay between fragile hydrogeology and intensive anthropogenic activity. This study investigated the concentrations, sources, and health risks of characteristic contaminants in groundwater from a karst coal mining area, aiming to provide a scientific basis for groundwater pollution control. Thirty-two groundwater samples were analyzed for nine target contaminants. Principal component analysis (PCA) and a health risk assessment model were integrated to identify pollution sources and evaluate health risks. Results showed that the concentrations of Fe, Mn, Fluoride, Pb, and Sulfate exceeded the Class III limits of the "Standard for Groundwater Quality" (GB/T 14848-2017), with maximum exceedance multiples of 3.60, 1.51, 1.08, and 1.22 times the standard limits, respectively. The maximum concentrations of Mn, Fluoride, and Pb exceeded the WHO guidelines by factors of 4.75, 0.67, and 1.00, respectively. Furthermore, the Pb concentration also surpassed the USEPA standard by a factor of 0.33. PCA identified three principal components, which together explained 71.065% of the total variance and were attributed to mining activities (PC1), mixed natural and anthropogenic sources (PC2), and natural geological processes (PC3), respectively. The health risk assessment reveals significant risks: arsenic poses a carcinogenic risk ($CR > 10^{-4}$), while both arsenic and Fluoride contribute to non-carcinogenic risks ($HI > 1$). The cumulative exposure from these contaminants demands immediate attention.

Keywords: cause analysis; coal mines; distribution characteristics; groundwater pollution; health risk assessment

1. Introduction

Coal resources are the cornerstone of China's energy structure [1]. Groundwater, representing the dominant freshwater resource (comprising over 95% of Earth's liquid freshwater), sustains drinking water supplies for approximately 2.5 billion people and irrigates more than 40% of global croplands [2–4]. However, there is always an opposition between coal mining and groundwater [5–8]. Under natural conditions, groundwater chemistry is primarily governed by atmospheric precipitation, evaporation, and water–rock interactions; however, intensive and prolonged coal mining activities increasingly alter the hydrochemical characteristics of aquifers in mining areas [9]. Empirical evidence from various coal mining regions underscores this impact. For instance, in the Zhongning area of Northwest China, Fluoride and nitrate concentrations in groundwater significantly exceeded drinking water standards [10]. Cheng et al. reported that long-term coal mining significantly disturbs groundwater systems, resulting in aquifer water characterized by elevated levels of Na^+, SO_4^{2-}, and total dissolved solids (TDS), thereby posing environmental health risks [10]. Bhuiyan et al. found that levels of Mn, Fe, Pb, Cu, etc., in most of the water samples in irrigation and drinking water systems in the vicinity of a coal mine area of northwestern Bangladesh exceeded the national and international standards [11]. Chandra et al. indicated that the average concentrations of Fe and Cd exceeded the limits by 1.43 times and 1.6 times, respectively, and Pb and Mn were close to reaching the permissible limits in groundwater samples near coalfield regions of India [12]. Furthermore, mining-induced aquifer damage can lead to a cascade of environmental issues, including groundwater level decline, water resource depletion, and the drying up of surface springs [13].

Guizhou Province is a representative karst region in southern China where approximately 75% of the terrain is underlain by karst formations [14]. Under natural conditions, karst groundwater often exhibits good quality and serves as a vital water source in many parts of China [15]. However, karst aquifers are commonly highly vulnerable to contamination, which is related to the fast recharge, discontinuous storage, and short residence time [16,17]. The release of toxic contaminants from tailings and waste rocks can pose a significant threat to human health [18]. Chronic exposure to Pb and Hg elements is associated with profound neurological and developmental impairments, as well as cardiovascular, renal, gastrointestinal, and hematological dysfunctions [19]. The general health effects of Fluoride on humans include affected or damaged soft tissues, muscles, erythrocytes, gastrointestinal mucosa, ligaments, spermatozoa and thyroid glands; brittle bones; inferior growth; Alzheimer disease; pain; or impeded movement due to bony and dental lesions [20]. Cd is a possible carcinogen and chronic exposure to elevated levels can lead to respiratory diseases [21]. Zn deteriorates human metabolism [21]. High-level inhalation exposure to Mn can cause central nervous system (CNS) damage and lead to various disorders, such as skeletal defects, infertility, heart disease, and hypertension [18].

Mining operations compromise groundwater quality through processes that are contingent upon hydrogeological conditions, coal seam geochemistry, and aquifer characteristics [22,23]. Mining activities can accelerate local groundwater circulation, enhance mixing effects, and intensify water–rock interactions, leading to more complex groundwater chemistry and damage to the hydrogeological structure [9,24]. Abandoned coal mines in well-developed karst areas especially pose significant hydrogeological challenges [25,26]. After the closure of a coal mine, rising water levels in mine shafts can lead to the seepage of water—carrying high loads of pollutants—into coal seams, roadways, and fractures, potentially contaminating nearby drinking water wells [27]. Additionally, large-scale excavation can induce secondary environmental challenges like ground subsidence and deformation [28]. Due to the concealment of groundwater systems, and their complex and rapidly mobile characteristics, once karst groundwater is contaminated it is extremely difficult to repair [29,30]. Hence, the study of the karst groundwater in mining areas is challenging.

Despite the acknowledged impacts of coal mining on groundwater quality, the source apportionment and integrated health risk assessment of contaminants in karst coal mining areas remain in the research stage. While health risks have been assessed in some mining areas, for the karst coal mining area, these assessments are rarely spatially linked to and informed by quantitative source apportionment results. This limits the ability to formulate

https://doi.org/10.3390/w18030351

targeted, source-specific risk management strategies for complex environments like karst coal mining zones.

To address these gaps, this study focuses on a typical karst coal mining concentration zone in Guizhou Province, China. We employ an integrated methodology combining principal component analysis (PCA) for source apportionment with a health risk assessment model. The primary objectives are to (1) characterize the spatial distribution of coal mining characteristic pollutants (CMCPs); (2) quantitatively identify and apportion their predominant sources (mining, mixed, and natural); and (3) evaluate the associated human health risks of CMCPs. This approach aims to provide a scientific basis not only for identifying pollution hotspots but also for prioritizing source-specific control measures, thereby contributing to the sustainable management of groundwater resources in vulnerable karst mining regions.

2. Study Area

2.1. Location and Climate

This investigation focuses on a region in northern Guizhou Province, China (Figure 1). The highest elevation is found in the southeastern part, while the lowest point lies in the western valley. The area is dominated by typical karst landforms, including corroded peak-cluster depressions, trough valleys, and low-mountain valleys [31]. Based on the on-site investigation, depressions and sinkholes are well-developed within the valleys, where the terrain is relatively gentle, with peak–valley height differences ranging from 20 m to 60 m. In contrast, the gullies exhibit significant incision, with relative elevation differences of approximately 300 m.

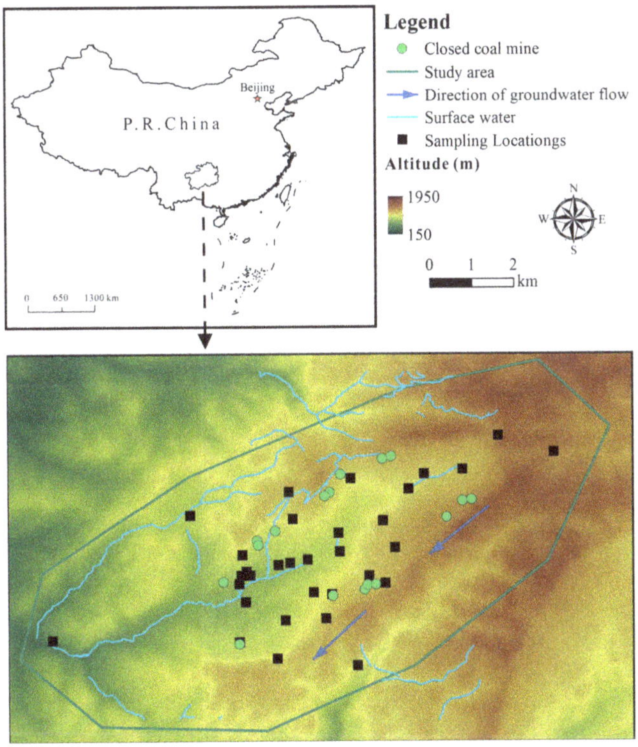

Figure 1. The distribution of sampling points.

https://doi.org/10.3390/ww18030351

The study area falls within the Cwa zone (temperate, dry winter, and hot summer) of the Köppen–Geiger climate classification system. This classification is derived from the high-resolution Köppen–Geiger climate maps (Version 2) [32]. We specifically used the dataset for the 1991—2020 historical baseline period at a 0.5° spatial resolution (approximately 56 km). According to data from the World Meteorological Organization (climatological information based on the 30-year monthly averages from 1971 to 2000), the highest daily average maximum temperature occurs in August at 30.1 °C, and the highest monthly average precipitation occurs in June at 199.4 mm [33].

2.2. Geology and Hydrogeology

The on-site investigation revealed that the oldest stratum in study area is the Permian System; strata mainly consist of limestone. The Triassic System is mainly composed of limestone, sandstone, mudstone and dolomite. The study area contains small perennial streams with low discharge, flowing from northeast to southwest. A river traverses the central part of the study area, where the valley banks are predominantly composed of limestone. This river receives inflows from gullies and springs along its course, stretching 2400 m within the study area and following a westbound flow path along the low-lying synclinal axis. Long-term monitoring data indicate an average discharge of approximately 330 L/s for this river, with a water chemistry type classified as HCO_3^-—SO_4^{2-}—Ca^{2+}.

Based on regional geological and coal mine data, the coal-bearing strata in the study area belong to the Upper Permian Longtan Formation (P_3l), with the main minable coal seam occurring at a burial depth of approximately 50–200 m. The underlying strata are Middle Permian Maokou Formation (P_2m) limestone, and the overlying strata are Upper Permian Changxing Formation (P_3c) limestone, both of which are regionally significant karst aquifers. The Longtan Formation itself is dominated by clastic rocks, acting as a relative aquitard to separate the upper and lower aquifers under natural conditions. However, mining activities can compromise its integrity, inducing fractures that thereby form a hydraulic connection conduit linking the goaf to the aquifers [34,35].

The following hydrogeological characterization is based on a comprehensive field investigation conducted for this study, including geological mapping and analysis of regional hydrogeological survey reports. It is primarily recharged by atmospheric precipitation concentrated between April and October [33]. The recharge mechanisms and intensity are jointly controlled by the lithology, geomorphology, and geological structure. Groundwater dynamics exhibit significant seasonal variations [36], with discharge occurring mainly through concentrated karst springs downstream. Based on regional hydrogeological characteristics, the aquifer system consists predominantly of limestone, with groundwater occurring in dissolution cavities, pores, fissures, and karst conduits [37].

2.3. Coal Mine Factors

According to historical investigations, small-scale coal mining in the study area began predominantly in the 1980s. Most of these mines were distributed near coal seam outcrops, characterized by small production scales and substantial water accumulation in the mining tunnels. To date, all such mines have been closed.

2.4. Selection of Sampling Locations

The groundwater sampling sites were selected based on a combination of scientific and practical considerations to ensure the data were representative of the study area's hydrogeochemical conditions. The selection criteria were as follows: (1) Points were distributed to cover the main groundwater flow paths, including presumed upstream (recharge) areas, within the mining-impacted zones, and downstream (discharge) areas near springs or rivers. (2) Samples were also collected from springs in nearby villages

https://doi.org/10.3390/ww18030351

and agricultural lands to evaluate the potential spread of contaminants to areas of human activity. (3) A limited number of samples were taken from areas presumed to be upgradient or distant from mining activities to provide a local geochemical background reference.

3. Materials and Methods

3.1. Sample and Analysis

Groundwater sampling was conducted from 1–16 April 2024, yielding a total of 32 samples for analysis. All sampling procedures strictly adhered to the Chinese national standards "standard for groundwater quality" (GB/T 14848-2017) [38] and "Technical specifications for environmental monitoring of groundwater" (HJ 164-2020) [39]. The spatial distribution of the sampling points is illustrated in Figure 1. Following collection, samples were promptly transported to a possessing CMA certification for chemical analysis. According to the standard "Emission Standard for Pollutants from Coal Industry" (GB 20426-2006) [40], this study analyzed 9 characteristic coal mine pollutants, including iron (Fe), manganese (Mn), zinc (Zn), Fluoride, mercury (Hg), arsenic (As), cadmium (Cd), lead (Pb) and Sulfate.

Analyses were performed using standardized methods with specified instrumentation: Fe and Mn were determined by flame atomic absorption spectrophotometry (WFX-200 instrument, Beijing Beifen-Ruili Analytical Instrument (Group) Co., Ltd., Beijing, China) following "Water quality—Determination of iron and manganese-Flame atomic absorption spectrophotometric method" (GB 11911-89) [41]. Cd and Pb were analyzed using the same atomic absorption spectrophotometer, according to "Standard examination methods for drinking water-Part 6: Metal and metalloid indices" (GB/T 5750.6-2023) [42]. Hg and As were measured by atomic fluorescence spectrometry (AFS-230E instrument, Beijing Haiguang Instrument Co., Ltd., Beijing, China) per "Water Quality—Determination of Mercury, Arsenic, Selenium, Bismuth and Antimony—Atomic Fluorescence Spectrometry" (HJ 694-2014) [43]. Zn was quantified via atomic absorption spectrometry (WFX-200 instrument) "Water quality—Determination of copper, zinc, lead and cadmium—Atomic absorption spectrometry" (GB 7475-87) [44]. Sulfate was detected by barium chromate spectrophotometry (VIS-7220N instrument, Beijing Beifen-Ruili Analytical Instrument (Group) Co., Ltd., Beijing, China) as per "Water quality—Determination of sulfate-barium chromate spectrophotometry" (HJ/T 342-2007) [45]. The method detection limits for Fe, Mn, Zn, Fluoride, Hg, As, Cd, Pb and Sulfate were 0.03, 0.01, 0.05, 0.05, 0.04×10^{-3}, 0.3×10^{-3}, 0.5×10^{-3}, 2.5×10^{-3} and 8 mg/L, respectively. For quality control, the standard deviation for all measured elements was maintained below 10%.

3.2. Evaluation Methodology

3.2.1. Spatial Distribution Analysis

The spatial distribution of CMCPs was constructed using the inverse distance weighting (IDW) interpolation technique within a Geographic Information System (GIS). This method estimates values at unsampled locations based on the principal of spatial proximity, assigning greater influence to known points that are closer to the prediction site [46,47]. IDW interpolation operates on the fundamental premise that an unknown value can be estimated from known neighboring points, with influence weighted inversely by distance. This distance decay effect ensures that the closest known point exerts the strongest influence, while the impact of farther points diminishes progressively; consequently, the method estimates values at unsampled locations by calculating a weighted average of nearby points [47,48].

https://doi.org/10.3390/ww18030351

3.2.2. Principal Component Analysis (PCA)

We employed principal component analysis (PCA), a dimensionality reduction method, to simplify complex datasets. PCA derives a small number of principal components that capture most of the variance present in the original multiple variables, allowing for streamlined analysis without substantial information loss [49–51]. In their method, the authors noted that implementation in SPSS software enabled the analysis of inter-indicator correlations and the formation of linearly independent composite indicators from related variables [52].

3.2.3. Health Risk Assessment

The human body's exposure to heavy metals in groundwater mainly occurs through oral intake and skin contact, and these two routes account for approximately 90% of the total intake of pollutants [51,53]. Scholars have indicated that human exposure to heavy metals and As is mainly occurring through oral intake and, to a lesser extent, through several other pathways, including dermal contact, inhalation and the food chain [26]. At the same time, given the observed use of some water bodies for irrigation or indirect use purposes, a conservative direct-ingestion scenario was adopted in this study to facilitate a precautionary risk assessment (i.e., the exposure level of residents to chemical components in water can be estimated solely based on their intake of drinking water). The health risk assessment model recommended by the standard "Technical guidelines for risk assessment of soil contamination of land for construction" (HJ 25.3-2019) [54] was used to evaluate the health risk of toxic substances (Cd, Pb, As and Fluoride) using the standard "Standards for drinking water quality" (GB 5749-2022) [55] in the groundwater of the study area. The Chemical-specific Groundwater Exposure Risk for Carcinogenic Effects ($CGWER_{ca}$) and Chemical-specific Groundwater Exposure Risk for Non-carcinogenic Effects ($CGWER_{nc}$) are determined with Equation (1) and Equation (2), respectively.

$$CGWER_{ca} = \frac{GWCR_a \times EF_a \times ED_a}{BW_a \times AT_{ca}} \tag{1}$$

$$CGWER_{nc} = \frac{GWCR_a \times EF_a \times ED_a}{BW_a \times AT_{nc}} \tag{2}$$

In the formula, $GWCR_a$ is the daily groundwater consumption rate of adults (1.8 L/d); EF_a is the exposure frequency of adults (250 d/a); ED_a is the exposure duration of adults (25 a); BW_a is the average body weight of adults (61.8 kg); AT_{ca} is the average time for carcinogenic effect (27,740 d); AT_{nc} is the average time for non-carcinogenic effects (9125 d).

Health risks from groundwater contaminants are defined by two primary metrics: the non-carcinogenic Hazard Index (HI) and the carcinogenic Cancer Risk (CR). These are quantified by the following formulae:

$$CR = CGWER_{ca} \times C_{gw} \times SF_o \tag{3}$$

$$HI = \frac{CGWER_{nc} \times C_{gw}}{RfD_o \times WAF} \tag{4}$$

where CR denotes carcinogenic health risks (mg/L); SF_o denotes the Oral Ingestion Carcinogenic Slope Factor (Table 1); RfD_o denotes the Oral Reference Dose (Table 1); WAF denotes the groundwater allocation factor (0.2).

According to the standard "Technical guidelines for risk assessment of soil contamination of land for construction" (HJ 25.3-2019) [54] and referenced USEPA guidelines, the acceptable CR level for regulatory purposes ranges from 10^{-4} to 10^{-6}. For non-carcinogenic risks, the HI exceeding 1 suggests potential adverse effects on human health [61].

https://doi.org/10.3390/w18030351

Table 1. SF_O and RfD_O of the CMCPs.

Element	SF_O [a]	RfD_O [b]	Source
Cd	- [c]	5.00×10^{-4}	I
Pb	8.50×10^{-3}	4.00×10^{-3}	II
As	32	6.00×10^{-5}	III
Fluoride	-	4.00×10^{-2}	IV

Note: [a]. per mg/kg-day. [b]. mg/kg-d. [c]. The absence of value. "I": [56]; "II": [57,58]; "III": [59]; "IV": [60].

The average population density was derived from the 2020 1 km gridded population dataset for China [62]. Using ArcGIS 10.8, the study area boundary was overlaid on the raster layer. The mean population count per grid cell within the boundary was calculated. This value, directly equivalent to persons per square kilometer (persons/km^2), represents the area-weighted average density that accounts for the spatial heterogeneity of settlement, including uninhabited zones.

Field surveys confirm that water from these sources is utilized by the local community through a variety of means (e.g., springs) and serves as a primary source for agricultural irrigation. Consequently, contaminants detected in these groundwater systems present plausible contamination risks. Given this context, and in line with a precautionary screening approach, a health risk assessment based on a conservative direct-ingestion scenario is applied to evaluate the potential baseline risks.

3.2.4. Uncertainty Analysis

This study focused solely on the health risks posed by drinking water, and the assessed risk values may be underestimated compared to real-world scenarios. The risk assessment was conducted based on fixed health analysis parameters for adults, whereas actual conditions involve inherent parameter uncertainties rather than deterministic values. Additionally, the limited number of groundwater samples collected in this study may introduce some degree of randomness in the evaluation results. Although all sampling and analytical procedures strictly adhered to national standards, potential interference factors during sample collection and testing could still influence the measured outcomes. These uncertainties may affect the accuracy of the health risk assessment; however, they do not invalidate the findings.

3.2.5. Data Analysis

The relationship among CMCPs in groundwater was determined using the Pearson correlation and abundance heatmap by Origin Pro (version 2024). We interpreted the results as follows: $p < 0.05$ was taken to indicate a significant correlation, and $p < 0.01$ was interpreted as indicating a highly significant one. The PCA and standardization of original data were also performed by IBM SPSS version 22 [63]. Sampling points and groundwater quality were mapped using ArcGIS Desktop 10.8 [64].

4. Results and Discussion

4.1. Concentration and Spatial Distribution Characteristics of CMCPs

In Table 2, Fe, Mn, Fluoride, Pb and Sulfate exceeded the limit values of Class III water in the "Standard for Groundwater Quality" (GB 14848-2017) [38]. The maximum Mn concentration detected in groundwater was 0.46 mg/L. This value exceeds the Class III water in the "Standard for Groundwater Quality" (GB 14848-2017) (0.1 mg/L) and the WHO permissible limit (0.08 mg/L), with max exceedance factors of 3.6 and 4.75, respectively. The maximum concentration of Fluoride detected in groundwater was 2.51 mg/L, exceeding the Class III limit (1.0 mg/L) of GB/T 14848-2017 and the WHO permissible limit (1.5 mg/L),

https://doi.org/10.3390/ww18030351

with max exceedance factors of 1.51 and 0.67, respectively. However, it did not surpass the standard set by the USEPA. The concentration of Pb in groundwater was 0.02, exceeding the standard values of the "Standard for Groundwater Quality" (GB 14848-2017) (0.01 mg/L), the WHO (0.01 mg/L), and also crossing the standard limit of the USEPA (0.015 mg/L), with max exceedance factors of 1.00, 1.00 and 0.33. The maximum concentrations of Fe and Sulfate in groundwater reached 0.49 mg/L and 556 mg/L, respectively. These values exceeded the Class III thresholds specified in GB/T 14848-2017, with max exceedance factors of 0.63 and 1.22. In contrast, the detected concentrations of Zn, Hg, As and Cd did not surpass any applicable standards or limits.

Table 2. The content of CMCPs in groundwater.

Element	Min	Max	Mean	SD [b]	C.V. [b]	Detection Rate [a]	Exceedance Rate of Monitoring Sites [a]	Class III [c]	WHO [d]	USEPA [e]
Fe	ND [a]	0.49	0.05	0.11	2.29	25.00	6.25	0.3	-	-
Mn	ND	0.46	0.04	0.11	2.53	21.88	12.50	0.1	0.08	-
Zn	ND	0.27	0.01	0.05	5.66	3.13	0.00	1	-	-
Fluoride	0.09	2.51	0.41	0.52	1.26	100.00	6.25	1	1.5	4.0
Hg	ND	4.00×10^{-5}	1.25×10^{-6}	7.07×10^{-6}	5.66	3.13	0.00	0.001	0.006	0.002
As	ND	1.90×10^{-3}	4.53×10^{-4}	5.27×10^{-4}	1.16	53.13	0.00	0.01	0.01	0.01
Cd	ND	2.80×10^{-3}	5.91×10^{-4}	8.64×10^{-4}	1.46	37.50	0.00	0.005	0.003	0.005
Pb	ND	0.02	3.57×10^{-3}	4.18×10^{-3}	1.17	78.13	3.13	0.01	0.01	0.015
Sulfate	9.33	556	162.26	160.64	0.99	100.00	18.75	250	-	-

Notes: [a]. "ND" means not detected; "-" means no guideline value or standards; detection rate (%); exceedance rate of monitoring sites (%) (compared with GB 14848-2017); except for C.V., which is dimensionless, others were mg/L. [b]. SD denotes the standard deviation; C.V. represents the coefficient of variation. [c]. The "Standard for Groundwater Quality" (GB 14848-2017) [38]. [d]. The "World Health Organization" [65]. [e]. The USEPA values listed are the Maximum Contaminant Level (MCL)—enforceable regulatory limits set under the National Primary Drinking Water Regulations (NPDWRs) [66].

Among the tested CMCPs, the coefficient of variation (C.V.) exceeded 1.00 in the following order: Zn = Hg > Mn > Fe > Cd > Fluoride > Pb > As >1.00, indicating there is a certain variation in water quality among the groundwater samples [10]. The overall spatial distribution characteristics of the elements Mn and Zn (Figure 2b,c) are opposite to those of Cd, Pb, and Sulfate (Figure 2g–i), which show higher concentrations in the southwestern portions of the area. The spatial map also shows that discrete zones of high concentration for Fe were observed (Figure 2a); the central region shows a high concentration of Fluoride and As (Figure 2d,f).

4.2. Source Apportionment of CMCPs

4.2.1. Correlation Analysis

Figure 3 shows the Pearson correlation analysis results of groundwater. The results show that the correlation coefficients between Fe and Zn, indicating significant positive correlations between them ($p < 0.05$). The correlation coefficients between Mn and Zn, Mn and Sulfate, Fluoride and As, Cd and Pb, Cd and Sulfate and Pb and Sulfate indicate significant strong positive correlations between them ($p < 0.01$).

4.2.2. PCA-Based Source Apportionment

The standardized data were deemed suitable for PCA, as supported by factorability tests. The Kaiser–Meyer–Olkin (KMO) measure yielded a value of 0.62 (exceeding the 0.50 threshold), and Bartlett's test of sphericity reached significance ($p < 0.001$). PCA extracted three significant principal components (eigenvalues > 1), which cumulatively capture 71.065% of the total dataset variance (Table 3). The first component (PC1), explaining

https://doi.org/10.3390/w18030351

34.470% of the variance, showed high loadings for Cd, Pb, and Sulfate. Subsequently, PC2 (19.253%) was characterized by Zn, Mn, and Fe, and PC3 (17.342%) by Fluoride and As.

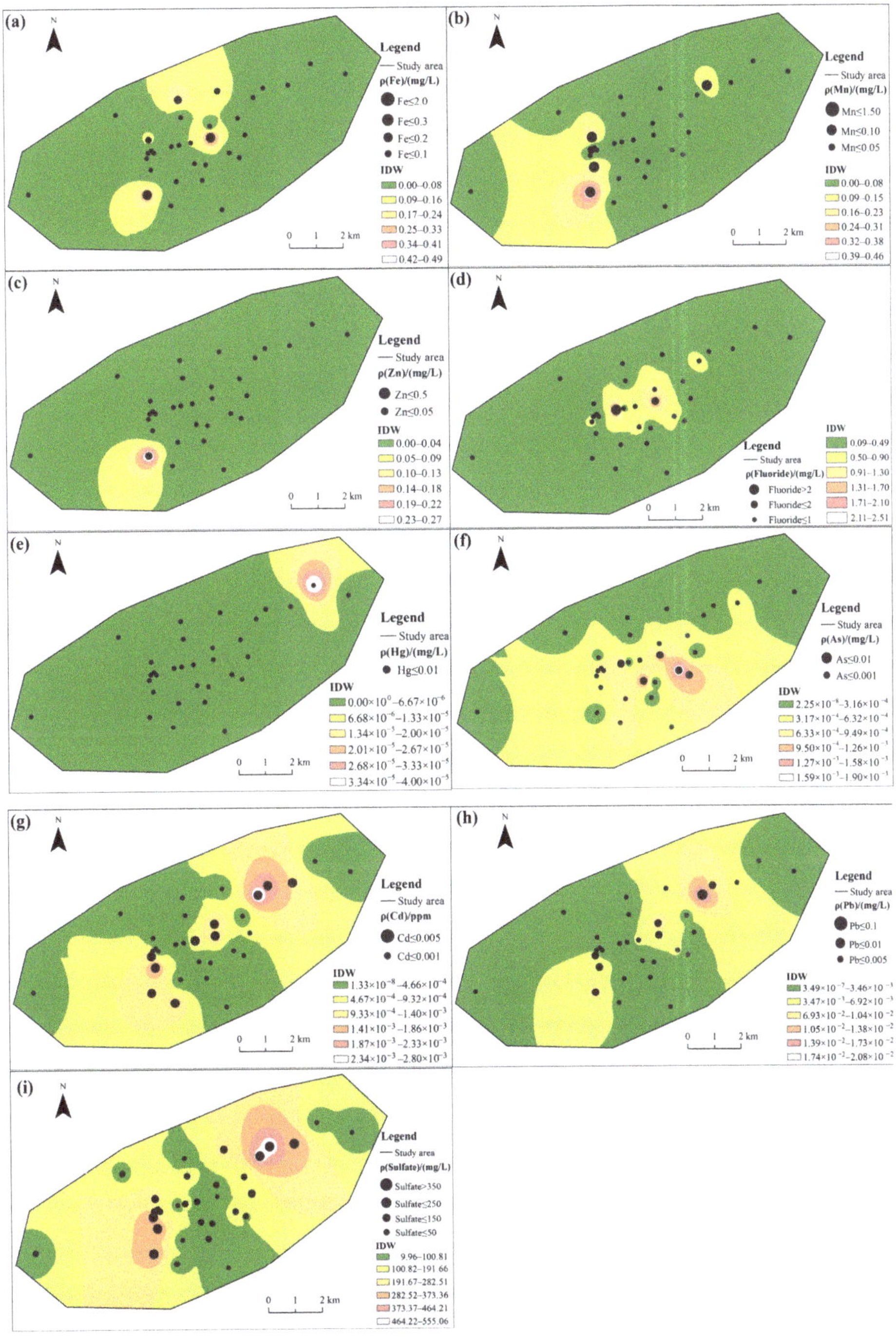

Figure 2. Spatial distribution of groundwater quality: (**a**) Fe, (**b**) Mn, (**c**) Zn, (**d**) Fluoride, (**e**) Hg, (**f**) As, (**g**) Cd, (**h**) Pb, and (**i**) Sulfate.

https://doi.org/10.3390/w18030351

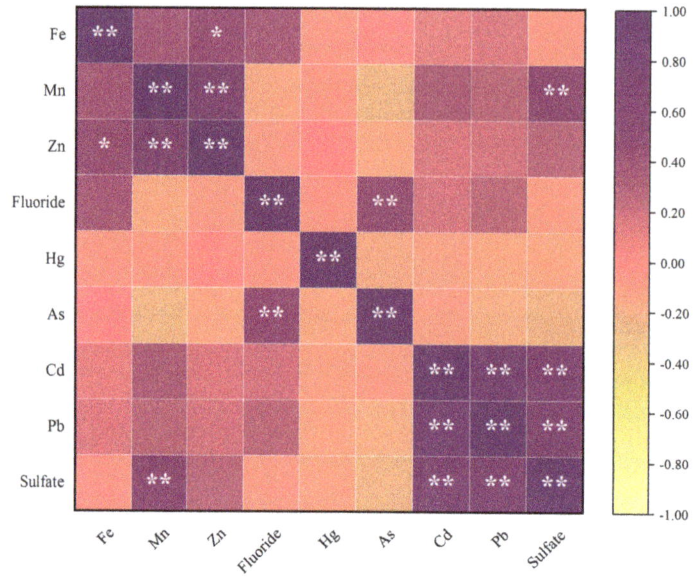

Figure 3. Analysis results of groundwater correlation. * means $p < 0.05$, ** means $p < 0.01$.

Table 3. Rotating load matrix.

Variable	PC1	PC2	PC3
Cd	**0.927**	0.098	0.115
Pb	**0.909**	0.098	0.131
Sulfate	**0.869**	0.164	−0.223
Zn	0.116	**0.872**	−0.125
Mn	0.324	**0.790**	−0.254
Fe	−0.051	**0.744**	0.418
Fluoride	0.123	0.034	**0.878**
As	−0.179	−0.174	**0.737**
Hg	−0.222	−0.043	−0.266
Eigenvalue	3.102	1.733	1.561
Explained variance (%)	34.470	19.253	17.342
Cumulative variance (%)	34.470	53.724	71.065

Note: Bold values represent the loadings > 0.7 in absolute value.

According to Table 3, PC1 exhibits high loadings for Cd, Pb, and Sulfate, primarily reflecting their contamination characteristics. The results in Figure 3 indicate significant positive correlations ($p < 0.01$) between Cd and Pb as well as Sulfate, suggesting that they may share similar sources. Cheng et al. indicates that the damage to underground aquifers during coal mining results in cross-layer pollution between aquifers at different depths, and the high concentration of Na^+ and SO_4^{2-} in the deep aquifer migrates to the shallow aquifer through fractured geological structures, thereby polluting the shallow groundwater source [10]. Kim et al. identified SO_4^{2-} in six metalliferous mines and one coal mine in South Korea, which has been shown to be the most reliable indicator of the mining impact on groundwater [67]. Bhuiyan et al. found that the Pb concentration was comparatively higher, with 80% samples exceeding the Bangladesh (Govt.) maximum admissible limit standard for underground mine water, which may pose serious ecological and human health hazards [68]. Dheeraj et al. used the PCA method to study a coalfield region in India, showing large positive loadings of Cd, suggesting an association with nearby mining

https://doi.org/10.3390/w18030351

activities, thus aligning with our study [47]. The combined evidence indicates that PC1 likely originates from mining activities.

PC2 shows high loadings of Mn, Zn and Fe, indicating their contamination characteristics. Mining activities generate substantial quantities of waste rock containing sulfide minerals such as pyrite (FeS_2) [69]. Bharat et al. observed that, in coal formations, iron is primarily hosted in sulfide minerals like pyrite (FeS_2), chalcopyrite ($CuFeS_2$), and arsenopyrite (AsFeS). Upon weathering, this increases iron and Sulfate environmental concentrations [70]. Zhou et al. considered that the natural weathering of sulfide minerals represented by pyrite was related to the Fe release process in abandoned coal mines [71]. Hao et al. found that seasonal variations in Mn contribute to the dynamic nature of geochemical processes [72]. This means that the natural geological process is dominated by Mn release from the weathering of carbonate rocks and Mn-rich mineral dissolution [72–74]. Mining activities can lead to elevated concentrations of trace elements (e.g., Mn, Ni, Si, and Sr) through the acidic leaching and release of these constituents from sulfide, silicate, and carbonate minerals within the strata [75]. Lin et al. previously published a study that found Zn and Fe showed a close correlation [76], similar to our study. Moreover, Lin et al. thought Zn could come from schistose rocks with sulfide seams [76]. Bharat et al. identified atmospheric dust, coal burning, and mining waste disposal as key anthropogenic contributors to Zn levels in aquatic systems. [70]. So, these results suggest that PC2 may represent both natural actions and anthropogenic sources.

PC3 exhibits high loadings on Fluoride and As, primarily reflecting their contamination characteristics. As indicated in Figure 3, there is a significant positive correlation between Fluoride and As. The arsenic and Fluoride ions affecting groundwater quality are primarily released from minerals such as arsenopyrite (FeAsS), fluorite (CaF_2), and micas, which are hosted within the sedimentary and metamorphic rocks and coal seams of the region [77–79]. At the abandoned Kettara mine in Morocco, hydrogeological and geochemical investigations attributed the As contamination in groundwater to a geological origin (mafic rocks) rather than mining activities [80]. According to Rashid et al., "the Middle Jurassic strata are chiefly composed of sedimentary rocks, including limestone, shale, and sandstone. Notably, the presence of minerals like arsenopyrite and fluorite within these strata is a key contributor to aquifer contamination, introducing arsenic and fluoride, respectively" [79]. The main lithologies in this study were limestone and sandstone, similar to Rashid's study. In the Mahendragarh district, Haryana, India, its high-Fluoride groundwater was likewise attributed to silicate weathering, highlighting the dominance of natural geochemical processes [81]. Wang et al. attributed the enrichment of arsenic and Fluoride in confined aquifers to organic matter degradation. They proposed that elevated temperatures exacerbate this release through a thermally driven process encompassing organic matter mineralization, arsenic desorption, and the dissolution of Fluoride-bearing minerals [82], which shows that PC3 likely originates from natural actions.

4.3. *Water Quality Evaluation of Human Drinking Water Health*

The average population density was derived from the 2020 1 km gridded population dataset for China, and the analysis yielded an average population density of 195.69 persons per square kilometer for the study area. This value reflects the rural character of the region, where settlements are dispersed amidst larger tracts of agricultural or natural land. Figure 4 shows the CR and HI statistical table. For cancer risks (CRs), according to the standard "Technical guidelines for risk assessment of soil contamination of land for construction" (HJ 25.3-2019) [54], and the reference USPEA standard, if the CR level in groundwater is between 10^{-4} and 10^{-6}, we consider that there is no considerable risk regarding detrimental health effects. The Cd and Fluoride in the groundwater were negligible for adults (the absence of value) (Figure 4a,c). Additionally, CR values between 10^{-4} and 10^{-6} were observed at other points, including one sampling point associated with lead in the northeastern part (Figure 4d) and two sampling points associated with As in the central part (Figure 4b). However, As carcinogenic risk (CR greater than 10^{-4}) was identified at 15 sampling points (Figure 4b), and the potential risks at these locations merit attention. These areas warrant priority consideration in monitoring and management strategies.

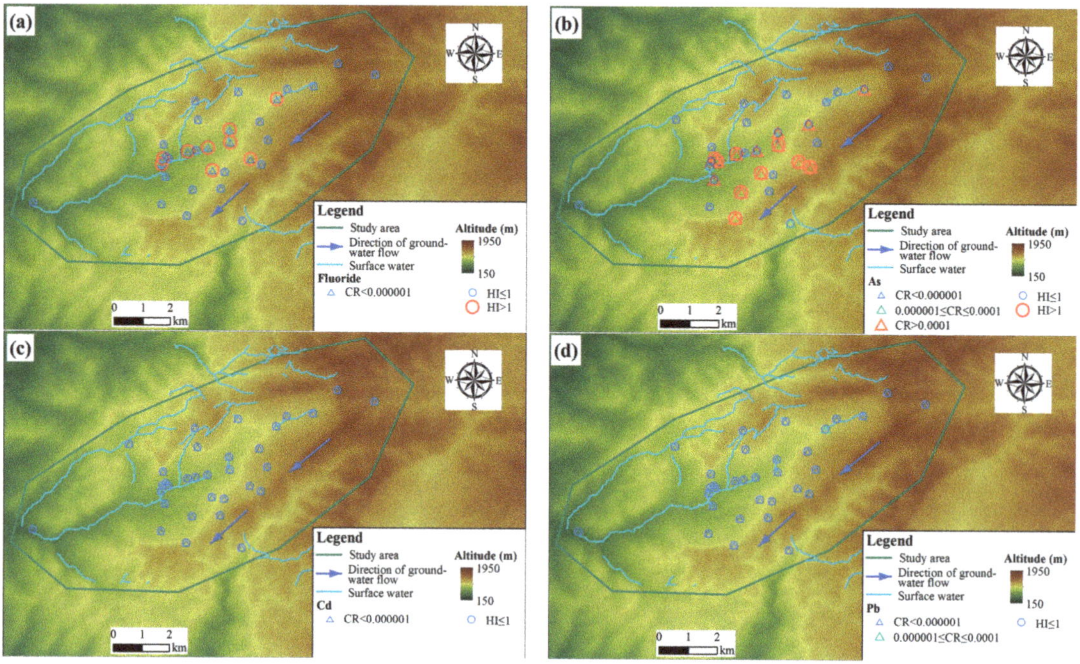

Figure 4. Cancer Risk (CR) and Hazard Index (HI) health risk assessment results: (**a**) Fluoride, (**b**) As, (**c**) Cd, and (**d**) Pb.

Groundwater samples were assessed for non-carcinogenic health risks using the Hazard Index (HI), as per HJ 25.3-2019. HI values of less than 1 are interpreted as indicating no significant risk. Any exceedance of 1, however, is deemed to reflect a potential for adverse health effects and thus represents an unacceptable risk level. The results showed that the health risk assessment identified nine sampling points each for Fluoride and As with a HI greater than 1, indicating an unacceptable non-carcinogenic risk. For both elements, these high-risk points are predominantly concentrated in the central part of the study area (Figure 4a,b).

4.4. Synthesis of Key Findings

In summary, the integrated analysis of this karst coal mining basin reveals the following groundwater contamination profile:

(1) The concentrations of Fe, Mn, Fluoride, Pb, and Sulfate in the groundwater of the study area exceeded the Class III limits specified in China's "Standard for Groundwater Quality" (GB/T 14848-2017), with maximum exceedance multiples of 0.63, 3.60, 1.51, 1.00, and 1.22 times the respective standards. The maximum concentrations exceeded WHO guidelines, with Mn, Fluoride, and Pb being higher by factors of 4.75, 0.67, and 1.00, respectively. Additionally, the Pb level exceeded the USEPA standard by a factor of 0.33. Analysis of the coefficient of variation (C.V.) indicated spatially heterogeneous distributions

https://doi.org/10.3390/w18030351

of Zn, Hg, Mn, Fe, Cd, Fluoride, Pb, and As. Elevated concentrations of Mn and Zn were observed in the southwestern part of the area, contrasting with the spatial pattern of Cd, Pb, and Sulfate, which showed higher levels in the northeast. Fe exhibited discrete high-concentration patches, while Fluoride and As had high contents in the central region.

(2) Correlation analysis revealed a highly significant positive relationship ($p < 0.01$) between Cd and Pb as well as Sulfate, suggesting their potential common origin. Principal component analysis (PCA) further supported this finding, with PC1 showing high loadings of Cd, Pb, and Sulfate, indicating that mining activities may serve as their predominant pollution source. Correlation analysis revealed a highly significant positive relationship ($p < 0.01$) between Mn and Zn, as well as the correlation coefficients between Fe and Zn, indicating significant positive correlations between them ($p < 0.05$). PC2 may therefore represent both natural actions and anthropogenic sources. A significant positive correlation ($p < 0.05$) was observed between Fluoride and As, and the PCA results indicated that PC3 was primarily loaded with Fluoride and As. This implies that natural geological processes may be the dominant source of these contaminants.

(3) Health risk assessment indicated As posed a carcinogenic risk ($CR > 10^{-4}$) at 15 sampling points. This finding is consistent with the study by Liu et al. in an industrial hub, which also identified arsenic as a primary contributor to carcinogenic risk [83]. Fluoride and As resulted in a non-carcinogenic risk ($HI > 1$) at nine sampling points, indicating that there is a possibility for adverse effects on human health. Similarly, research at the abandoned Kettara mine in Morocco indicated that As in groundwater poses a potential threat to public health [80]. In the Mahendragarh district, Haryana, India, Fluoride exposure from groundwater also resulted in a high non-carcinogenic risk [81]. Therefore, for our study area, it is necessary to incorporate regular groundwater quality monitoring into the regional environmental management framework. These threats highlight that it is necessary to follow the evolution of As and Fluoride concentrations, as well as perform microbiological and clinical studies, which are an essential preventive measure.

5. Conclusions

The three PCA-derived sources collectively delineate a "triple-overlay" pollution framework that is likely characteristic of complex mining districts in vulnerable karst terrains. This framework synthesizes three concurrent and interactive processes: (i) direct anthropogenic injection from mining activities (PC1: Cd, Pb, and Sulfate), which is widely recognized as the primary impact; (ii) mining-enhanced geochemical leaching (PC2: Mn, Zn and Fe), wherein the natural weathering of minerals is accelerated by anthropogenic disturbances; and (iii) pristine geogenic release (PC3: Fluoride and As) from the local aquifer matrix, governed by natural water–rock interactions. The co-existence of these processes challenges the simplistic binary attribution of pollution to either "anthropogenic" or "natural" sources and provides a more nuanced conceptual model for environmental assessment in similar settings.

A critical and pragmatic insight from this study is the necessity to further study the health significance of geogenically sourced contaminants in mining-impacted areas. Although Fluoride and As (PC3) were attributed to natural geological processes, their associated health risks were found to be non-negligible. This underscores a key principle: a natural origin does not equate to an acceptable risk where sustained human exposure occurs. Consequently, water quality management strategies in such regions must expand beyond a sole focus on conventional industrial pollutants. Rigorous and long-term monitoring of inherent geogenic contaminants, informed by local geological knowledge, should be integrated into standard practice to ensure comprehensive public health protection.

https://doi.org/10.3390/w18030351

Supplementary Materials: The following supporting information can be downloaded at: https://www.mdpi.com/article/10.3390/w18030351/s1, Table S1: Information on the sampled wells.

Author Contributions: Conceptualization, X.F.; methodology, X.F. and C.Y.; software, Z.W. and B.H.; validation, H.G. and C.C.; formal analysis, H.G., L.J. and L.N.; investigation, C.Y., H.G. and X.F.; resources, X.F.; data curation, L.J., C.Y. and Z.T.; writing—original draft preparation, H.G. and C.Y.; writing—review and editing, X.F., H.G. and C.Y.; visualization, X.L. and C.Y.; supervision, X.F.; project administration, X.F.; funding acquisition, X.F. All authors have read and agreed to the published version of the manuscript.

Funding: This work was supported by Bonanza and Precision Mining, Guizhou Provincial Academician Expert Workstation (KXJZ [2024]003), Anhui University of Science and Technology's high-level talent team launch funding project (YJ20250001), and the National Key R&D Program of China (2021YFC2902100).

Data Availability Statement: The original contributions presented in this study are included in the article/Supplementary Materials. Further inquiries can be directed to the corresponding author.

Conflicts of Interest: The authors declare no conflicts of interest.

References

1. Jiang, C.; Gao, X.; Hou, B.; Zhang, S.; Zhang, J.; Li, C.; Wang, W. Occurrence and environmental impact of coal mine goaf water in karst areas in China. *J. Clean. Prod.* **2020**, *275*, 123813. [CrossRef]
2. Lall, U.; Josset, L.; Russo, T. A snapshot of the world's groundwater challenges. *Annu. Rev. Environ. Resour.* **2020**, *45*, 171–194. [CrossRef]
3. Wang, Y.; Yuan, S.; Shi, J.; Ma, T.; Xie, X.; Deng, Y.; Du, Y.; Gan, Y.; Guo, Z.; Dong, Y.; et al. Groundwater Quality and Health: Making the Invisible Visible. *Environ. Sci. Technol.* **2023**, *57*, 5125–5136. [CrossRef]
4. Li, Y.; Zhang, X. Characterization of the health and irrigation risks and hydrochemical properties of groundwater: A case study of the Selian coal mine area, Ordos, Inner Mongolia. *Appl. Water Sci.* **2022**, *12*, 236. [CrossRef]
5. Qiao, W.; Li, W.; Zhang, S.; Niu, Y. Effects of coal mining on the evolution of groundwater hydrogeochemistry. *Hydrogeol. J.* **2019**, *27*, 2245–2262. [CrossRef]
6. Jarvis, A.P.; Younger, P.L. Broadening the scope of mine water environmental impact assessment: A UK perspective. *Environ. Impact Assess. Rev.* **2000**, *20*, 85–96. [CrossRef]
7. Kallioras, A.; Ruzinski, N. Special Issue: Sustainable Development of Energy, Water and Environment Systems. *Water Resour. Manag.* **2011**, *25*, 2917–2918. [CrossRef]
8. Johnson, K.L.; Younger, P.L. The co-treatment of sewage and mine waters in aerobic wetlands. *Eng. Geol.* **2006**, *85*, 53–61. [CrossRef]
9. Hou, Z.; Huang, L.; Zhang, S.; Han, X.; Xu, J.; Li, Y. Identification of groundwater hydrogeochemistry and the hydraulic connections of aquifers in a complex coal mine. *J. Hydrol.* **2024**, *628*, 130496. [CrossRef]
10. Cheng, W.; Yin, H.; Dong, F.; Li, Y.; Guo, Q.; Wang, Y. Hydrochemical characteristics, cross-layer pollution and environmental health risk of groundwater system in coal mine area: A case study of Jiangzhuang coal mine. *Environ. Geochem. Health* **2024**, *46*, 523. [CrossRef]
11. Bhuiyan, M.A.H.; Islam, M.A.; Dampare, S.B.; Parvez, L.; Suzuki, S. Evaluation of hazardous metal pollution in irrigation and drinking water systems in the vicinity of a coal mine area of northwestern Bangladesh. *J. Hazard. Mater.* **2010**, *179*, 1065–1077. [CrossRef]
12. Chandra, G.V.; Ghosh, P.K. Systematic review on groundwater quality and its controlling mechanisms in coal mining regions: Health risk assessment and strategies for sustainability. *Environ. Dev. Sustain.* **2024**, 1–25. [CrossRef]
13. Sun, K.; Fan, L.; Xia, Y.; Li, C.; Chen, J.; Gao, S.; Wu, B.; Peng, J.; Ji, Y. Impact of coal mining on groundwater of Luohe Formation in Binchang mining area. *Int. J. Coal Sci. Technol.* **2021**, *8*, 88–102. [CrossRef]
14. Wang, Z.; Torres, M.; Paudel, P.; Hu, L.; Yang, G.; Chu, X. Assessing the Karst Groundwater Quality and Hydrogeochemical Characteristics of a Prominent Dolomite Aquifer in Guizhou, China. *Water* **2020**, *12*, 2584. [CrossRef]
15. Tian, B.; Lu, C.; Zhao, D.; Li, J.; Zhu, H.; Dong, F. Hydrochemical Evolution of Karst Groundwater and Mining-Induced Activation Effects in a Coal Mining Area: A Case Study from the Tengxian Coalfield, China. *ACS Omega* **2025**, *10*, 42548–42560. [CrossRef] [PubMed]
16. Li, J.; Yang, G.; Zhu, D.; Xie, H.; Zhao, Y.; Fan, L.; Zou, S. Hydrogeochemistry of karst groundwater for the environmental and health risk assessment: The case of the suburban area of Chongqing (Southwest China). *Geochemistry* **2022**, *82*, 125866. [CrossRef]

17. Su, C.; Zhang, X.; Sun, Y.; Meng, S.; Cui, X.; Fei, Y. Hydrochemical characteristics and evolution processes of karst groundwater in Pingyin karst groundwater system, North China. *Environ. Earth Sci.* **2023**, *82*, 67. [CrossRef]
18. Dey, S.; Tripathy, B.; Kumar, M.S.; Das, A.P. Ecotoxicological consequences of manganese mining pollutants and their biological remediation. *Environ. Chem. Ecotoxicol.* **2023**, *5*, 55–61. [CrossRef]
19. Pedron, T.; Cristine Augusto, C.; Carvalho Silva, G.; Cezar Valeriano, M.; Benicia Mamián-López, M.; Slaveykova, V.I.; Lemos Batista, B. Human health risk assessment and concentration of Al, Mn, Fe, Cu, Zn, Se, As, Cd, Pb, Hg, Rb, and REEs in chocolate. *Food. Chem. Toxicol.* **2025**, *206*, 115770. [CrossRef]
20. Kabir, H.; Gupta, A.K.; Tripathy, S. Fluoride and human health: Systematic appraisal of sources, exposures, metabolism, and toxicity. *Crit. Rev. Environ. Sci. Technol.* **2020**, *50*, 1116–1193. [CrossRef]
21. Karar, K.; Gupta, A.K.; Kumar, A.; Biswas, A.K. Characterization and Identification of the Sources of Chromium, Zinc, Lead, Cadmium, Nickel, Manganese and Iron in Pm10 Particulates at the Two Sites of Kolkata, India. *Environ. Monit. Assess.* **2006**, *120*, 347–360. [CrossRef]
22. Ju, Q.; Hu, Y.; Liu, Q. Hydrogeochemical impacts of coal mining on multi-aquifer systems: A case study in the Linhuan Mining District, China. *Ecotoxicol. Environ. Saf.* **2025**, *308*, 119492. [CrossRef]
23. Burri, N.M.; Weatherl, R.; Moeck, C.; Schirmer, M. A review of threats to groundwater quality in the anthropocene. *Sci. Total Environ.* **2019**, *684*, 136–154. [CrossRef]
24. Huang, P.; Hu, Y.; Gao, H.; Su, Q. Dynamic identification and radium—Radon response mechanism of floor mixed water source in high ground temperature coal mine. *J. Hydrol.* **2021**, *603*, 126942. [CrossRef]
25. Qin, W.; Han, D.; Song, X.; Liu, S. Environmental isotopes (δ18O, δ2H, 222Rn) and hydrochemical evidence for understanding rainfall-surface water-groundwater transformations in a polluted karst area. *J. Hydrol.* **2021**, *592*, 125748. [CrossRef]
26. Hao, H.; Hao, X.; Wu, P.; Li, X.; Han, Z.; Cao, X.; Yang, B.; Zhang, R.; Zhang, K.; Chen, M.; et al. Different hydrogeological characteristics of springs associated with abandoned coal mines in well-developed karst area. *J. Hydrol.* **2025**, *653*, 132683. [CrossRef]
27. Zhu, M.; Li, B.; Liu, G. Groundwater risk assessment of abandoned mines based on pressure-state-response—The example of an abandoned mine in southwest China. *Energy Rep.* **2022**, *8*, 10728–10740. [CrossRef]
28. Wang, L.; Jia, B.; Yang, F.; Huang, Q.; Peng, Q.; Wu, R.; Xie, Z. Anthropogenic coal mining reducing groundwater storage in the Yellow River Basin. *Sci. Total Environ.* **2025**, *958*, 178120. [CrossRef] [PubMed]
29. Cao, H.; Dong, W.; Chen, H.; Wang, R. Groundwater vulnerability assessment of typical covered karst areas in northern China based on an improved COPK method. *J. Hydrol.* **2023**, *624*, 129904. [CrossRef]
30. Zhang, W.; Xin, C.; Yu, S. A Review of Heavy Metal Migration and Its Influencing Factors in Karst Groundwater, Northern and Southern China. *Water* **2023**, *15*, 3690. [CrossRef]
31. Ya, L.; Zhang, R.; Xue, X.; Liu, M.; Wang, J.; Lou, J. Research on karst landform classification based on positive and negative terrainsTaking the karst region of Guizhou Province as an example. *J. Guizhou Norm. Univ. (Nat. Sci.)* **2024**, *42*, 9–19. (In Chinese)
32. Beck, H.E.; McVicar, T.R.; Vergopolan, N.; Berg, A.; Lutsko, N.J.; Dufour, A.; Zeng, Z.; Jiang, X.; van Dijk, A.I.J.M.; Miralles, D.G. High-resolution (1 km) Köppen-Geiger maps for 1901–2099 based on constrained CMIP6 projections. *Sci. Data* **2023**, *10*, 724. [CrossRef]
33. World Meteorological Organization (WMO). Climatological Information for Zunyi. Available online: https://worldweather.wmo.int/zh/city.html?cityId=1874 (accessed on 24 December 2025).
34. Xu, S.; Zhang, Y.; Shi, H.; Wang, K.; Geng, Y.; Chen, J. Physical Simulation of Strata Failure and Its Impact on Overlying Unconsolidated Aquifer at Various Mining Depths. *Water* **2018**, *10*, 650. [CrossRef]
35. Sun, Q.; Zhang, J.; Li, M.; Zhou, N. Experimental evaluation of physical, mechanical, and permeability parameters of key aquiclude strata in a typical mining area of China. *J. Clean. Prod.* **2020**, *267*, 122109. [CrossRef]
36. Ding, K.; Li, Y.; Wang, L.; Xue, P.; Li, J.; Han, D.; Yu, L. Research on Groundwater Drought Characteristics in Guizhou Province Based on GRACE Data. *J. China Hydrol.* **2025**, 1–10. (In Chinese) [CrossRef]
37. Zhang, J.; Zhu, J.; He, C.; Jia, Z.; Zhu, L.; Yi, Y.; Gong, J. Research progress on the water-carbon coupling processes at different scales in karst regions. *Carsologica Sin.* **2025**, *44*, 261–273. (In Chinese)
38. *GB/T 14848-2017*; Standard for Groundwater Quality; General Administration of Quality Supervision. Inspection and Quarantine of the People's Republic of China: Beijing, China, 2017.
39. *HJ 164-2020*; Technical Specifications for Environmental Monitoring of Groundwater. Ministry of Ecology and Environment of the People's Republic of China: Beijing, China, 2020.
40. *GB 20426-2006*; Emission Standard for Pollutants from Coal Industry. State Environmental Protection Administration of China: Beijing, China, 2006.
41. *GB 11911-89*; Water Quality-Determination of Iron and Manganese-Flame Atomic Absorption Spectrophotometric Method. State Environmental Protection Administration of China: Beijing, China, 1989.

https://doi.org/10.3390/w18030351

42. *GB/T 5750.6-2023*; Standard Examination Methods for Drinking Water-Part 6: Metal and Metalloid Indices. State Administration for Market Regulation: Beijing, China, 2023.
43. *HJ 694-2014*; Water Quality—Determination of Mercury, Arsenic, Selenium, Bismuth and Antimony—Atomic Fluorescence Spectrometry. Ministry of Ecology and Environment of the People's Republic of China: Beijing, China, 2014.
44. *GB 7475-87*; Water Quality—Determination of Copper, Zinc, Lead and Cadmium—Atomlc Absorption Spectrometry. State Environmental Protection Administration of China: Beijing, China, 1987.
45. *HJ/T 342-2007*; Water Quality-Determination of Sulfate-Barium Chromate Spectrophotometry. State Environmental Protection Administration of China: Beijing, China, 2007.
46. Choudhury, T.R.; Ferdous, J.; Haque, M.M.; Rahman, M.M.; Quraishi, S.B.; Rahman, M.S. Assessment of heavy metals and radionuclides in groundwater and associated human health risk appraisal in the vicinity of Rooppur nuclear power plant, Bangladesh. *J. Contam. Hydrol.* **2022**, *251*, 104072. [CrossRef]
47. Dheeraj, V.P.; Singh, C.S.; Sonkar, A.K.; Kishore, N. Heavy metal pollution indices estimation and principal component analysis to evaluate the groundwater quality for drinking purposes in coalfield region, India. *Sustain. Water Resour. Manag.* **2024**, *10*, 31. [CrossRef]
48. Uddin, M.G.; Diganta, M.T.M.; Sajib, A.M.; Hasan, M.A.; Moniruzzaman, M.; Rahman, A.; Olbert, A.I.; Moniruzzaman, M. Assessment of hydrogeochemistry in groundwater using water quality index model and indices approaches. *Heliyon* **2023**, *9*, e19668. [CrossRef]
49. Costa, J.C.; Alves, M.M.; Ferreira, E.C. Principal component analysis and quantitative image analysis to predict effects of toxics in anaerobic granular sludge. *Bioresour. Technol.* **2009**, *100*, 1180–1185. [CrossRef]
50. Qu, S.; Liang, X.; Liao, F.; Mao, H.; Xiao, B.; Duan, L.; Shi, Z.; Wang, G.; Yu, R. Geochemical fingerprint and spatial pattern of mine water quality in the Shaanxi-Inner Mongolia Coal Mine Base, Northwest China. *Sci. Total Environ.* **2023**, *854*, 158812. [CrossRef]
51. Yan, B.; Wei, Q.; Li, X.; Song, X.; Gao, Z.; Liu, J.; Zhang, R.; Wang, M. Heavy Metal Content Characteristics and Pollution Source Analysis of Shallow Groundwater in Tengzhou Coal Mining Area. *Water* **2023**, *15*, 4091. [CrossRef]
52. Yao, R.W.; Yan, Y.T.; Wei, C.L.; Luo, M.; Xiao, Y.; Zhang, Y.H. Hydrochemical Characteristics and Groundwater Quality Assessment Using an Integrated Approach of the PCA, SOM, and Fuzzy c-Means Clustering: A Case Study in the Northern Sichuan Basin. *Front. Environ. Sci.* **2022**, *10*, 907872. [CrossRef]
53. Giri, S.; Singh, A.K. Risk assessment, statistical source identification and seasonal fluctuation of dissolved metals in the Subarnarekha River, India. *J. Hazard. Mater.* **2014**, *265*, 305–314. [CrossRef]
54. *HJ 25.3-2019*; Technical Guidelines for Risk Assessment of Soil Contamination of Land for Construction. Ministry of Ecology and Environment of the People's Republic of China: Beijing, China, 2019.
55. *GB 5749-2022*; Standards for Drinking Water Quality. State Administration for Market Regulation: Beijing, China, 2022.
56. U.S. Environmental Protection Agency. Cadmium CASRN 7440-43-9 | DTXSID1023940. Available online: https://iris.epa.gov/ChemicalLanding/&substance_nmbr=141 (accessed on 13 January 2026).
57. Nag, R.; Cummins, E. Human health risk assessment of lead (Pb) through the environmental-food pathway. *Sci. Total Environ.* **2022**, *810*, 151168. [CrossRef] [PubMed]
58. Liu, Y.; Liu, D.; Zhang, W.; Chen, X.; Zhao, Q.; Chen, X.; Zou, C. Health risk assessment of heavy metals (Zn, Cu, Cd, Pb, As and Cr) in wheat grain receiving repeated Zn fertilizers. *Environ. Pollut.* **2020**, *257*, 113581. [CrossRef]
59. U.S. Environmental Protection Agency. Arsenic, Inorganic CASRN 7440-38-2 | DTXSID4023886. Available online: https://iris.epa.gov/ChemicalLanding/&substance_nmbr=278 (accessed on 13 January 2026).
60. U.S. Environmental Protection Agency. Regional Screening Levels (RSLs)—Generic Tables. Available online: https://www.epa.gov/risk/regional-screening-levels-rsls-generic-tables (accessed on 13 January 2026).
61. Abedi Sarvestani, R.; Aghasi, M. Health risk assessment of heavy metals exposure (lead, cadmium, and copper) through drinking water consumption in Kerman city, Iran. *Environ. Earth Sci.* **2019**, *78*, 714. [CrossRef]
62. Xu, X. Resource and Environmental Science Data Registration and Publishing System. China Population Spatial Distribution Kilometer Grid Dataset. 2017. Available online: https://www.resdc.cn/DOI/doi.aspx?DOIid=32 (accessed on 24 December 2025).
63. Wen, X.; Lu, J.; Wu, J.; Lin, Y.; Luo, Y. Influence of coastal groundwater salinization on the distribution and risks of heavy metals. *Sci. Total Environ.* **2019**, *652*, 267–277. [CrossRef]
64. Ababakr, F.A.; Ahmed, K.O.; Amini, A.; Moghadam, M.K.; Gökçekuş, H. Spatio-temporal variations of groundwater quality index using geostatistical methods and GIS. *Appl. Water Sci.* **2023**, *13*, 206. [CrossRef]
65. World Health Organization. *Guidelines for Drinking-Water Quality: Fourth Edition Incorporating the First and Second Addenda*; World Health Organization: Geneva, Switzerland, 2022; ISBN 978-92-4-004506-4.
66. U.S. Environmental Protection Agency. National Primary Drinking Water Standards. Available online: https://permanent.access.gpo.gov/websites/epagov/www.epa.gov/OGWDW/mcl.html#7 (accessed on 24 December 2025).

https://doi.org/10.3390/w18030351

67. Kim, D.; Yun, S.; Cho, Y.; Hong, J.; Batsaikhan, B.; Oh, J. Hydrochemical assessment of environmental status of surface and ground water in mine areas in South Korea: Emphasis on geochemical behaviors of metals and sulfate in ground water. *J. Geochem. Explor.* **2017**, *183*, 33–45. [CrossRef]
68. Bhuiyan, M.A.H.; Bodrud-Doza, M.; Rakib, M.A.; Saha, B.B.; Islam, S.M.D. Appraisal of pollution scenario, sources and public health risk of harmful metals in mine water of Barapukuria coal mine industry in Bangladesh. *Environ. Sci. Pollut. Res.* **2021**, *28*, 22105–22122. [CrossRef]
69. Sun, J.; Tang, C.; Wu, P.; Strosnider, W.H.J.; Han, Z. Hydrogeochemical characteristics of streams with and without acid mine drainage impacts: A paired catchment study in karst geology, SW China. *J. Hydrol.* **2013**, *504*, 115–124. [CrossRef]
70. Bharat, A.P.; Singh, A.K.; Mahato, M.K. Heavy metal geochemistry and toxicity assessment of water environment from Ib valley coalfield, India: Implications to contaminant source apportionment and human health risks. *Chemosphere* **2024**, *352*, 141452. [CrossRef]
71. Zhou, L.; Zhang, J.; Zhu, X.; Xu, D.; Zheng, S. The pollution characteristics and causes of dual sources—Iron (Fe) in abandoned coal mines: A literature review. *J. Clean. Prod.* **2024**, *471*, 143358. [CrossRef]
72. Hao, X.; Hao, H.; Bu, C.; Long, E.; Wang, H.; Wu, P.; Li, X. Temporal and spatial distribution, sources and health risk assessment of trace elements in a typical karst river basin in Southwest China: Influence of acid mine drainage from abandoned coal mines. *Ecotoxicol. Environ. Saf.* **2026**, *309*, 119524. [CrossRef]
73. Yang, X.; Du, J.; Jia, C.; Yang, T.; Shao, S. Unravelling integrated groundwater management in pollution-prone agricultural cities: A synergistic approach combining probabilistic risk, source apportionment and artificial intelligence. *J. Hazard. Mater.* **2025**, *481*, 136514. [CrossRef]
74. Zhou, J.; Wu, Q.; Gao, S.; Zhang, X.; Wang, Z.; Wu, P.; Zeng, J. Coupled controls of the infiltration of rivers, urban activities and carbonate on trace elements in a karst groundwater system from Guiyang, Southwest China. *Ecotoxicol. Environ. Saf.* **2023**, *249*, 114424. [CrossRef]
75. Zhang, X.; Li, X.; Gao, X. Hydrochemistry and coal mining activity induced karst water quality degradation in the Niangziguan karst water system, China. *Environ. Sci. Pollut. Res.* **2016**, *23*, 6286–6299. [CrossRef] [PubMed]
76. Lin, M.; Peng, W.; Gui, H. Heavy metals in deep groundwater within coal mining area, northern Anhui province, China: Concentration, relationship, and source apportionment. *Arab. J. Geosci.* **2016**, *9*, 319. [CrossRef]
77. Rashid, A.; Farooqi, A.; Gao, X.; Zahir, S.; Noor, S.; Khattak, J.A. Geochemical modeling, source apportionment, health risk exposure and control of higher fluoride in groundwater of sub-district Dargai, Pakistan. *Chemosphere* **2020**, *243*, 125409. [CrossRef]
78. Rashid, A.; Ayub, M.; Bundschuh, J.; Gao, X.; Ullah, Z.; Ali, L.; Li, C.; Ahmad, A.; Khan, S.; Rinklebe, J.; et al. Geochemical control, water quality indexing, source distribution, and potential health risk of fluoride and arsenic in groundwater: Occurrence, sources apportionment, and positive matrix factorization model. *J. Hazard. Mater.* **2023**, *460*, 132443. [CrossRef] [PubMed]
79. Rashid, A.; Ayub, M.; Gao, X.; Xu, Y.; Ullah, Z.; Zhu, Y.G.; Ali, L.; Li, C.; Ahmad, A.; Rinklebe, J.; et al. Unraveling the impact of high arsenic, fluoride and microbial population in community tubewell water around coal mines in a semiarid region: Insight from health hazards, and geographic information systems. *J. Hazard. Mater.* **2024**, *480*, 136064. [CrossRef] [PubMed]
80. Moyé, J.; Picard-Lesteven, T.; Zouhri, L.; El Amari, K.; Hibti, M.; Benkaddour, A. Groundwater assessment and environmental impact in the abandoned mine of Kettara (Morocco). *Environ. Pollut.* **2017**, *231*, 899–907. [CrossRef]
81. Ambastha, S.K.; Haritash, A.K. Prevalence and risk analysis of fluoride in groundwater around sandstone mine in Haryana, India. *Rend. Lincei. Sci. Fis. Nat.* **2021**, *32*, 577–584. [CrossRef]
82. Wang, Z.; Guo, H.; Adimalla, N.; Pei, J.; Zhang, Z.; Liu, H. Co-occurrence of arsenic and fluoride in groundwater of Guide basin in China: Genesis, mobility and enrichment mechanism. *Environ. Res.* **2024**, *244*, 117920. [CrossRef]
83. Liu, S.; Yi, B.; Liu, F.; Liu, C.; Yang, S.; Zhang, H.; Kang, W.; Jiang, K. Groundwater metal pollution and health risk assessment in river valley heavy industrial cities of arid regions in China. *China Geol.* **2025**, *8*, 526–539. [CrossRef]

Disclaimer/Publisher's Note: The statements, opinions and data contained in all publications are solely those of the individual author(s) and contributor(s) and not of MDPI and/or the editor(s). MDPI and/or the editor(s) disclaim responsibility for any injury to people or property resulting from any ideas, methods, instructions or products referred to in the content.

https://doi.org/10.3390/ww18030351

water

MDPI

Article

Behavioral Responses of *Unio tumidus* Freshwater Mussels to Neonicotinoid Pesticide Contamination

Marta Szostak [1,*], Krzysztof Szoszkiewicz [1], Krzysztof Achtenberg [1] and Dariusz Drożdżyński [2]

1 Department of Ecology & Environmental Protection, Faculty of Environmental and Mechanical Engineering, University of Life Sciences, Wojska Polskiego 28 Street, 60-637 Poznan, Poland; krzysztof.szoszkiewicz@up.poznan.pl (K.S.); krzysztof.achtenberg@up.poznan.pl (K.A.)
2 Institute of Plant Protection in Poznan, Węgorka 20 Street, 60-318 Poznan, Poland; d.drozdzynski@iorpib.poznan.pl
* Correspondence: marta.szostak@up.poznan.pl; Tel.: +48-061-846-65-17

Abstract: This investigation examined the behavioral responses of freshwater mussels to neonicotinoid pesticide exposure, a widely implemented agricultural crop protection agent. The study systematically evaluated the behavioral modifications of *U. tumidus* under controlled laboratory conditions, with particular emphasis on shell dynamics and activity patterns during both acute (2.5 h) and prolonged (20 h) exposure to imidacloprid at 50 μg/L concentration. The experimental findings revealed consistent and statistically significant behavioral alterations. Upon initial pesticide exposure, specimens exhibited an immediate reduction in shell aperture, followed by a sustained period of enhanced opening amplitude. Activity metrics demonstrated acute elevation immediately post-exposure, indicative of stress-induced responses, subsequently achieving homeostatic equilibrium before declining in later temporal phases. These behavioral modifications demonstrated statistical significance across all four experimental iterations, indicating a robust and reproducible stress response pattern. This study demonstrated that exposure to high concentrations of the neonicotinoid pesticide imidacloprid significantly affects the freshwater mussel causing significant, repeatable changes in mussel behavior: initial shell closure followed by prolonged opening and increased activity, indicating stress and subsequent toxic effects.

Keywords: bivalves; *Unio tumidus*; biomonitoring; pesticides; neonicotinoids; imidacloprid

1. Introduction

The widespread use of pesticides in modern agriculture has undoubtedly increased crop yields and food production efficiency. These substances are widely used to control pests and increase yields, but their excessive and irrational application has led to significant contamination of both terrestrial and aquatic ecosystems [1]. The use of pesticides in agriculture has been increasing from year to year, particularly in more economically developed countries. Pesticides have brought many benefits to agriculture and industry, but their harmful effects on humans and animals have always been a source of concern. The overuse of pesticides and their long-term impact on the environment continues to pose significant challenges [2]. Pesticides can be found everywhere in honey, fruits, vegetables, cereals, and drinking water, but also in human hair, which is a particular threat to human health. They are also a threat to aquatic ecosystems and biodiversity. When pesticides enter water bodies through runoff, leaching, or atmospheric deposition, they pose a serious threat to non-target organisms. Excessive concentrations of these chemicals can disrupt fundamental ecological processes, leading to changes in community structures and, in severe cases, the collapse of aquatic populations [3,4]. The impact of pesticides extends to various trophic levels, affecting not only primary consumers such as invertebrates but also the higher trophic levels that depend on them. Therefore, understanding the effects of pesticide exposure on sensitive aquatic species is essential for assessing ecosystem health and developing sustainable management strategies [5].

Neonicotinoid pesticides are extensively used directly on crops to protect plants, as well as for seed treatments, leaf spraying, and soil drenching. Due to the high water solubility of neonicotinoids, concerns have arisen about the risk of their entry into water bodies and the environment, posing threats to aquatic organisms [6,7]. Under simulated environmental conditions, neonicotinoids are readily leached from soil, plants, and seeds into water [8,9]. It has been established that neonicotinoids infiltrate watercourses through direct leaching into groundwater, which then enters surface waters through the decomposition of treated plants in waterways, direct contact with dust stirred up during the sowing of treated seeds, and drift of sprays into water bodies [10,11]. The main effects resulting from exposure to imidacloprid include negative impacts on the central nervous system by altering levels of metabolites such as nucleotides and amino acids, reducing filtration and the exchange of choline and acetylcholine [12], as well as decreasing heart rate and gill ventilation, which lead to death. This impact varies and depends on the species of the contaminated animal but also on the concentration of this pesticide [13].

Molluscs, representing one of the most species-rich phyla in the animal kingdom (over 130,000 known extant species), play an important role in environmental biomonitoring. Particularly bivalves, constituting about 15% of all mollusc species (20,500 species), and gastropods (110,000 species) demonstrate exceptional qualities as indicator organisms. The sedentary lifestyle of adult specimens enables local monitoring and also limits their ability to escape from contaminated areas. It should also be noted that bivalves not only accumulate substances from food but also from water, which is their living environment, resulting in accumulation through absorption by their entire body. The biology and ecology of these organisms are well understood, and their monitoring has a long history, notably including the "Mussel Watch" program, which was initiated in 1986 [14,15]. Benthic macroinvertebrates, particularly bivalve mollusks, have proven to be valuable bioindicators due to their sensitivity to a range of pollutants, including pesticides, heavy metals, microplastic, wastewater, and nutrient enrichment [16–23]. Mussels respond to water quality changes through observable behaviors such as shell opening and closing, which can be monitored in real time using valvometry [24]. As filter feeders, mussels are particularly susceptible to various organic and inorganic pollutants dissolved in the water [25].

Unio tumidus, a widespread species in Europe's freshwater bodies, is particularly well suited for biomonitoring [26–29]. Its sedentary lifestyle and filter-feeding behavior make it highly responsive to environmental disturbances, while its abundance allows for its use in ecotoxicological studies without impacting local populations. Although *U. tumidus* has been shown to be sensitive to various pollutants [17–19,28,30] and it is regarded as a valuable bioindicator of environmental pollution [31], its reaction to pesticides remains relatively understudied [18,28,30]. The majority of research on neonicotinoid toxicity has focused on pollinators, particularly bees [32]. However, in recent years, there has been growing attention to the effects of neonicotinoids on aquatic organisms [32–35]. Despite this increased interest, data on both acute and chronic toxicity of neonicotinoids for aquatic organisms remain limited, with imidacloprid being the most extensively studied compound in this context [34,36].

Given the increasing detection of neonicotinoids in aquatic environments and potential adverse effects on non-target organisms, it is necessary to investigate how *U. tumidus* responds behaviorally to imidacloprid exposure. Such studies may provide insights into the sublethal effects of neonicotinoids on freshwater mussels and enhance the utility of *U. tumidus* as a bioindicator species.

This study investigates the impact of pesticides on aquatic organisms. Specifically, we aimed to evaluate the effects of imidacloprid, a neonicotinoid pesticide, on the behavior of freshwater mussels (*Unio tumidus*). We examined both short- and long-term behavioral responses to pesticide exposure, hypothesizing that (1) imidacloprid would significantly alter bivalve behavior and (2) behavioral responses would evolve over time following exposure. We hypothesize that: imidacloprid exposure will significantly alter the behavior of *U. tumidus* by affecting shell movement dynamics and activity levels due to its neurotoxic effects. By investigating these hypotheses, the study seeks to fill the existing research gap and contribute valuable data on the sublethal effects of neonicotinoids on freshwater bivalves. The findings will have implications for environmental monitoring and risk assessment protocols, emphasizing the role of *U. tumidus* as a reliable bioindicator species for detecting pesticide contamination in aquatic ecosystems.

2. Materials and Methods

2.1. Materials and Model Organism

Mussels (*Unio tumidus*) were collected from Lake Wolsztyńskie in the Wolsztyn community, approximately 80 km from the biomonitoring laboratory at Poznań University of Life Sciences, Faculty of Environmental and Mechanical Engineering, Department of Ecology and Environmental Protection. Samples were gathered using rakes at depths of 0.5 to 1.0 m. Mature mussels (3–6 cm in length) were selected using vernier calipers. After collection, the mussels were manually cleaned and stored in a 10 L mobile fridge filled with 8 L of lake water to maintain optimal conditions during transport.

Upon arrival at the laboratory, mussels were transferred to an acclimatization aquarium for two weeks to minimize stress and allow adaptation to laboratory conditions before use in experimental studies. This step ensured the organisms' well-being, as mussels are highly sensitive to environmental changes.

2.2. Construction of the Monitoring System

A standard Biological Early Warning System (SYMBIO) was employed in this study, with its design illustrated in Figure 1. Shell movements were monitored using Hall sensors, which measure magnetic field changes caused by the movement of a ferrite magnet (8 mm diameter) attached to one shell half. The sensors recorded continuous data on shell opening levels, which were processed and visualized by dedicated software. Calibration established the minimum and maximum shell-opening ranges for each individual mussel.

For experiments, eight mussels were placed in a 150 × 25 × 25 cm aquarium filled with 60 L of aerated commercially distributed spring water. Water parameters were maintained at a temperature of 14–16 °C, pH of 8.0–8.4, and dissolved oxygen (DO) levels of 7.8–9.1 mg/L. An air pump ensured adequate aeration, and an aquarium chiller regulated the water temperature. Control conditions were defined as five days in clean water at 10–12 ± 1 °C, during which mussel behavior remained stable.

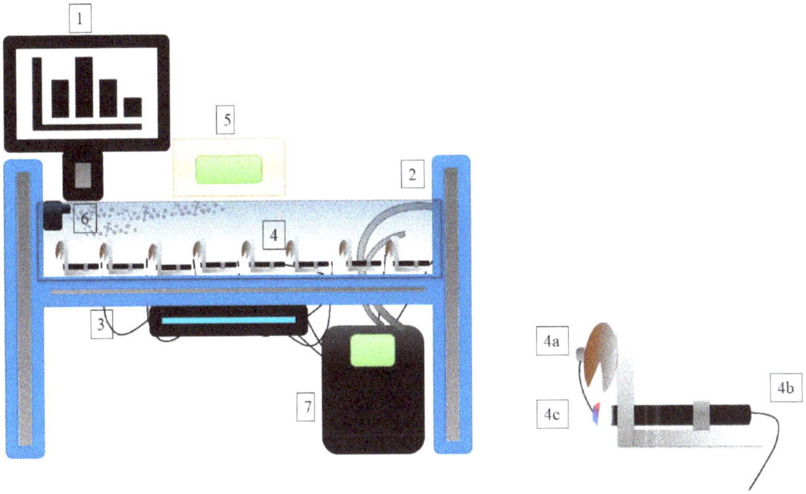

Figure 1. Construction of biological early warning system: (1) computer with dedicated software; (2) aquarium/tank; (3) connection of sensors with the controller; (4) organisms connected to the system; (4a) bivalve; (4b) hall sensor; (4c) magnet; (5) controller; (6) air pump; (7) aquarium chiller.

2.3. Pesticides Tested

This study utilized imidacloprid, a neonicotinoid pesticide, to assess behavioral responses in mussels. Four experimental series were conducted:

Experiment 1: 26 January 2023
Experiment 2: 7 July 2023
Experiment 3: 29 August 2023
Experiment 4: 3 January 2024

During experiments, imidacloprid was introduced at a concentration of 50 µg/L. The system continuously monitored shell movements, with data recorded 50 times per minute, amounting to approximately 72,000 records daily per individual. Short-term reactions were analyzed over 2.5 h, with a 0.5 h control period and 2 h following pesticide exposure. Long-term reactions were evaluated over 20 h, with a 5 h control period and 15 h of pesticide exposure.

2.4. Experimental Setup and Behavioral Assessment

Each mussel was mounted on a methacrylate holder and equipped with a magnet and Hall sensor, which detected shell movements by measuring changes in magnetic field strength. The collected data were processed to calculate the degree of shell opening (average percentage) and activity levels (average standard deviation in 1 min intervals). Behavioral responses were compared between control conditions and pesticide exposure periods.

The setup ensured precise monitoring and analysis of mussel behavior under controlled conditions, enabling detailed assessments of the effects of imidacloprid. No mortality was observed during the acclimatization or experimental periods, and all mussels were of uniform size (6 ± 1 cm length, 3.5 ± 0.5 cm width) and estimated age (6–9 years).

The significance of differences between the calculated means was tested based on ANOVA, and differences between individual intervals were assessed using Tukey's HSD post hoc test. STATISTICA 13.0 software was used for statistical analyses.

3. Results

3.1. Analysis for 5 min Intervals

In the experiment, the change in mussel behavior under the influence of a high concentration of the pesticide (imidacloprid at a concentration of 50 µg/L) was analyzed in 5 min intervals. Changes in the average level of mussel shell opening and the intensity of these changes were observed in control conditions and after the pesticide was administered (Table 1).

Table 1. Average shell opening (%) and mussel activity (average standard deviation in 1 min intervals) for 5 min intervals; control in time intervals 1–6, pesticide application in intervals 7–30.

Interval Number	Experiment 1		Experiment 2		Experiment 3		Experiment 4	
	Opening (%)	Activity	Opening (%)	Activity	Opening (%)	Activity	Opening (%)	Activity
1	43.16	0.09	43.75	0.09	43.16	0.00	43.09	0.14
2	43.00	0.15	43.72	0.14	42.82	0.30	43.16	0.12
3	43.43	0.08	43.43	0.16	43.13	0.07	42.79	0.27
4	43.32	0.06	42.97	0.13	43.38	0.03	42.46	0.16
5	42.54	0.12	43.03	0.16	43.45	0.10	42.62	0.13
6	41.72	0.23	42.78	0.53	43.58	0.08	43.07	0.08
7	39.17	0.35	38.76	0.38	35.64	1.22	42.69	0.26
8	37.55	0.35	38.35	0.60	34.43	0.54	41.33	0.27
9	44.09	0.77	40.48	2.44	41.89	1.81	46.51	2.37
10	51.22	0.93	55.10	1.97	58.68	2.63	53.62	1.53
11	69.68	1.10	67.76	1.10	75.88	1.36	61.74	1.05
12	73.32	0.96	71.83	0.90	77.10	1.06	66.17	0.75
13	73.72	0.95	72.06	0.81	76.29	0.85	69.03	1.05
14	73.81	0.79	78.48	0.49	77.11	0.39	71.65	0.80
15	74.75	0.69	79.22	0.36	77.83	0.62	71.45	0.41
16	76.37	0.78	79.08	0.27	77.04	1.11	70.92	0.61
17	77.67	0.80	80.14	0.23	79.16	0.45	70.50	0.56
18	78.61	0.50	81.77	0.33	78.81	0.48	69.32	0.60
19	79.20	0.53	81.85	0.19	79.66	0.33	65.97	1.14
20	80.76	0.18	82.97	0.30	77.78	1.35	65.94	0.52
21	81.17	0.40	83.29	0.17	79.49	0.37	66.08	0.35
22	81.84	0.18	84.03	0.23	79.09	0.18	67.14	0.19
23	82.26	0.17	84.24	0.26	78.36	0.26	67.05	0.27
24	82.30	0.25	84.50	0.28	77.91	0.62	67.26	0.55
25	82.44	0.13	85.05	0.29	77.66	0.56	67.10	0.69
26	82.55	0.21	85.14	0.33	77.89	0.18	67.94	0.18
27	81.94	0.29	85.59	0.22	77.04	0.36	68.11	0.25
28	81.82	0.29	86.32	0.36	76.70	0.34	68.18	0.07
29	81.89	0.33	86.10	0.24	76.99	0.32	68.36	0.04
30	81.34	0.23	85.53	0.22	76.45	0.22	67.98	0.31

The significance of differences between the calculated means was tested based on ANOVA (Table 2). Significant effects were demonstrated for each experiment with respect to both average opening and activity ($p < 0.05$)

Table 2. Analysis of variance for average shell opening and average clam activity in 5 min intervals; significant effects at $p < 0.01$ are marked in red, MS—the mean sum of squares, SS—the sum of squares.

Experiment	Variable	SS	MS	SS	MS	F	p
1	Opening	45,125	1556.1	235.9	1.97	791.6	<0.01
	Activity	14	0.5	14.3	0.12	4.2	<0.01
2	Opening	49,838	1718.6	358.2	2.98	575.7	<0.01
	Activity	40	1.4	11.5	0.10	14.7	<0.01
3	Opening	42,122	1452.57	554.5	4.62	314.3	<0.01
	Activity	51	1.77	53.3	0.44	4.0	<0.01
4	Opening	19,899	686.2	116.2	0.97	708.6	<0.01
	Activity	37	1.3	15.8	0.13	9.7	<0.01

In Experiment 1, before the application of the pesticide (imidacloprid 50 µg/L), the average shell opening level ranged from 41.72% to 43.43% (intervals 1–6 in Figure 2 and Table 1). After the application of imidacloprid, the average shell opening level decreased to 37.55% within the first 10 min, and then strongly increased over the next 20 min, reaching 73.32%. Later, the opening level continued to slightly increase, reaching 82.55% in the 19th interval after pesticide application, which is after 1 h and 35 min. In subsequent intervals, the shell opening level remained high, only slightly decreasing to 81.34%.

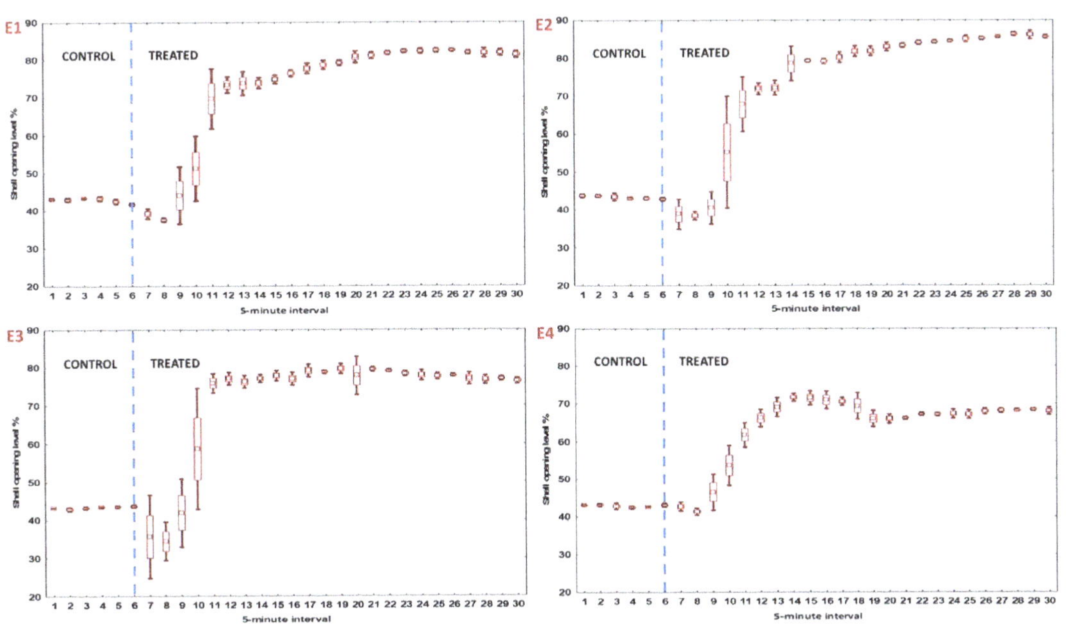

Figure 2. Average shell opening for 5 min intervals in Experiments 1–4 (E1–E4); control in time intervals 1–6, pesticide application in intervals 7–30. Values presented on graphs: mean SD, SD, and 1.96 SD.

In Experiment 2, before the application of the pesticide, the average shell opening level ranged from 42.78% to 43.75% (intervals 1–6 in Figure 2 and Table 1). After the application of imidacloprid, the average shell opening level decreased to 38.35% within the first 5 min and then began to strongly increase over the next 20 min, reaching 71.83%. The opening level continued to increase but to a much lesser extent, reaching 85.14% in the 19th interval after pesticide application, which is 1 h and 35 min later. In subsequent intervals, the shell opening level remained high at 86.32%, and then decreased to 85.53% in interval 30.

Experiment 3, before the application of the pesticide, the average shell opening level was very stable and ranged from 42.82% to 43.58% (intervals 1–6 in Figure 2 and Table 1). After the application of imidacloprid, the average shell opening level decreased to 34.43% within the first 10 min and then began to strongly increase over the next 20 min, reaching 77.10%. The opening level continued to increase but to a much lesser extent, reaching 79.66% in the 19th interval after pesticide application, which is 1 h and 35 min later. In subsequent intervals, the shell opening level remained high, ranging from 76.70% to 79.49%, and then decreased to 76.45% over an interval of 30 min.

In Experiment 4, before the application of the pesticide, the average shell opening level ranged from 42.62% to 43.16% (intervals 1–6 in Figure 2 and Table 1). After the application of imidacloprid, the average shell opening level decreased to 41.33% within the first 5 min and then began to strongly increase over the next 30 min, reaching 71.65%. In subsequent intervals, the shell opening level remained high, gently oscillating above 65.94%.

The post hoc analysis using Tukey's test allowed us to verify the significance of the average clam opening level for 5 min time intervals. The change in shell opening level after pesticide application was significant in each experiment for both shell opening and consequent activity ($p < 0.01$).

In Experiment 1, during the control period, clams exhibited consistent behavioral activity, with an average minute standard deviation ranging from 0.06 to 0.23 (Figure 3 and Table 1). Following the introduction of the pesticide, there was a noticeable increase in activity, with the standard deviation rising to 1.10 after 25 min. This elevated activity persisted above 0.5 for the next 40 min, eventually settling into a gentle oscillation slightly higher than the control period levels.

In Experiment 2, clams showed steady behavioral activity during the control period, with measurements ranging from 0.09 to 0.53 (Figure 3 and Table 1). Post-pesticide introduction, activity significantly increased, and the average minute standard deviation climbed to 2.44 within 15 min. It remained high, between 0.23 and 1.97, for the following 45 min before stabilizing at a lower range of 0.17 to 0.36, still above pre-pesticide levels.

In Experiment 3, during the control period, a consistent behavioral activity of clams was observed (Figure 3 and Table 1), determined based on the average 5 min standard deviation, which ranged from 0.00 to 0.30. After the introduction of the pesticide, a marked increase in activity was observed, and the average minute standard deviation rose, reaching 2.63 after 20 min. For the next 50 min, it remained within a broad range of 0.39 to 1.36, and then gently oscillated at a lower level but higher than during the control period.

In Experiment 4, during the control period, a consistent behavioral activity of clams was observed, ranging from 0.08 to 0.27 (Figure 3 and Table 1). After the introduction of the pesticide, a marked increase in activity was observed, and the average minute standard deviation rose, reaching 2.37 after 15 min. For the next 50 min, it remained at a very high level with fluctuations exceeding 1.0, and then decreased but oscillated within a broad range from 0.04 to 0.69, most often exceeding the level before the pesticide was administered.

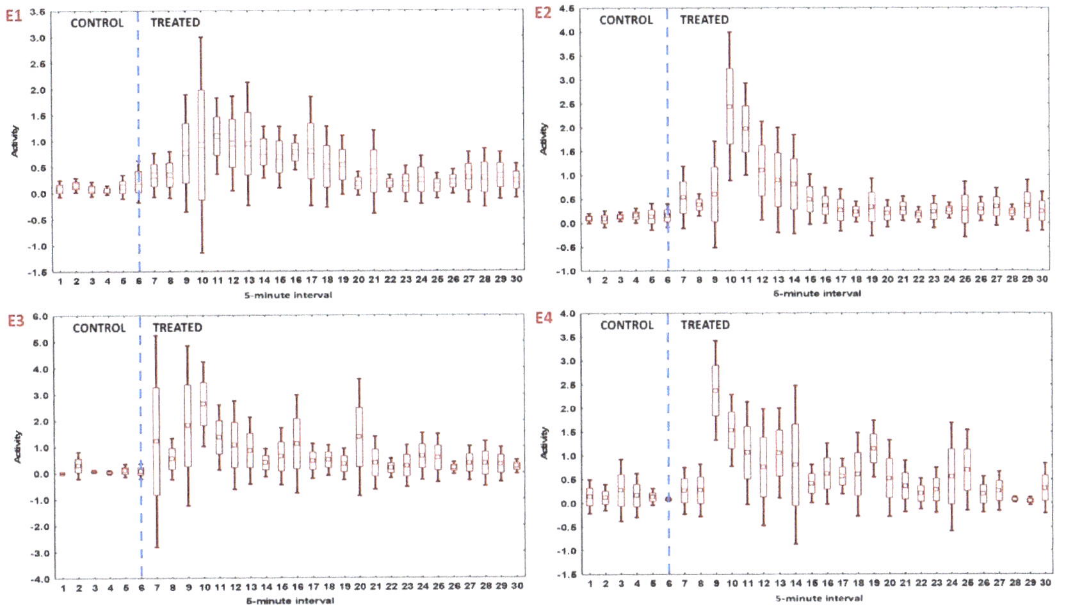

Figure 3. Bivalves' activity in 5 min intervals in Experiments 1–4 (E1–E4); control in time intervals 1–6, pesticide application in intervals 7–30. Values presented on graphs: mean, SD, and 1.96 SD.

Post hoc analysis using Tukey's test allowed us to verify the significance of the mean activity of mussels in 5 min time intervals. The change in mussel activity after pesticide application was significant in each experiment ($p < 0.01$) with respect to the first time interval, but further oscillations did not show significance anymore.

The behavioral responses of *Unio tumidus* to imidacloprid exposure reveal a sophisticated and reproducible stress response pattern, characterized by an initial protective phase followed by prolonged compensatory behavior, thereby validating this species' potential as a sensitive bioindicator of aquatic ecosystem perturbation.

3.2. Analysis for Hourly Intervals

In the experiment, the change in clam behavior under the influence of a high concentration of pesticide (imidacloprid at a concentration of 50 µg/L) was analyzed in 60 min intervals. Changes in the average level of clam opening and the intensity of these changes were observed under control conditions and after pesticide administration (Table 3).

Table 3. Average shell opening (%) and mussel activity (average standard deviation in 1 min intervals); 1 h intervals; control in time intervals 1–5, pesticide application in intervals 6–21.

Interval Number	Experiment 1		Experiment 2		Experiment 3		Experiment 4	
	Opening (%)	Activity	Opening (%)	Activity	Opening (%)	Activity	Opening (%)	Activity
1	30.79	0.23	33.90	0.40	33.33	0.24	31.46	0.18
2	35.59	0.16	42.48	0.22	29.79	0.26	33.41	0.15
3	42.18	0.12	42.05	0.18	26.58	0.04	36.20	0.26
4	41.76	0.13	41.51	0.15	26.84	0.05	35.41	0.12
5	42.53	0.11	40.62	0.17	30.49	0.15	35.66	0.14

Table 3. *Cont.*

Interval Number	Experiment 1		Experiment 2		Experiment 3		Experiment 4	
	Opening (%)	Activity	Opening (%)	Activity	Opening (%)	Activity	Opening (%)	Activity
6	64.16	0.75	41.89	0.57	41.94	0.77	55.97	0.93
7	81.63	0.27	76.20	0.45	71.55	0.58	60.16	0.29
8	79.42	0.32	81.22	0.30	70.31	0.28	59.82	0.22
9	79.26	0.34	79.13	0.74	69.08	0.25	60.35	0.27
10	74.61	0.29	82.59	0.32	65.16	0.31	61.75	0.17
11	66.01	0.22	82.25	0.15	59.32	0.28	61.28	0.18
12	58.98	0.23	80.99	0.14	57.27	0.34	60.74	0.18
13	52.17	0.23	79.86	0.13	64.10	0.48	60.23	0.11
14	44.48	0.16	78.65	0.11	63.14	0.54	56.14	0.18
15	37.70	0.31	74.77	0.23	52.50	0.47	44.86	0.20
16	27.53	0.14	70.69	0.08	45.01	0.40	44.38	0.13
17	25.67	0.19	72.40	0.16	35.72	0.33	37.32	0.11
18	18.69	0.09	72.32	0.37	27.89	0.24	37.49	0.11
19	28.03	0.19	62.85	0.11	24.05	0.00	37.10	0.19
20	27.79	0.18	28.10	0.16	24.12	0.00	36.93	0.20
21	27.44	0.18	34.28	0.14	24.15	0.00	35.05	0.27

The significance of differences between the calculated means was tested based on ANOVA (Table 4). Significant effects were demonstrated for each experiment concerning both average opening and activity ($p < 0.05$).

Table 4. Analysis of variance for average shell opening and average clam activity in 1 h intervals; significant effects at $p < 0.01$ are marked in red, MS—the mean sum of squares, SS—the sum of squares.

Experiment	Variable	SS	MS	SS	MS	F	p
1	Opening	490,951	24,547.6	21,404.9	17.28	1420.9	<0.01
	Activity	23	1.1	64.1	0.05	22.0	<0.01
2	Opening	472,585	23,629.3	15,618.5	12.61	1874.5	<0.01
	Activity	35	1.8	164.1	0.132	13.3	<0.01
3	Opening	382,888	19,144.4	17,771.9	14.34	1334.7	<0.01
	Activity	52	2.6	185.7	0.15	17.2	<0.01
4	Opening	168,440	8422.0	8140.9	6.57	1281.8	<0.01
	Activity	35	1.8	104.3	0.08	21.0	<0.01

In Experiment 1, before the pesticide application, the average shell opening level ranged from 30.79% to 42.53% (intervals 1–5 in Table 3 and Figure 4). After the application of imidacloprid, the average shell opening level began to increase significantly, reaching a maximum value of 81.63% in the second hour. In the following hours, a decrease in the shell opening level was observed initially, the decrease was slight for three hours, then it became more pronounced, reaching a minimum shell opening level of 18.69% 13 h after the pesticide application (interval 18, Figure 4 and Table 4). In the subsequent hours, the shell opening level remained low, fluctuating between 27.44% and 28.03%

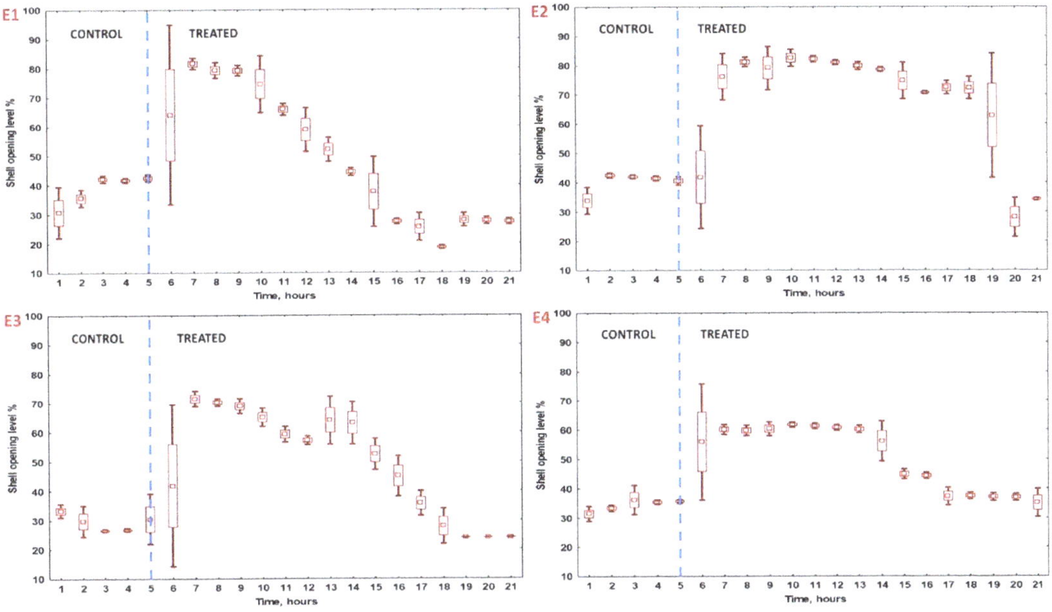

Figure 4. Average shell opening for 1 h intervals in Experiments 1–4 (E1–E4); control in time intervals 1–5, pesticide application in intervals 6–21. Values presented on graphs: mean, SD, and 1.96 SD.

In Experiment 2, before the pesticide application, the average shell opening level ranged from 33.90% to 42.48% (intervals 1–5 in Table 3 and Figure 4). After the application of imidacloprid, the behavioral response was immediate, but the average shell opening level only significantly increased in the second hour. The maximum shell opening value, 82.59%, was reached in the fifth hour after the pesticide application. In the following hours, a gradual decrease in the shell opening level was observed initially, the decrease was slight for eight hours, then it dropped sharply over the next two hours to 28.10% (interval 20, Table 3 and Figure 4). In the next hour, it slightly increased again.

In Experiment 3, before the pesticide application, the average shell opening level ranged from 26.58% to 33.33% (intervals 1–5 in Table 3 and Figure 4). After the application of imidacloprid, the behavioral response was immediate, and the average shell opening level increased, reaching a maximum value of 71.55% in the second hour. In the following hours, a gradual decrease in the shell opening level was observed, with one fluctuation, ultimately reaching a level of 24.05% (interval 19, Table 3 and Figure 4). In the subsequent hours, this level remained very similar.

In Experiment 4, before the pesticide application, the average shell opening level ranged from 31.46% to 36.20% (intervals 1–5 in Table 3 and Figure 4). During the first two hours after the application of imidacloprid, the average shell opening level of the mussels increased significantly, reaching 60.16%. Over the next six hours, the shell opening level remained similarly high, reaching its maximum of 61.75% in the fifth hour after the pesticide application. Then, over the next eight hours, the shell opening level decreased to 37.32% (interval 17, Table 3 and Figure 4). In the subsequent hours, the shell opening level remained at a similarly low level.

The post hoc analysis using Tukey's test allowed for the verification of the significance of the average activity shell opening level for 1 h time intervals. The change in shell opening level after pesticide application was significant in each experiment ($p < 0.01$). The later oscillations also showed significance in most of the following intervals.

In Experiment 1, during the control period, the behavioral activity of mussels ranged between 0.11 and 0.23 (Table 3 and Figure 5). Following the introduction of the pesticide, a significant increase in activity was observed, reaching 0.75 within the first hour and remaining above 0.2 for the next 7 h. In subsequent intervals, mussel activity decreased and stayed below 0.20, with the exception of a sporadic increase to 0.31 in the tenth hour after pesticide application.

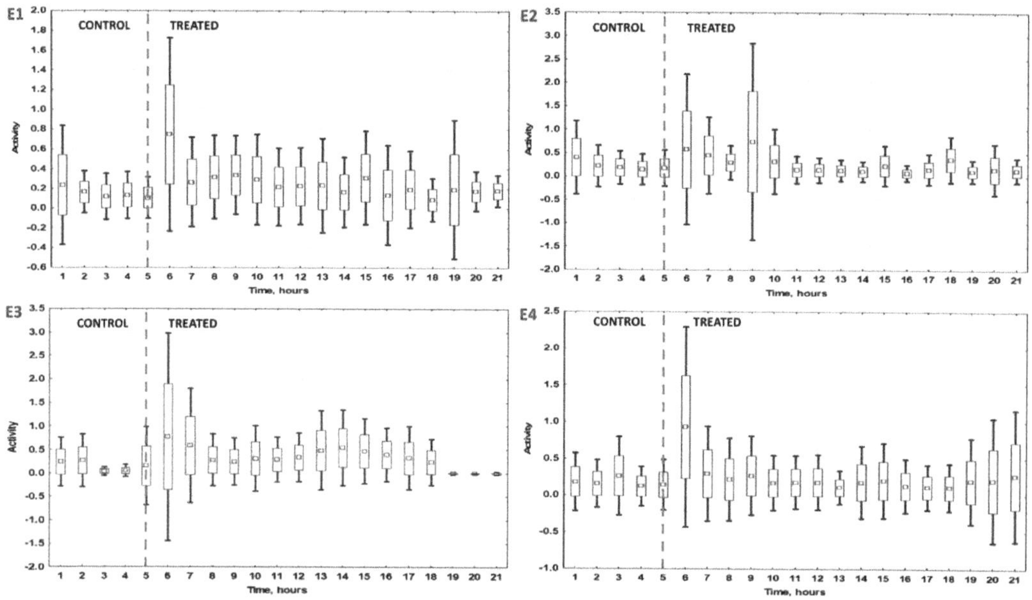

Figure 5. Mussel activity in 1 h intervals in Experiments 1–4 (E1–E4); control period in intervals 1–5, pesticide application in intervals 6–21. Values presented on the graphs: mean, SD, and 1.96 SD.

In Experiment 2, during the control period, mussel behavioral activity varied between 0.15 and 0.40 (Table 3 and Figure 5). After the pesticide was introduced, a marked increase in activity was observed, peaking at 0.74 in the fourth hour, and subsequently remaining at a low level below 0.2, apart from three peaks (0.32, 0.23, and 0.37).

In Experiment 3, during the control period, the visible behavioral activity of mussels varied within relatively wide ranges of 0.04–0.26 (Table 3 and Figure 5). After the introduction of the pesticide, a significant increase in activity was observed, with the average minute standard deviation rising to 0.77 within the first hour and remaining high for the next hour (0.58). In the subsequent intervals, mussel activity fluctuated widely, but 14 h after the pesticide application, it decreased sharply, and the activity of all individuals almost ceased.

In Experiment 4, during the control period, with slight changes in the average shell opening level of mussels, the average standard deviation ranged from 0.12 to 0.26 (Table 3 and Figure 5). After the introduction of the pesticide, a one-hour increase in activity to a level of 0.93 was observed, which then clearly decreased to 0.29 in the next hour. In the subsequent intervals, this activity gently oscillated at the level observed before the pesticide application.

A post hoc analysis using Tukey's test was conducted to assess the significance of the average activity of mussels for 1 h time intervals. Pesticide application caused a significant change in activity ($p < 0.01$) with respect to the following interval. In the subsequent interval, a significant decrease in activity was also observed (Experiments 1, 2, and 4). Further oscillations were no longer significant.

The data revealed that exposure to imidacloprid induced significant alterations in both the magnitude and temporal dynamics of shell movements. These modifications manifested as complex, time-dependent responses that reflected both immediate defensive mechanisms and longer-term physiological adaptations to chemical stress.

4. Discussion

The conducted experimental research allowed us to achieve the aim of the study; to demonstrate changes in the behavioral reaction of the molluscan under the influence of pesticides. Studies have shown changes in mussel behavior at different time intervals. Statistical analysis confirmed the relevance of the observations made. Both hypotheses were verified positively by proving that the inflow of pesticides has a significant effect on bivalves' behaviors.

In four experiments examining the short-term response of mussels to imidacloprid (50 μg/L), distinct, repeatable patterns of behavioral changes were observed, including both shell opening degree and activity level. In all experiments, immediately after pesticide application, a sharp decrease in shell opening degree was observed. This demonstrates that pesticides trigger an immediate biological response, and the reaction changed over time, confirming mussel stress responses to various types of pollution [16,18,27]. These organisms react this way to stress or attempt to protect themselves from a new environmental factor. This decrease was short-lived, typically lasting several minutes before the next stage of response occurred. Healthy and strong mussels open their shells for feeding and gas exchange. When exposed to pollutants, their typical response is closure [18,23], which protects their soft tissues from harmful substances. However, exposure to highly toxic substances can cause abnormal functioning of the adductor muscles responsible for shell closure. Substances such as dichlorvos (concentrations 0.1 and 1 μg/L) [22] and DDVP (concentrations 10 and 100 μg/L) caused paralysis of these muscles, which may be due to their prolonged contraction, but this process is not fully understood and requires further research.

After the initial closing phase, the degree of shell opening in the conducted experiments increased rapidly. Within 20–30 min after the application of imidacloprid, the shell opening level reached significantly higher values than those observed under control conditions. This increase can be considered a specific stress response caused by the presence of the pesticide. Subsequently, the degree of opening stabilized at a high level, higher than before the pesticide application. The increase in shell opening reached maximum values within the first 2–5 h. The opening level increased to about 60–80%, which was significantly higher than the control period values (about 30–40%). After the growth period, a stabilization of the elevated shell opening was observed, with the highest values maintained in various experiments within 3–8 h. After the stabilization period, a gradual decrease in shell opening was noted in all experiments. After 10–13 h from the pesticide application, the opening level decreased to values close to or even lower than the control period, oscillating at a low level (about 18–37%) for the last hours of the study. This phenomenon indicates the cessation of stress after exposure to the pesticide.

For some mussel species, such as mussels (Mytilidae), which produce byssal threads, shell opening and the associated formation of byssal threads are considered good indicators for monitoring poor mussel health, which can be linked to reduced vitality and negative impacts on their survival [31]. According to the literature, among the studied mussels, endosulfan was found to be the most toxic among various pesticides tested by Roberts [37], causing a reduction in the number of byssal threads by up to 50% after 24 h of exposure at a concentration of 0.45 mg/L. The reduction in the number of byssal threads is also confirmed by studies on imidacloprid. This example can be linked to the reaction of mussels in our experiment, which initially closed, and then opened more than under control conditions.

In terms of activity after pesticide application, a rapid increase in mussel activity was observed. Activity, measured as the standard deviation of the average opening measured for 1 min intervals, showed that mussels quickly opened and closed. In each experiment, behavioral activity increased rapidly, reaching its maximum within the first few to several minutes after pesticide application. After reaching the peak level, activity began to decrease just as quickly but remained slightly higher than during the control period for the next several minutes (about 0.2–0.3) within 2–4 h after pesticide application. Subsequently, activity oscillated at a level higher than during the control period for several hours but gradually decreased. In the later hours, especially after 10–14 h from application, mussel activity decreased significantly. This decrease indicates the cessation of stress after exposure to the pesticide.

The literature indicates that mussel responses to pesticides can vary depending on many factors, such as substance concentration, exposure time, or the specific characteristics of the studied species [25]. Nevertheless, there are studies that can help understand these reactions, especially in the context of increased mussel activity in the initial phase of pesticide exposure. Studies indicate that the initial phase of exposure to harmful chemical substances can cause increased mobility or escape attempts [38], before a decrease in activity due to the toxic effects of the harmful substance. Similar reactions regarding changes in mussel behavior, such as filtration, activity, and avoidance of harmful factors, were noted during studies on *Anodonta anatina* mussels exposed to salt. The mussels showed an initial increase in activity, trying to escape from the stressor, which can be an indicator of toxicity in ecological studies, especially in freshwater mussels [25].

During the study of the impact of neonicotinoids on the behavior of land mollusks *Helix aspersa* and squids, it was confirmed that this species exhibited a decrease in overall activity, which may indicate a characteristic response of mollusks to the administered poison and can be associated with nervous system damage and an increased risk of death [39].

Studies on the activity and increased movements of *Dreissena polymorpha* mussels have shown that chemical substances can cause an initial increase in activity before toxins lead to a decrease in overall motility and survival. This suggests that mussels may respond with a short-term increase in activity as a reaction to chemical stressors [40], before the effects of pollution become apparent [41,42].

Statistical analysis—both ANOVA and Tukey's post hoc analyses showed that short-term and long-term changes in both shell opening levels and activity were statistically significant ($p < 0.01$) in all experiments. All four studies demonstrated that the mussel response is highly repeatable. The significance of these results suggests that the observed reactions are repeatable and consistent, confirming the impact of imidacloprid as a factor causing changes in mussel behavior. Some behavioral traits indeed differed; for instance, the percentage of shell opening in Experiment 3 was noticeably distinct. However, the response to the pesticide remained repeatable across all experiments.

Behavioral responses of *Unio tumidus* in the presence of the pesticide confirmed observations by other researchers regarding the impact of neonicotinoids on the activity of the enzyme acetylcholinesterase present in mussel gills [43]. Acetylcholinesterase is the most important cholinesterase, breaking down acetylcholine, a substance that causes muscle contraction and transmits impulses between neurons in the nervous system. Exposure of the mussel *Mytilus galloprovincialis* to high concentrations of imidacloprid significantly reduced the activity of acetylcholinesterase [44]. The study showed that imidacloprid lowers AChE activity, which can disrupt nervous functions by causing improper regulation of the neurotransmitter acetylcholine. This effect would also confirm the behavioral response of the swollen river mussel *Unio tumidus* to high concentrations of imidacloprid.

Exposure of the mussel *Mytilus galloprovincialis* to a mixture of neonicotinoids impacts mussel health, with an emphasis on molecular and biochemical changes in these organisms. The study uses transcriptomic (gene expression analysis) and proteomic (protein analysis) approaches to identify changes in response to chemical stress induced by pesticides [44].

In the laboratory experiments conducted, we studied the response of mussels to imidacloprid at a concentration of 50 μg/L, which is high but can be found in the environment. The concentration of neonicotinoid pesticides in natural aquatic ecosystems varies and ranges from 0.1 to 100 μg/L, and their presence has been documented across five continents [7]. It is important to note that in some agricultural catchments, this concentration may even exceed levels considered lethal [38]. The highest values reported in the literature are 320 μg/L [7]. Thus, the results of our studies demonstrate the real-world implications of pesticide pollution.

Our laboratory findings reveal significant implications for real-world mussel populations in pesticide-contaminated water bodies. Mussels exposed to imidacloprid exhibit stress and behavioral changes that disrupt essential functions such as feeding and respiration, potentially reducing their survival rates. Prolonged stress responses suggest energy depletion and physiological damage, which can impair growth and immune function, particularly in areas with recurring contamination. Given the ecological importance of mussels, their decline could destabilize aquatic ecosystems by disrupting nutrient cycling, water filtration, and food webs, leading to broader impacts on biodiversity.

The research we conducted was based on a single species of mussel. This introduces a certain limitation in the interpretation of the obtained conclusions when referring to the threats of the tested insecticide to existing ecosystems. The justification for using *Unio tumidus* is that it is common, is not protected, occurs in clean water reservoirs, is sensitive to changes in the quality of environmental water [17–19,28,30], and its collection does not threaten biodiversity. The species *Unio tumidus* is widely used in biomonitoring, and its use is scientifically supported [26–29]. In the future, we plan to conduct similar comparative experiments with other species of mussels, but the scale of these experiments may be smaller because we will only be looking for confirmation of the observed phenomena.

The risk of imidacloprid contamination [45] indicates the importance of incorporating mussels into monitoring programs as bioindicators for assessing water quality and the ecological impact of agricultural practices and beyond, as the application of neonicotinoids is much broader. Other studies also show that these compounds affect many aquatic organisms, including amphibians [46], aquatic arthropods [47], aquatic vertebrates [48], and fish [49], and also impact entire ecosystems [50]. There is also a lack of research on the effects of these pesticides on humans and higher vertebrates over a longer period of time. Regulators should consider setting restrictions on the use of neonicotinoids, as their ability to bioaccumulate can reveal long-term ecological consequences for entire trophic chains, including humans [51].

5. Conclusions

Neonicotinoids like imidacloprid significantly impact aquatic ecosystems, with *Unio tumidus* serving as a reliable bioindicator due to its measurable behavioral responses.

This study demonstrated that exposure to high concentrations of the neonicotinoid pesticide imidacloprid significantly affects the freshwater mussel. The introduction of imidacloprid caused significant, repeatable changes in mussel behavior: initial shell closure followed by prolonged opening and increased activity, indicating stress and subsequent toxic effects.

Experiment findings were consistent across multiple experiments, showing minimal seasonal variation and reinforcing the reliability of *U. tumidus* as a test organism.

This study underscores the importance of behavioral monitoring for assessing pesticide risks and highlights the need for further research on long-term and multi-pesticide impacts. Future research should focus on studies of long-term exposure and recovery to gain a deeper understanding of the chronic effects and potential resilience of *Unio tumidus*.

The observed changes in shell opening and activity levels indicate a clear stress response, which highlights the potential ecological risks posed by such contaminants in aquatic environments. The consistent patterns across multiple experiments underscore the reliability of these behavioral indicators in assessing water quality and the health of aquatic ecosystems.

Author Contributions: Conceptualization, M.S., K.S. and D.D.; Methodology, M.S. and K.S.; Formal analysis, M.S., K.S. and D.D.; Resources, M.S.; Data curation, M.S.; Writing—original draft, M.S. and K.S.; Writing—review & editing, K.S.; Supervision, K.A.; Funding acquisition, K.A. All authors have read and agreed to the published version of the manuscript.

Funding: The publication was financed by the Polish Minister of Science and Higher Education as part of the Strategy of the Poznan University of Life Sciences for 2024–2026 in the field of improving scientific research and development work in priority research areas.

Data Availability Statement: Data are contained within the article.

Acknowledgments: We sincerely thank PROTE Technologies for the Environment for their collaboration. Our gratitude also goes to the students attending the Biotechnology and the Environmental Protection courses and the Erasmus program who supported the laboratory work in the Department of Ecology and Environmental Protection.

Conflicts of Interest: The authors declare no conflict of interest.

References

1. Goh, M.S.; Lam, S.D.; Yang, Y.; Naqiuddin, M.; Addis, S.N.K.; Yong, W.T.L.; Luang-InMa, V.; Sonne, C.N.L. Omics technologies used in pesticide residue detection and mitigation in crop. *J. Hazard. Mater.* **2021**, *420*, 126624. [CrossRef] [PubMed]
2. Kumar, V.; Kumar, P. Pesticides in agriculture and environment: Impacts on human health. In *Contaminants in Agriculture and Environment: Health Risks and Remediation*; Agro Environ Media: Haridwar, India, 2019; pp. 77–91. [CrossRef]
3. Pathiratne, A.; Kroon, F.J. Using species sensitivity distribution approach to assess the risks of commonly detected agricultural pesticides to Australia's tropical freshwater ecosystems. *Environ. Toxicol. Chem.* **2016**, *35*, 419–428. [CrossRef] [PubMed]
4. Nowell, L.H.; Moran, P.W.; Schmidt, T.S.; Norman, J.E.; Nakagaki, N.; Shoda, M.E.; Mahler, B.J.; Van Metre, P.C.; Stone, W.W.; Sandstrom, M.W.; et al. Complex mixtures of dissolved pesticides show potential aquatic toxicity in a synoptic study of Midwestern U.S. streams. *Sci. Total Environ.* **2014**, *613–614*, 1469–1488. [CrossRef] [PubMed]
5. Hassaan, M.A.; El Nemr, A. Pesticides pollution: Classifications, human health impact, extraction and treatment techniques. *Egypt. J. Aquat. Res.* **2020**, *46*, 207–220. [CrossRef]
6. Bonmatin, J.M.; Giorio, C.; Girolami, V.; Goulson, D.; Kreutzweiser, D.P.; Krupke, C.; Liess, M.; Long, E.; Marzaro, M.; Mitchell, E.A.D.; et al. Environmental fate and exposure; neonicotinoids and fipronil. *Environ. Sci. Pollut. Res.* **2015**, *22*, 35–67. [CrossRef]
7. Morrissey, C.A.; Mineau, P.; Devries, J.H.; Sanchez-Bayo, F.; Liess, M.; Cavallaro, M.C.; Liber, K. Neonicotinoid contamination of global surface waters and associated risk to aquatic invertebrates: A review. *Environ. Int.* **2015**, *74*, 291–303. [CrossRef]
8. Gupta, S.; Gajbhiye, V.T.; Gupta, R.K. Soil Dissipation and Leaching Behavior of a Neonicotinoid Insecticide Thiamethoxam. *Bull. Environ. Contam. Toxicol.* **2008**, *80*, 431–437. [CrossRef]
9. Tišler, T.; Jemec, A.; Mozetic, B.; Trebše, P. Hazard identification of imidacloprid to aquatic environment. *Chemosphere* **2009**, *76*, 907–914. [CrossRef]
10. Krupke, C.H.; Hunt, G.J.; Eitzer, B.D.; Andino, G.; Given, K. Multiple Routes of Pesticide Exposure for Honey Bees Living Near Agricultural Fields. *PLoS ONE* **2012**, *7*, e29268. [CrossRef]
11. Nuyttens, D.; Devarrewaere, W.; Verbovenb, P.; Foqu'ea, D. Pesticide-laden dust emission and drift from treated seeds during seed drilling: A review. *Pest Manag. Sci.* **2013**, *69*, 564–575. [CrossRef]

12. Tufi, S.; Stel, J.M.; de Boer, J.; Lamoree, M.H.; Leonards, P.E. Metabolomics to explore imidacloprid-induced toxicity in the central nervous system of the freshwater snail Lymnaea stagnalis. *Environ. Sci. Technol.* **2015**, *49*, 14529–14536. [CrossRef] [PubMed]
13. Mandal, A.H.; Sadhu, A.; Ghosh, S.; Saha, N.C.; Mossotto, C.; Pastorino, P.; Saha, S.; Faggio, C. Evaluating the Impact of Neonicotinoids on Aquatic Non-target Species: A comprehensive Review. *Environ. Toxicol. Pharmacol.* **2025**, *113*, 1382–6689. [CrossRef] [PubMed]
14. Oehlmann, J.; Schulte-Oehlmann, U. Chapter 17 Molluscs as bioindicators. In *Trace Metals and Other Contaminants in the Environment*; Bioindicators and Biomonitors; Elsevier: Amsterdam, The Netherlands, 2003; Volume 6, pp. 577–635. [CrossRef]
15. Beiras, R. Chapter 17—Marine Pollution Monitoring Programs. In *Marine Pollution: Sources, Fate and Effects of Pollutants in Coastal Ecosystems*; Elsevier: Amsterdam, The Netherlands, 2018; pp. 293–311. [CrossRef]
16. Bae, M.J.; Park, Y.S. Biological early warning system based on the responses of aquatic organisms to disturbances: A review. *Sci. Total Environ.* **2014**, *466*, 635–649. [CrossRef] [PubMed]
17. Chmist, J.; Szoszkiewicz, K. Choice of the most useful biological early warning system, based on ahp and rembrandt analysis. *Acta Sci. Pol. Form. Circumiectus* **2018**, *17*, 95–102. [CrossRef]
18. Chmist, J.; Szoszkiewicz, K.; Drożdżyński, D. Behavioural responses of Unio tumidus freshwater mussel to pesticide contamination. *Arch. Environ. Contam. Toxicol.* **2019**, *77*, 432–442. [CrossRef]
19. Mbah, J.T.; Chmist-Sikorska, J.; Szoszkiewicz, K.; Czekała, W. The effects of inflow of agricultural biogas digestate on bivalves' behavior. *Environ. Sci. Pollut. Res.* **2021**, *28*, 67385–67393. [CrossRef]
20. Salánki, J.; Farkas, A.; Kamardina, T.; Rózsa, K.S. Molluscs in biological monitoring of water quality. *Toxicol. Lett.* **2006**, *140*, 403–410. [CrossRef]
21. Savorelli, F.; Manfra, L.; Croppo, M.; Tornambè, A.; Palazzi, D.; Canepa, S.; Trentini, P.L.; Cicero, A.M.; Faggio, C. Fitness Evaluation of Ruditapes philippinarum Exposed to Ni. *Biol. Trace Elem. Res.* **2017**, *177*, 384–393. [CrossRef]
22. Le Bris, H.; Maffart, P.; Bocquené, G.; Buchet, V.; Galgani, F.; Blanc, G. Laboratory study on the effect of dichlorvos on two commercial bivalves. *Aquaculture* **1995**, *138*, 139–144. [CrossRef]
23. Almeshal, W.; Takács, A.; Aradi, L.; Sandil, S.; Dobosy, P.; Záray, G. Comparison of Freshwater Mussels Unio tumidus and Unio crassus as Biomonitors of Microplastic Contamination of Tisza River (Hungary). *Environments* **2022**, *9*, 122. [CrossRef]
24. Sow, M.; Durrieu, G.; Briollais, L.; Ciret, P.; Massabuau, J.C. Water quality assessment by means of HFNI valvometry and high-frequency data modeling. *Environ. Monit. Assess.* **2011**, *182*, 155–170. [CrossRef] [PubMed]
25. Hartmann, J.T.; Beggel, S.; Auerswald, K.; Stoeckle, B.C.; Geist, J. Establishing mussel behavior as a biomarker in ecotoxicology. *Aquat. Toxicol.* **2016**, *170*, 279–288. [CrossRef] [PubMed]
26. Lopes-Lima, M.; Sousa, R.; Geist, J.; Aldridge, D.C.; Araujo, R.; Bergengren, J.; Bespalaya, Y.; Bódis, E.; Burlakova, L.; Van Damme, D.; et al. Conservation status of freshwater mussels in Europe: State of the art and future challenges. *Biol. Rev.* **2017**, *92*, 572–607. [CrossRef] [PubMed]
27. Jonsson, A.; Bertilsson, A.; Rydgard, M. Spatial distribution and age structure of the freshwater unionid mussels Anodonta anatina and Unio tumidus: Implications for environmental monitoring. *Hydrobiologia* **2013**, *711*, 61–70. [CrossRef]
28. Drewek, A.; Lubawy, J.; Domek, P.; Polak, J.; Słocińska, M.; Dzięgelewska, A.; Klimaszyk, P. Behavioral and Biochemical Effects of Glyphosate-Based Herbicide Roundup on Unionid Mussels: Are Mussels Good Indicators of Water Pollution with Glyphosate-Based Pesticides? *Water* **2024**, *16*, 1882. [CrossRef]
29. Chmist, J.; Szoszkiewicz, K. Attempt at assessment of Unio tumidus bivalve mollusks suitability for monitoring water iron content. *Environ. Prot.* **2017**, *39*, 39–43.
30. Alhassan, A.B.; Aljahdali, M.O. Behavioural and Biochemical Responses of Freshwater Bivalve Anodonta marginata Exposed to Dichlorvos. *Water* **2024**, *16*, 3572. [CrossRef]
31. Stara, A.; Pagano, M.; Albano, M.; Savoca, S.; Di Bella, G.; Albergamo, A.; Koutkova, Z.; Sandova, M.; Velisek, J.; Fabrello, J.; et al. Effects of long-term exposure of Mytilus galloprovincialis to thiacloprid: A multibiomarker approach. *Environ. Pollut.* **2021**, *289*, 117892. [CrossRef]
32. Pisa, L.; Goulson, D.; Yang, E.-C.; Gibbons, D.; Sánchez-Bayo, F.; Mitchell, E.; Aebi, A.; Sluijs, J.; MacQuarrie, C.J.K.; Giorio, C.; et al. An update of the Worldwide Integrated Assessment (WIA) on systemic insecticides. Part 2: Impacts on organisms and ecosystems. *Environ. Sci. Pollut. Res.* **2021**, *28*, 11749–11797. [CrossRef]
33. Raby, M.; Nowierski, M.; Perlov, D.; Zhao, X.; Hao, C.; Poirier, D.G.; Sibleya, P.K. Acute Toxicity of 6 Neonicotinoid Insecticides to Freshwater Invertebrates. *Environ. Toxicol. Chem.* **2018**, *37*, 1430–1445. [CrossRef]
34. Sánchez-Bayo, F.; Goka, K.; Hayasaka, D. Contamination of the Aquatic Environment with Neonicotinoids and its Implication for Ecosystems. *Front. Environ. Sci.* **2016**, *4*, 71. [CrossRef]
35. Schmidt, T.S.; Miller, J.L.; Mahler, B.J.; Metre, P.C.; Nowell, L.H.; Sandstrom, M.W.; Carlisle, D.M.; Moran, P.W.; Bradley, P.M. Ecological consequences of neonicotinoid mixtures in streams. *Sci. Adv.* **2022**, *8*, eabj8182. [CrossRef]
36. Hladik, M.L.; Main, A.R.; Goulson, D. Environmental Risks and Challenges Associated with Neonicotinoid Insecticides. *Environ. Sci. Technol.* **2018**, *52*, 3329–3335. [CrossRef] [PubMed]

37. Roberts, D. Sub-lethal effects of chlorinated hydrocarbons on bivalve. *Mar. Pollut. Bull.* **1975**, *6*, 20–24. [CrossRef]
38. Wang, W.-X.; Rainbow, R.S. Subcellular Partitioning and the Prediction of Cadmium Toxicity to Aquatic Organisms. *Environ. Chem.* **2006**, *3*, 395–399. [CrossRef]
39. Hamlet, S.A.; Bensoltane, S.; Djekoun, M.; Yassi, F.; Berrebbah, H. Histological changes and biochemical parameters in the hepatopancreas of terrestrial gastropod Helix aspersa as biomarkers of neonicotinoid insecticide exposure. *Afr. J. Biotechnol.* **2012**, *11*, 16277–16283. [CrossRef]
40. Wang, N.; Ivey, C.D.; Ingersoll, C.G.; Brumbaugh, W.G.; Alvarez, D.; Hammer, E.J.; Bauer, C.R.; Augspurger, T.; Raimondo, S.; Barnhart, M.C. Acute sensitivity of a broad range of freshwater mussels to chemicals with different modes of toxic action. *Environ. Toxicol. Chemistry* **2016**, *36*, 786–796. [CrossRef]
41. Dzierżyńska, A.; Białończyk, Ł.; Jermacz, Ł.; Zielska, J.; Kobak, J. What scares a mussel? Changes in valve movement pattern as an immediate response of a byssate bivalve to biotic factors. *Hydrobiologia* **2019**, *841*, 65–77. [CrossRef]
42. Robson, A.; Wilson, R.; de Leaniz, C.G. Mussels Xexing their muscles: A new method for quantifying bivalve behaviour. *Mar. Biol.* **2007**, *151*, 1195–1204. [CrossRef]
43. Reynolds, A.; Cody, E.; Giltrap, M.; Chambers, G. Toxicological and Biomarker Assessment of Freshwater Zebra Mussels (Dreissena polymorpha) Exposed to Nano-Polystyrene. *Toxics* **2024**, *12*, 774. [CrossRef]
44. Dondero, F.; Negri, A.; Boatti, L.; Marsano, F.; Mignone, F.; Viarengo, A. Transcriptomic and proteomic effects of a neonicotinoid insecticide mixture in the marine mussel (*Mytilus galloprovincialis*, Lam.). *Sci. Total Environ.* **2010**, *408*, 3775–3786. [CrossRef] [PubMed]
45. Yamamoto, A.; Terao, T.; Hisatomi, H.; Kawasaki, H.; Arakawa, R. Evaluation of river pollution of neonicotinoids in Osaka City (Japan) by LC/MS with dopant-assisted photoionization. *J. Environ. Monit.* **2012**, *14*, 2189–2194. [CrossRef] [PubMed]
46. Kerner, M.; Flach, H.; Dietmann, P.; Kühl, M.; Kühl, S.J. The impact of the insecticide acetamiprid on the embryogenesis of the aquatic model organism Xenopus laevis. *Environ. Toxicol. Pharmacol.* **2023**, *103*, 104278. [CrossRef] [PubMed]
47. Gacem, H.; Bendali-Saoudi, F.; Serradj, N.; Houmani, M.; Soltani, N. Risk assessment of the neonicotinoid insecticide acetamiprid on two non-target species, *Daphnia magna* Straus, 1820 (Crustacea, Cladocera) and *Plea minutissima* Leach, 1817 (Insecta, Heteroptera). *Appl. Ecol. Environ. Res.* **2023**, *21*, 2. [CrossRef]
48. Vajargah, M.F.; Namin, J.I.; Mohsenpour, R.; Yalsuyi, A.M.; Prokić, M.D.; Faggio, C. Histological effects of sublethal concentrations of insecticide Lindane on intestinal tissue of grass carp (*Ctenopharyngodon idella*). *Vet. Res. Commun.* **2021**, *45*, 373–380. [CrossRef]
49. Velisek, J.; Stara, A. Effect of thiacloprid on early life stages of common carp (*Cyprinus carpio*). *Chemosphere* **2018**, *194*, 481–487. [CrossRef]
50. Singh, S.; Gupta, A.; Waswani, H.; Prasad, M.; Ranjan, R. Impact of pesticides on the ecosystem. In *Agrochemicals in Soil and Environment: Impacts and Remediation*; Springer Nature: Singapore, 2022; pp. 157–181. [CrossRef]
51. Stehle, S.; Ovcharova, V.; Wolfram, J.; Bub, S.; Herrmann, L.Z.; Petschick, L.L.; Schulz, R. Neonicotinoid Insecticides in Global Agricultural Surface Waters—Exposure, Risks and Regulatory Challenges. *Sci. Total Environ.* **2023**, *867*, 161383. [CrossRef]

Disclaimer/Publisher's Note: The statements, opinions and data contained in all publications are solely those of the individual author(s) and contributor(s) and not of MDPI and/or the editor(s). MDPI and/or the editor(s) disclaim responsibility for any injury to people or property resulting from any ideas, methods, instructions or products referred to in the content.

water

MDPI

Article

Investigation of Used Water Sediments from Ceramic Tile Fabrication

Simona Elena Avram [1,*], Bianca Violeta Birle [1], Lucian Barbu Tudoran [2,3], Gheorghe Borodi [3] and Ioan Petean [4,*]

1 Faculty of Materials and Environmental Engineering, Technical University of Cluj-Napoca, 103–105 Muncii Bd., 400641 Cluj-Napoca, Romania; birle.ma.bianca@student.utcluj.ro
2 Faculty of Biology and Geology, Babes-Bolyai University, 44 Gheorghe Bilaşcu Street, 400015 Cluj-Napoca, Romania; lucian.barbu@ubbcluj.ro
3 National Institute for Research and Development of Isotopic and Molecular Technologies, 65-103 Donath Street, 400293 Cluj-Napoca, Romania; borodi@itim-cj.ro
4 Faculty of Chemistry and Chemical Engineering, Babes-Bolyai University, 11 Arany Janos Street, 400028 Cluj-Napoca, Romania
* Correspondence: simona.avram@imadd.utcluj.ro (S.E.A.); ioan.petean@ubbcluj.ro (I.P.)

Abstract: Used water treatment is one of the most important aspects of environmental protection regarding industrial processes. Particulate matter dispersions affect water parameters; for example, increased pH values such as 10.21 are found for used floor tile water, and values of 10.84 are found for used wall tile water. However, pH decreases to about 9.42 after the sediment filtration process. This influences water turbidity, which is higher for used wall tile water due to its finer suspensions, and it is considerably decreased after the filtration process. Thus, the main aim of the present research is to investigate particulate matter dispersion into the water flows that are involved in ceramic tile technological processes before and after treatment at used water treatment facilities. X-ray diffraction (XRD) coupled with mineralogical optical microscopy (MOM) reveals that waters from wall tiles and floor tiles have similar mineral dispersions, containing mineral particles of quartz (5–50 μm), kaolinite (1–30 μm), and mullite (5–125 μm). Glass particles (having a dark appearance at MOM investigation) were also found in both samples in a size range of 20–55 μm. High-resolution SEM imaging coupled with the EDS elemental analysis confirms the XRD and MOM observations. Water samples collected after treatment at the treatment facility reveal a significant reduction in the particulate matter MOM, evidencing only small traces of quartz, kaolinite, and mullite in a size range of 1–15 μm, with most of the particles being attached to the filters, as confirmed by XRD. Atomic force microscopy (AFM) effectuated on this sample reveals the presence of kaolinite nanoparticles with a tabular–lamellar aspect and sizes ranging from 40 to 90 nm. The obtained results prove the efficacy of the filtering system regarding targeted particulate matters, ensuring water recirculation into the technological processes. The sludge resulting from the filtration process presents with a dense grainy structure of sediment particles containing quartz, mullite, and kaolinite, along with traces of iron hydroxide crystallized as goethite. Therefore, it cannot be reused in the technological flux, as the iron causes glaze staining; but the observed microstructure, along with the mineralogical composition, indicates that it could be used for other applications, such as ecological bricks or plasters, which will be further investigated.

Keywords: used water; ceramic tiles; sediments; pressed slurry; kaolinite ultra-structure

1. Introduction

Ceramic tile fabrication involves various mineral powders such as quartz and clay, and vitreous materials such as broken glass [1,2]. Some of these particulate matters pass into the water used during the technological process and must be properly filtered out at water treatment facilities [3].

Used water is collected from all working places on the floor and wall tile facilities of ceramic factories. Therefore, particulate matters originating in both the raw material and the intermediary sub-products might be found in suspension in the used water [4,5].

Both floor and wall tiles are produced by molding a conditioned ceramic paste into metallic dies to obtain crude ceramic biscuits. Ceramic paste preparation and the washing of metallic dies induce a lot of raw material particles. Biscuit firing and their manipulation can also generate particulate matters that gather in the water used, and this can occur after the glazing process and the final firing. Tile-production facilities have tanks that collect used water from all the mentioned sources. Therefore, it was decided to establish an analysis of water within these tanks in floor tile and wall tile facilities. The sampling procedures and sample preservation and transport approaches used in this study are further described in the Materials and Methods Section.

The most important raw materials in tile production are clays with a controlled purity grade. The purity grade is related to the prevalence of kaolinite and possibly muscovite particles in the clay, which often are found as small fractions [6,7]. Some research regarding clay particles has indicated that kaolinite has nanostructured fractions [8,9].

Another important raw material is quartz sand, which also has a controlled purity grade; this strongly depends on the exploitation point. Generally, local sources are preferred for economic reasons. Thus, the sand deposits situated in the Transylvanian Basin are of great interest; here, large halite deposits [10,11] which are intercalated with sedimentary areas might alter the properties of the raw materials. However, this possibility will lead to the need for Cl management within the ceramic mass, which should be investigated through the elemental analysis of collected samples.

Besides clays and quartz, Ziman et al. show that the raw materials involved in this process might contain various amounts of calcite and feldspars, such as orthose, which also might be found as suspended matters in the used water [4]. Some of the ceramic tiles' technological processes use broken glass as a melting mediator, identifying particles that could contain small traces of chloride [12,13]. Tile biscuits which have defects after firing are ground and recirculated within the raw material in different proportions according to charge prescriptions [14,15]. This includes mullite particles among the particulate matter, which originates from the raw materials, and ceramic paste which is used for molding.

Used water, resulting from the ceramic industry's technological processes, is mainly generated by the wet cleaning of the molding dies with water jets (without cleaning agents) rather than from raw material preparation. The clay material, along with glaze and pigment particles, is washed away, becoming suspended particles in the used water. Such dispersed materials might be valuable if they are recycled in a smart manner. However, certain challenges arise due to the details of the technological processes.

Particulate matter suspensions within used water are required, as they allow fast sedimentation to occur during the decantation process in the collecting tanks; however, the current practice proves that this process requires a very long time to be carried out, which is not economically feasible. Therefore, these particles are removed from the used water through a physical filtration system, generating ceramic sludge. Theoretically, this contains most of the raw material particles, and would thus be great for recirculation. Unfortunately, it has been observed that the collected sludge contains various amounts of iron contamination, which compromises its potential for use as a recirculation material. Data in the literature show that iron contamination in the ceramic raw material induces brown stains over the glaze layer [16,17].

The aim of the present study is to investigate the sediment formed by dispersed particles within the used water collected from floor and wall tile facilities and to compare it with the sludge collected after the filtration process at a water treatment station; this will enable us to figure out the best ecological strategy for the management of such waste. The technological process requirements regarding raw materials and processing parameters are significantly different for floor and wall tiles. Thus, we expect to find significant variations regarding used water parameters and the subsequent sediments that might influence the formation of slurry. Water pretreatment for suspended particle filtration is of great interest in the current study for two reasons: water flow recirculation into the technological process and pressed slurry formation.

There is a lack of studies in the literature addressing the sediment found in used water from ceramic tile production and there is an urgent need to diminish the amount of waste produced in these processes; this lack represents an important gap in the research surrounding the operation of the ceramic industry. The novelty of the current research lies in the presented advanced approach that enables slurry to be transformed from a potentially dangerous and harmful waste into a raw material that is potentially useful for certain suitable applications, such as ecological bricks or plasters. The presented investigations of water properties and detailed physicochemical characterizations of the pressed slurry (made possible through mineralogical investigations coupled with advanced scanning electron microscopy (SEM) and ultra-structural observations, performed using atomic force microscopy (AFM)) are expected to provide practitioners in the field with important and useful information regarding the possibilities of slurry recycling.

2. Materials and Methods

2.1. Samples Collection and Storage

The current investigation was conducted using samples collected from an industrial facility that produces floor and wall ceramic tiles in Cluj County, Romania (the company name is kept anonymous for economic reasons). The water circuit for the technological processes of the floor and wall tiles was analyzed, and the sample collection points are identified in Figure 1. The on-site monitoring was effectuated by weekly measurements for two months. The water parameters were measured and mean values were established.

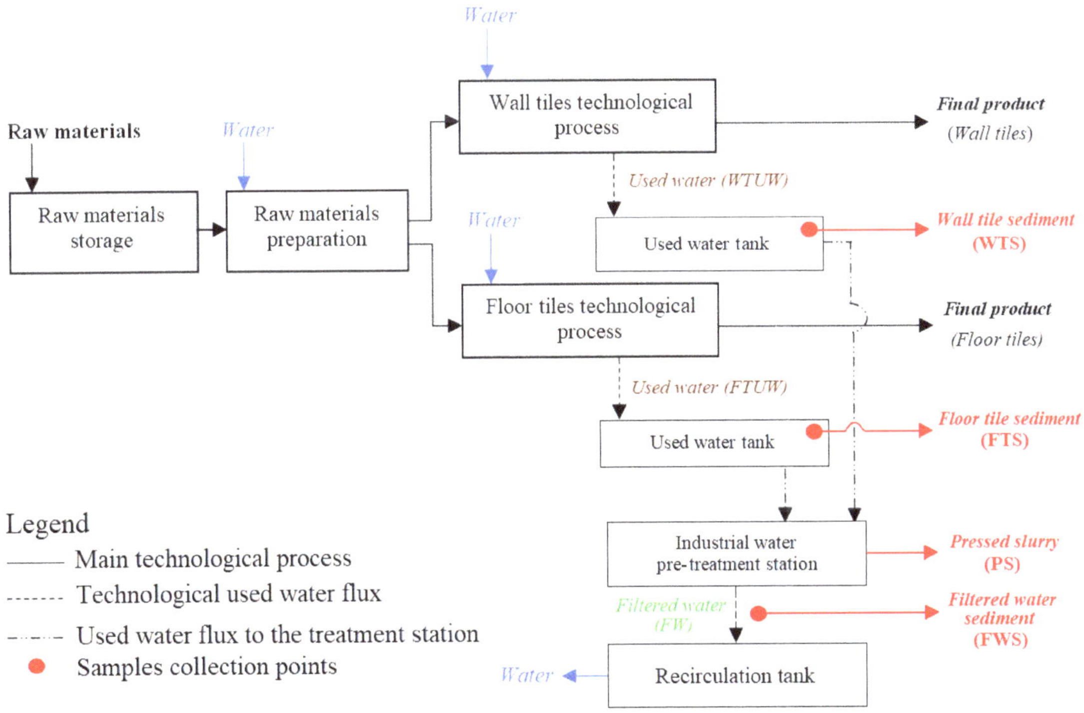

Figure 1. Ceramic tiles facility technological scheme with used water circuit and sampling sites.

The used water resulting from each technological process was collected in a storage tank and then was directed to the pretreatment station for the filtration process, as described in Figure 1. We collected samples from the used water tanks from the floor tile (FTUW) process, from the tanks from the wall tile (WTUW) process, and from the treatment facility after the filtration (FW).

The water samples were extracted from the tanks with a manual pump and deposited into glass vials, which were properly labeled for easy identification. The sample vials were transported to the laboratory in a cooled container at 4 °C and further stored in a refrigerator at the same temperature. The measurements were effectuated within 24 hours of collection to prevent sample alterations. The sediment samples were collected from the bottom of each used water storage vial and merged onto an average representative sample for each of the following: floor tile sediment (FTS), wall tile sediment (WTS), and filtered water sediment (FWS).

The procedures were repeated weekly for two months, and the samples were analyzed in the laboratory; the obtained results were centralized and statistically processed.

2.2. *Experimental Methods*

2.2.1. Water Properties Measurement Method

The water properties—such as pH, electrical conductivity, resistivity, total dissolved solids, salinity, and temperature—were measured using a multi-parameter device, HANNA HI 9829, with a GPS positioning system, produced by Hanna Instruments Co., Bedforshire, UK. The measurements were effectuated according to the device instructions and the obtained values were properly centralized.

Their evolution was statistically analyzed by an ANOVA test with a significance level of $p = 0.05$; this was followed by a Tukey post hoc test using Origin software version 2019b, produced by Microcal Company, Northampton, MA, USA.

2.2.2. Mineralogical Investigations

Mineralogical microscopy was effectuated in cross-polarized light using a Laboval 2 Microscope, Carl Zeiss, Jena, Germany, equipped with digital photo system, Samsung 10 megapixels. The sediment particles were spread on the glass lamella to reveal both small and bigger particle sizes.

X-ray diffraction (XRD) was effectuated with a Bruker D8 Advance diffractometer using monochromatic radiation Cu Kα1 (λ = 1.54056 Å) at room temperature. All patterns were registered in the 2 theta range of 10–80 degrees at a speed of 1 degree/minute. The mineral phase identification was performed with Match 1.0 software and PDF 2.0 database, Crystal Impact Company, Bonn, Germany. The PDF files used for the minerals' identification are as follows: quartz—PDF 89-8936; kaolinite—PDF 01-0527; mullite—PDF 02-0428; lepidocrocite—PDF 76-2301.

2.2.3. Microstructural Investigations

The sediments' morphologies and elemental compositions were investigated with a Hitachi SU8230 scanning electron microscope (SEM) equipped with an energy-dispersive spectroscopy (EDS) detector, X-Max 1160 EDX (Oxford Instruments). The investigation was effectuated in the high-vacuum mode with the electron beam acceleration voltage of 40 kV. All samples were metalized with a thin gold layer to assure their electrical conductivity, which was necessary for the high-resolution view. The gold component was subtracted from the elemental analysis.

The atomic force microscopy investigation was effectuated with a Jeol JSPM 4210 Scanning Probe Microscope (JEOL, Tokyo, Japan), operated in tapping mode. The sample surface was probed with an NSC 15 cantilever produced by MikroMasch Co., Sofia, Bulgaria, with a resonant frequency of 325 kHz and a force constant of 40 N/m. The scanning rate was in the range of 1–3 Hz, depending on the scanned area. The obtained images were processed with Win SPM 2.0 processing soft (JEOL, Tokyo, Japan), measuring the particle diameters and local heights. At least 3 different macroscopic sites on the sample surface were investigated for proper statistical relevance.

3. Results

The current section is organized starting with the investigation of the collected water samples and the measurement of their physical values, followed by an investigation of the dispersed particles up to the final slurry formation.

3.1. Water Samples Properties

The properties of used water are strongly influenced by the particularities of a given technological process and, especially, by the induced dispersed particles. The micro dispersoids might release ions and might be partly dissolved into used water, altering its properties. The mean values of water properties, with their standard deviations, are presented in Figure 2.

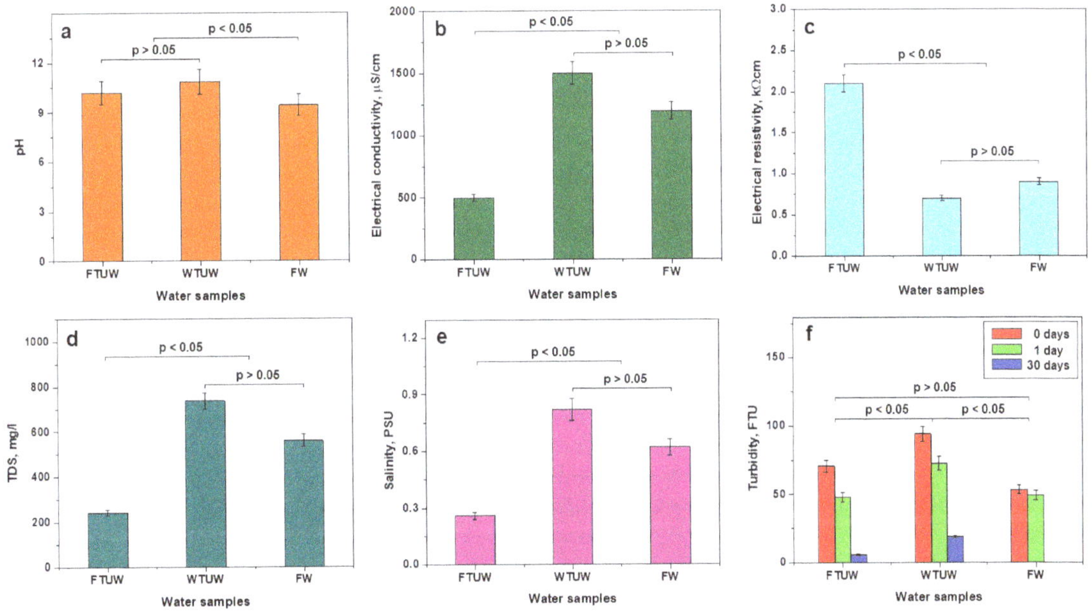

Figure 2. Mean values of water samples properties: (**a**) pH, (**b**) electrical conductivity, (**c**) electrical resistivity, (**d**) total dissolved solids, (**e**) salinity, and (**f**) turbidity. The error bars within the graphs represent the standard deviation.

The pH variation in Figure 2a shows that all water samples are alkaline, with mean values of 10.21 for the used floor tile water (FTUW) and 10.84 for the used wall tile water (WTUW). There are strong statistical similarities between these samples, so they form a statistical group. The pretreated water has a mean pH of about 9.42 after the filtration process, evidencing a significant decrease in alkalinity; this fact is supported by the statistical analysis. The p value is lower than the significance level of 0.05, indicating that the filtered water (FW) forms a different statistical group than the used water. The pH value did not decrease to the NTPA 002 prescriptions (i.e., Romanian regulations regarding used water quality) of 6.5–8.5 due to the removal of dispersoids through the filtration process; thus, it requires further water neutralization to be discharged, or it could be collected in the recirculation tank and reused in the technological process. The pH values are strongly influenced by the dissolved solids and salts and less influenced by the mineral dispersions. The significant decrease in pH in the FW sample indicates that some salt traces might be kept on the slurry after the pressing filtration process. It will be useful to find these traces in the elemental analysis.

The electrical properties of the water samples are represented by the conductivity (see Figure 2b) and the resistivity (see Figure 2c). These two properties are antagonistic and strongly depend on the dispersed particles and dissolved solids. The results show that FTUW presents with low conductivity and high resistivity; meanwhile, WTUV has high conductivity and low resistivity. The statistical analysis reveals significant differences between FTUW and WTUW, indicating that they are distinct statistical populations. The statistical results show that the conductivity and resistivity of the FW sample belong to the WTUW statistical population. This means that FW is more influenced by the WTUW than it is by the FTUW.

This fact is explained by the low amount of total dissolved solids (TDS) and salinity of FTUW (see Figure 2d,e); meanwhile, WTUW has a significant level of TDS and increased salinity. The gross difference between the used water samples is supported by the statistical analysis ($p < 0.05$). FTUW and WTUW, when collecting in the used water tanks, hold mixtures of both used water and TDS; salinity slightly decreases due to the partial dilution. This is the situation that was observed for the FW sample. The statistical analysis shows that WTUW and FW form a statistical population ($p > 0.05$) and proves that WTUW has a greater influence on FW than FTUW does ($p < 0.05$).

The turbidity directly reflects the number of dispersed particles in the used water. The WTUW sample is more turbid immediately after sampling; this is followed by FTUW. Meanwhile, the filtered water has low turbidity as a consequence of the removal of micro-dispersions in the filtration process (see Figure 2f). Turbidity significantly decreases after 1 day, as a consequence of micro-particle sedimentation; this is visible mostly in FTUW and FW, and is less evident in WTUW (after 1 day, its turbidity is slightly higher than the initial value for FTUW). The turbidity after 30 days is very low for FTUW and 0 for FW; meanwhile, for WTUW, it is still higher. The statistical analysis reveals good similarities in the turbidity behavior of FTUW and FW, with a significant difference shown by WTUW. This indicates the presence of ultrafine particles in the WTUW sample, which requires further advanced investigations.

3.2. Mineral Constituents Assessment

The variation in the properties of the water samples indicates a strong dependence on mineral dispersions; this fact is supported by the XRD patterns found for the investigated sediments (see Figure 3). All the obtained patterns present with crystalline features but they have particularities depending on the sample.

The floor tile sediment (FTS) sample presents with an XRD pattern with narrow and very intense peaks, proving that the sample has high crystallinity (see Figure 3a). The tile floor water sediment contains quartz as a dominant mineral, which is followed by mullite and kaolinite. We note the heterogeneity in the kaolinite peaks: some of them are intense and narrow at 29.46 degrees; meanwhile, the peak at 17.92 is broadened and less intense. This means that the FTS sample contains some large kaolinite particles which are preferentially oriented; meanwhile, smaller fractions are randomly oriented. Applying the Scherrer formula on the broadened peak of kaolinite results in an average crystallite size of 41 nm.

The wall tile sediment (WTS) sample, shown in Figure 3b, has an XRD pattern with narrow and intense peaks, which are related to the high crystallinity of the sample; but the patterns also show wide baseline noise, which is related to the presence of very small particles. This fact is in good agreement with the increased values of the turbidity, which is related to the dispersion of the finest particles. The composition is dominated by quartz and kaolinite, with only some mullite. Some of the kaolinite peaks in the WTS sample are broadened (e.g., for example to 15.62 degrees) due to the presence of ultrafine particles. Their average size, calculated with the Scherrer formula, is about 39 nm.

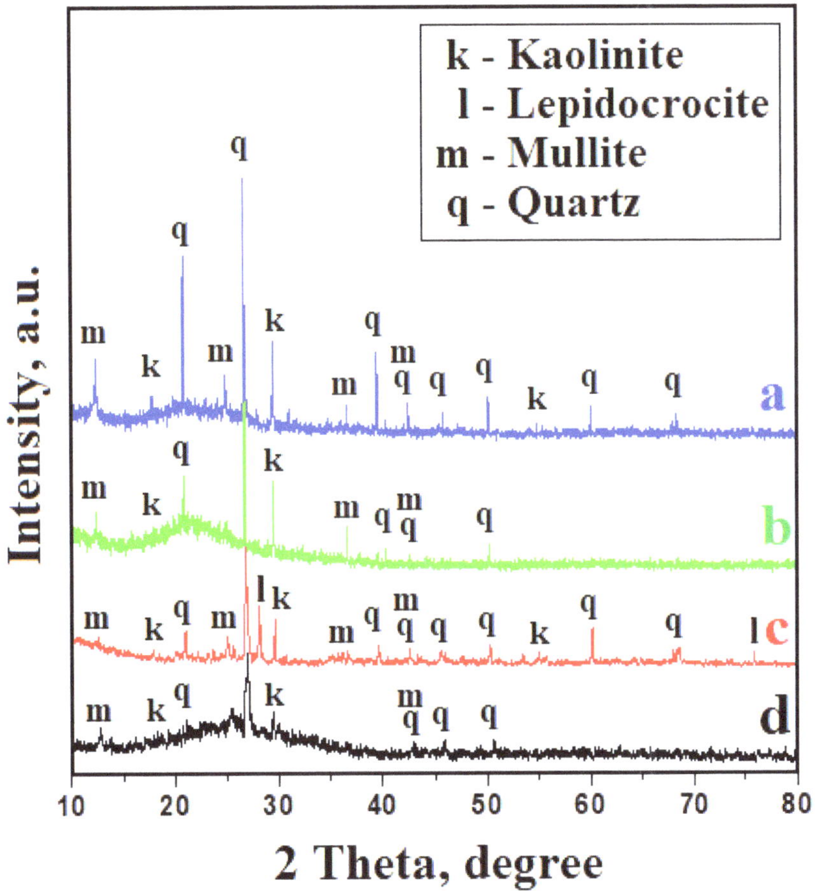

Figure 3. XRD patterns for the investigated samples: (a) FTS, (b) WTS, (c) PS, and (d) FWS.

The pressed slurry sample (PS) has increased crystallinity, as evidenced by the XRD pattern shown in Figure 3c; here, we can see well-developed diffraction peaks with a narrow allure. This result shows that quartz is the dominant mineral, followed closely by kaolinite and mullite. Both forms of kaolinite particles—large and small—are observed according to the height and full width at half maximum (FWHM) of the relevant peaks. A great discovery is the presence of lepidocrocite. It is an iron hydroxide γ-FeO(OH) crystallized in the orthorhombic system. It has been widely reported as component of environmental particulate matter samples [18,19].

The sediment collected from the filtered water (after the treatment facility) presents with an XRD pattern with a profound nanocrystalline characteristic (see Figure 3d); this is due to the amorphous baseline of the pattern and due to the presence of less intense and more broadened diffraction peaks of quartz, kaolinite, and mullite. The Scherer formula was applied to the most relevant peaks; as a consequence, we derive results showing an average crystallite of 35 nm for kaolinite, 75 nm for mullite, and 85 nm for quartz.

The weight amount of each mineral fraction in the samples was evaluated by the RIR (reference intensity ratio) method. This method assumes that, if there are two phases in the sample, then we have Equation (1):

$$\frac{Ia}{Ib} = \left(\frac{xa}{xb}\right) \cdot \left(\frac{fa}{fb}\right). \tag{1}$$

where *Ia* is the intensity for the most intense diffraction peak from phase *a*; *Ib* is the intensity for the most intense diffraction peak from phase *b*; *xa* and *xb* are the mass percentage concentrations for the two phases; *fa* and *fb* are the Corundum factors for the *a* and *b* phases.

If a third mineral fraction occurs, noted with *c*, then the relation can be written as follows:

$$\frac{Ia}{Ic} = \left(\frac{xa}{xc}\right) \cdot \left(\frac{fa}{fc}\right). \tag{2}$$

Considering that

$$xa + xb + xc = 100\%, \tag{3}$$

Thus, the mass concentrations of the three phases in the samples can be determined. The results of the phase analysis performed by this method are presented in Table 1.

Table 1. Minerals distribution within investigated samples as results of XRD and MOM investigation.

Component	Kaolinite	Mullite	Quartz	Lepidocrocite	Glass
Formula Color nuances	$Al_2Si_2O_5(OH)_4$ White-blue	$Al_6Si_2O_{13}$ Red-orange	SiO_2 Green-gray	$\gamma FeO(OH)$ Redish-brown	$6SiO_2 \cdot CaO \cdot Na_2O$ Black
			FTS		
Amount, wt. % Particle size, μm	14 1–100	32 10–180	54 5–70	0 -	* 10–30
			WTS		
Amount, wt. % Particle size, μm	37 1–30	21 3–35	43 1–35	0 -	* 20–55
			FS		
Amount, wt. % Particle size, μm	41 1–100	32 1–180	19 1–70	8 3–20	* 10–60
			FWS		
Amount, wt. % Particle size, μm	68 1–2.5	6 1–10	26 1–15	0 -	* -

Note: * undetectable due to the amorphous structure.

Mineralogical optical microscopy, shown in Figure 4, evidences the mineral particles within the sediment samples. Each of the identified minerals has its own nuanced color: quartz—green-gray; kaolinite—predominantly white; mullite—brownish-red; glass particles have dark nuance due to the lack of crystallinity.

The floor tile sediment is dominated by large mullite and quartz particles, surrounded by fine particles (see Figure 4a). Larger kaolinite particles are seen to have a polyhedral aspect with a tabular shape and sizes between 30 and 100 μm; the finest fractions appear as small white dots with sizes below 30 μm (see Figure 4a′). It results are in good agreement with the XRD observations. The high-magnification image shown in Figure 4a′ reveals the presence of small fractions of mullite (red, dotted particles). The particles size ranges measured from the MOM images are summarized in Table 1, along with their corresponding XRD results.

The WTS sample is more refined due to the atomization process that develops during the preparation of the raw materials; consequently, it contains less mullite. Thus, the bigger particle class is dominated by quartz (see Figure 4b). Fine kaolinite particles overwhelm the bigger ones (which do not exceed size of 30 μm) and surround the bigger particles of the WTS sample, giving a heterogeneous aspect. This is in correspondence with the aspect of the XRD pattern which showed the presence of fine particles in the sediment composition. The refinement of the particles is better evidenced in the higher-magnification image shown in Figure 4b′, because of the atomization process. All the finest fractions have a deeply milled aspect and small sizes; the values of these are shown in Table 1.

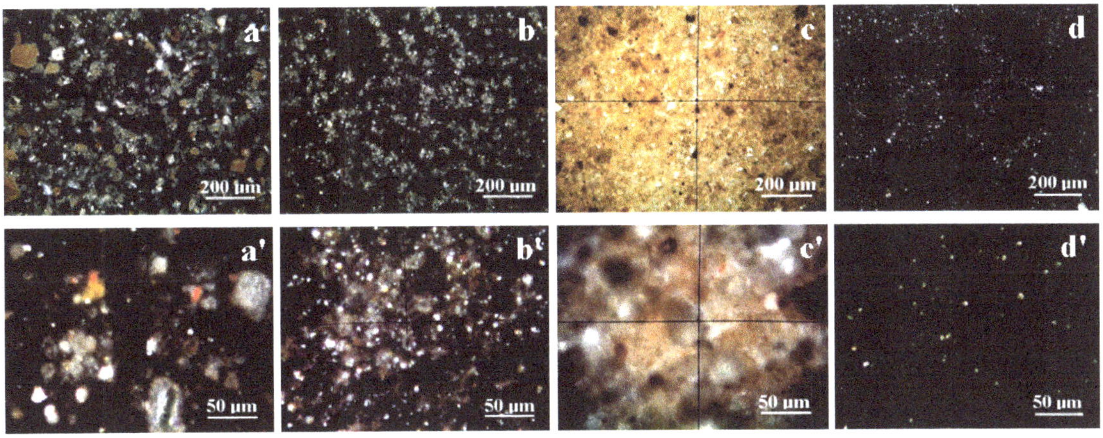

Figure 4. MOM images of the investigated samples observed at 300× magnification: (**a**) FTS, (**b**) WTS, (**c**) PS, and (**d**) FWS. At 1200× magnification: (**a′**) FTS, (**b′**) WTS, (**c′**) PS, and (**d′**) FWS.

All the dispersed particles within FTS and WTS were collected into the pressed slurry through the filtration process. Therefore, PS has a dense aspect under MOM observation, with larger particles of quartz and mullite that are strongly bonded onto the finer Kaolinite particle matrix. The kaolinite matrix is so dense that it has a yellow shade instead of bright white (see Figure 4c). However, the high-magnification images in Figure 4c′ clearly evidence some bigger kaolinite particles with a bright-white nuance. The lepidocrocite presence (reddish-brown nuance) in the FS sample is one of the most important MOM findings in this sample, and is in good agreement with the XRD observations shown in Figure 3c.

The sediment from the filtered water (FWS) contained only finer particles, which were equivalent to atmospheric PM 1, PM 2.5, and PM 10 (see Figure 4d). The higher-magnification images in Figure 4d′ allow a better view of the few fine quartz and mullite particles, surrounded by ultrafine kaolinite fractions.

3.3. *Microstructural Observation and Elemental Spectroscopy*

The sediment sample microstructures were investigated with high-resolution scanning electron microscopy imaging; a corresponding elements distribution map was made for each investigated site, along with the EDS spectrum (see Figure 5).

The SEM backscattered electron image (BSE) derived for the floor tile sample (Figure 5a) reveals a view that is complementary to the MOM image in Figure 4a. The corresponding elemental distribution map and the EDX spectrum are presented beside the EDS map. Oxygen is the dominant element due to the oxide nature of the mineral compounds within the sample; this is followed by Si, which is found in all involved minerals (see Table 2). The aluminum amount of 6.1 at. % belongs to both the kaolinite $Al_2Si_2O_5(OH)_4$ and the mullite $Al_6Si_2O_{13}$.

Glass is an amorphous material, described by the following chemical formula: $6SiO_2 \cdot CaO \cdot Na_2O$. Thus, the amounts of Ca and Na found in the elemental composition can be explained by the glass particles. On the other hand, the amount found of Na might be a trace influence from the water's total dissolved solids and salts. This fact is supported by the pink spots that are randomly spread out on the elemental maps (not only on the glass particles) shown in Figure 5a,b; additionally, see the pale-green spots on the map in Figure 5c.

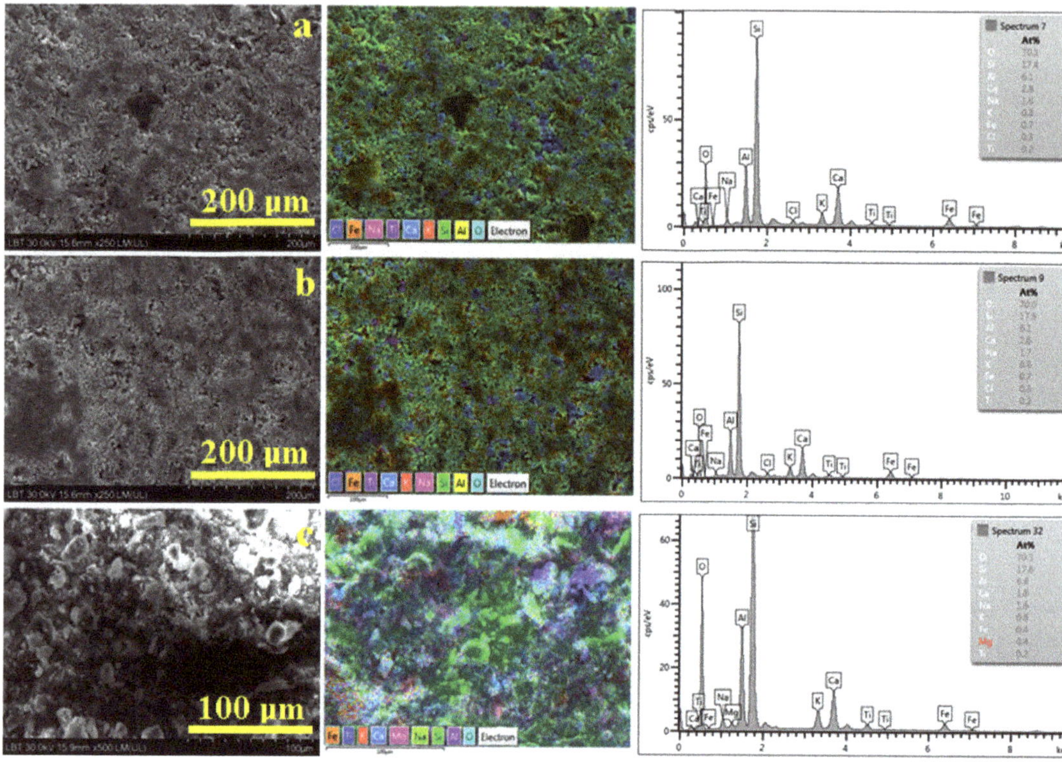

Figure 5. XRD patterns for the investigated samples: (**a**) FTS, (**b**) WTS, and (**c**) PS.

Table 2. Samples elemental composition.

Element	**Elemental Composition, At. %**		
	FTS	**WTS**	**PS**
O	70.1	70.0	70.3
Si	14.7	17.5	17.6
Al	6.1	6.1	6.8
Ca	2.8	2.6	1.8
Na	1.6	1.7	1.6
K	0.8	0.8	0.8
Fe	0.7	0.7	0.4
Cl	0.3	0.3	-
Mg	-	-	0.4
Ti	0.2	0.2	0.2

However, the data in the literature specify that glass might contain traces of Cl, which becomes more evident when one looks at recycled glass [20,21]. The other elements that were found, such as K, Fe, and Ti, are present in low amounts, which indicates their occurrence as natural traces from the used raw materials; the data in the literature show that clays might contain these elements [22,23]. Otherwise, it might be a mild contamination from the metallic parts of the devices in the technological flux. The correlation between the morphological aspect in the BSE image and the elemental map clearly evidences that quartz particles with rounded corners are present, with a size range of 5–80 μm; these have an intense green color. Additionally, mullite is present, with a light-green nuance (caused by the traces of K, Fe, and Ti), and is surrounded by fine fractions of kaolinite; these are situated between 1 and 10 μm and are also colored with a light-green nuance. The glass particles have a blue nuance due to the presence of Ca and Na with the chloride traces. Their shapes are irregular because they were broken during the technological process.

The SEM–BSE images obtained for the wall tile sediment (see Figure 5b) provide evidence for smaller particle sizes compared to those from the floor tile sediment, but the elemental compositions and distribution maps for the two are similar. Thus, the boulder-like particles with rounded corners around 5–45 μm belong to quartz; meanwhile, the mullite rounded shapes that are present have diameters of around 90 nm. They are surrounded by finer kaolinite particles with a size range of 1–30 μm. The glass particles are well distributed among the other constituents and have sizes of around 20–40 μm; this finding is in good agreement with the MOM investigation. Some minor differences were observed in the elemental map in Figure 5b: there are some brown segregations, which indicates that there are some kaolinite particles containing more Fe impurities than those observed for the floor tile sediment; some glass particles present with a violet nuance due to an increased presence of Na.

The SEM investigations of the solidified sludge samples (shown in Figure 5c) revealed the presence of well-defined particles ranging between 5 and 90 μm in size, with two typical morphologies: boulders with some blunted edges, related to quartz, and rounded boulders, related to mullite. These are surrounded by a compact mass corresponding to the kaolinite, which ensures the good cohesion of the microstructure due to its binding ability. Therefore, the fine kaolinite particles, ranging from 1 to 10 μm in size, are less visible in Figure 5c. The ceramic sludge elemental composition is dominated by 70.3% O; this is followed by 17.6% Si and 6.8% Al. These belong to the quartz, kaolinite, and mullite. Moderate amounts of 1.8% Ca and 0.8% K are related to the presence of the glass particles; these are in good agreement with the elemental compositions of the FTS and WTS sediments, but are significantly lower. The sludge accumulation from many different ceramic tiles charges explains this variation. The iron amount of 0.4% is related to the presence of lepidocrocite particles, confirming the XRD results. The elemental map within Figure 5c shows the quartz particles in bright green; this is due to the presence of O and Si; additionally, mullite particles have a violet shade due to their Al content. The kaolinite binding effect is proven in the elemental map by the formation of violet "clouds", which are caused by the fine distribution of Al atoms within the kaolinite particles. The lepidocrocite particles appear to be colored with an orange nuance, with some turquoise dots due to the presence of iron and oxygen. The disposal mode indicates that iron hydroxide crystallizes during the relative water evaporation from sludge over the deposit dump. This fact is supported by their small sizes of about 15–20 μm and confirms the data in the literature [16,17].

3.4. *Ultra-Structure Observation*

The samples' ultra-structures are formed by the submicron and nanodomains, requiring high-magnification, high-resolution imaging. Therefore, the ultra-structural aspects were investigated by atomic force microscopy (AFM). This approach presents the benefit of topographical investigation in addition to the simple visualization of the finest details.

The self-assembly of the FTS sample on the glass substrate generates a thin film of submicron particles with boulder shapes (see Figure 6a), ranging from 250 to 800 nm; these mostly belong to the quartz and mullite fine fractions. The upper-left section of Figure 6a reveals the end of a bigger kaolinite particle, evidencing parallel layers. The topographical arrangements of the particles lead to a local height of 425 nm, enabling the uniformity of the thin film. All submicron features are covered with kaolinite nanoplatelets, ensuring the base of the microstructural cohesion within the FTS sample.

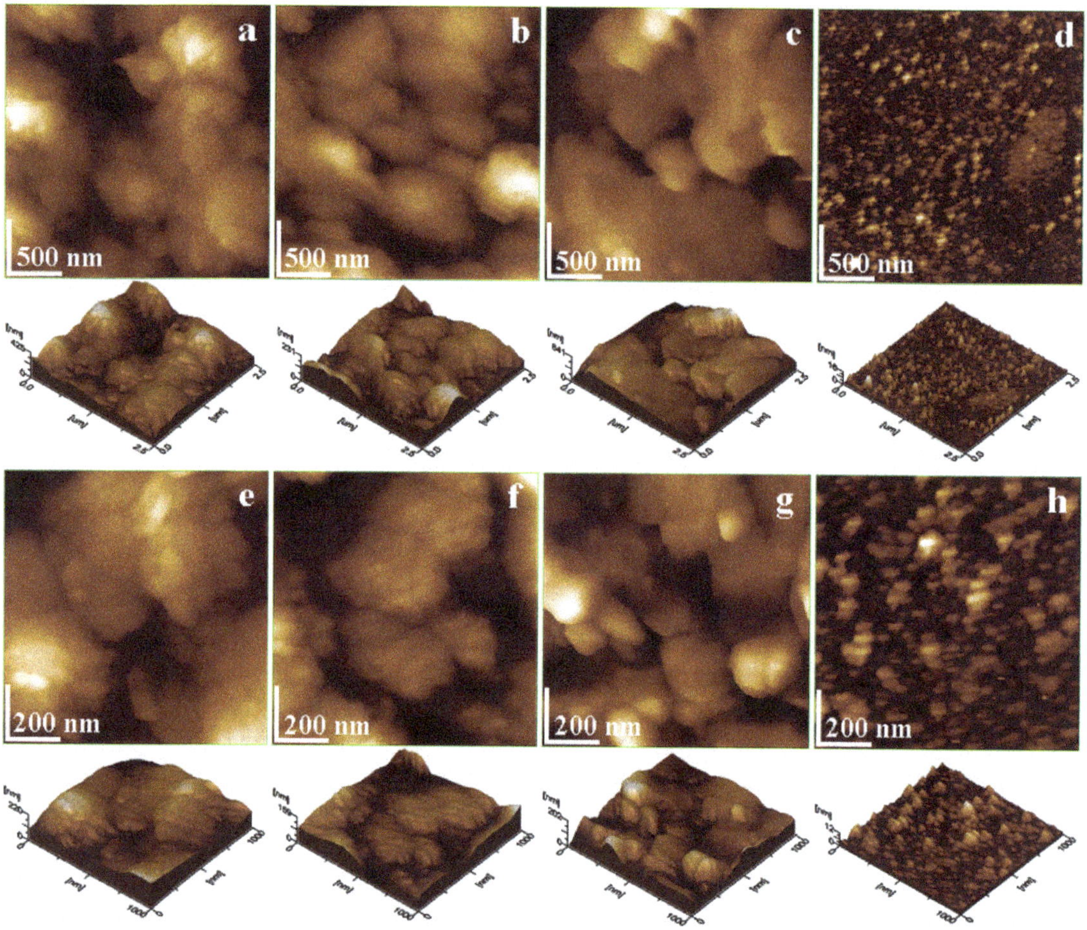

Figure 6. AFM topographic images the samples fine microstructure of: (**a**) FTS, (**b**) WTS, (**c**) PS, and (**d**) FWS. Nanostructures: (**e**) FTS, (**f**) WTS, (**g**) PS, and (**h**) FWS. Three-dimensional profiles are given below each topographic image.

Similar disposal of the boulder submicron particles is observed for the WTS sample, shown in Figure 6b, but the kaolinite nanoparticles are more numerous and lead to a more compact, uniform topography. This fact is supported by the local height of 231 nm.

The PS sample collects particles from FTS and WTS during the pressing filtration process. This technological process does not allow for the self-assembly of the particles to occur, but it forces the kaolinite nanoplatelets to exert a supplementary binding of the quartz and mullite submicron particles; this generates a compact, fine microstructure, as depicted in Figure 6c. The rough intercalation of the boulder-shaped submicron particles into the compact arrangements leads to an increased local height of 541 nm. After pressing filtration, water is completely free of submicron particles, but it features several nanoparticles that are self-assembled on a very thin film (see Figure 6d). There are two main kinds of nano-structural features: some bigger nanoparticles randomly spread out, with significant distances between them, surrounded by fine ones. These arrangements require further investigation.

The nano-structural aspect was observed with a scanned area of 1 μm^2, revealing the organization of kaolinite nanoplatelets. The FTS sample presents a significant amount of kaolinite nanoplatelets, which are adsorbed randomly onto the surface of the submicron particles (see Figure 6e). Their average size is about 60 nm; this is in good agreement with the crystallite size calculated by the Scherrer formula from the XRD pattern. The local height of the nano-arrangements is 220 nm, as observed in the three-dimensional profiles. The WTS sample features a denser adsorption of kaolinite nanoplatelets onto the submicron particle surface (see Figure 6f). The nanoplatelet size is about 40 nm and is disposed in parallel rows, ensuring the optimal and compact binding of the self-assembled thin film. This fact is also illustrated by the low local height of about 159 nm.

The pressed slurry nanostructure provide evidence for the presence of kaolinite nanoplatelets in the range of 40–60 nm; these are very well attached onto the small submicron particles, ensuring a dense formation (see Figure 6g). The borders of the nanoplatelets are more visible on the three-dimensional profile than they are on the topographic image, due to the spatial view. The compact nano-structuring of the pressed slurry leads to a local height of about 202 nm. Finally, the FWS topography in Figure 6h reveals a high-magnification view of the nanoparticles that are left in the water after filtration. The bigger particles have sizes of about 80–90 nm; this corroborates the XRD observations for mullite and quartz. These are surrounded by the finest nanoparticles, measuring about 40 nm, corresponding to the kaolinite crystallites.

4. Discussion

Used industrial water is a cause of great environmental concern due to the high associated pollution risk. The neutralization of this water requires complex physicochemical processes before the water can be released into the public water system [24,25]. Data in the literature shows that the used water from the ceramic industry is more affected by the mineral particle dispersions that are derived from raw materials and the additives used; the water is not affected by the contamination caused by organic compounds [26,27].

Our findings show that both the floor and wall tiles contain the same minerals as raw materials: quartz and kaolinite mixed up with mullite particles, originating from the recirculation of the broken fired biscuits. The main differences consist in the particle size range, which is partly coarse for the floor tile and very fine for the wall tile. The kaolinite/quartz ratio is also greater in the wall tile mixture than in the floor tile mixture. Therefore, the resulting used water from each has different physicochemical properties (see Figure 2). FTUW has fewer dissolved solids and lower salinity; these features are associated with a coarse particle distribution, implying a relatively lower electrical conductivity (increased electrical resistivity). This is reflected in the alkaline pH and the average turbidity. On the other hand, WTUW has a significant amount of dissolved solids and increased salinity, facilitating electrical conductivity (and low electrical resistivity).

As observed, the pH values are strongly influenced by the total dissolved solids in the used water. We found unusual sodium traces in both FTS and WTS, that were related mainly to the presence of glass particles. But the EDS maps clearly show some Na spots outside of the glass particles. Consequently, these Na traces are related to the dissolved salts in the water. The data in the literature reveal that sodium ions induce an alkaline pH that keeps fine kaolinite particles suspended for long periods of time [28,29]. Consequently, the high amount of dispersed fine particles is conducive to higher turbidity values. The sedimentation process is merely influenced by the dispersion of the finest particles; thus, the process takes longer, as was observed for the WTUW sample after 30 days (see Figure 2f).

The detailed ultra-structural observation shows that the filtration process eliminates all coarse particles and most of the fine fractions from the used water. Consequently, the filtered water (FW) is clear, with no turbidity after 30 days and a slightly reduced pH value. On the other hand, the pressed slurry (PS) is a complex mixture of coarse particles, containing quartz and mullite mixed up with fine fractions of kaolinite. The filtration process is optimally described by Equation (4):

$$FTUW_{(l)} + WTUW_{(l)} \xrightarrow{filtration} PS_{(S)} \downarrow + FW_{(l)} \uparrow \quad (4)$$

The FW sample is collected into the recirculation tank, where the FWS particles can become a sediment on the bottom of the tank, and the water is able to be recirculated in the technological process. This step is appropriate and in accordance with the concept of a circular economy [30]; moreover, it certainly reduces the neutralization costs and increases the efficiency of ceramic tiles production facilities regarding used water [31].

Lepidocrocite is one of the most important discoveries regarding pressed slurry. It did not appear in the used water sediments or in the filtered water sediments. Iron's fuzzy elemental distribution in the PS sample is related to the kaolinite particles. The literature mentions the occasional contamination of kaolinite with iron oxides [16,17], but our findings show that the raw kaolinite used here did not contain iron traces. The intense use of the water for the cleaning of the metallic molds leads to the removal of mineral particles; also, it might dissolve iron hydroxides from some rusted parts. Thus, the lepidocrocite (α FeOOH) is crystallized from the remnant liquid during the pressing procedure; this is in agreement with the data in the literature [9,32]. Its presence in the PS sample leads to its inability to be recirculated as raw material because of the glaze-staining risk [13,16]; this transforms the pressed slurry into a problematic waste as related to the production of ceramic tile; thus, it is supposed to be dumped [13,16]. It cannot be used for ceramic tile production, similarly to the red mud that results from the alumina Bayer process, because of the glaze requirement [33,34].

Lepidocrocite recrystallization, during the pressing process, might occur with increases in slurry concentration along with the water removal; on the other hand, the literature shows that iron hydroxides have a solubility of about 1×10^{-9} mol/L; this is independent of the pH value [35]. This suggests that small lepidocrocite nanoparticles, randomly dispersed in the used water, are kept trapped in the pressed slurry; here, they coalesces with each other, forming submicron clusters; this fact is supported by the EDS elemental distribution map shown in Figure 5c. We wonder whether a selective magnetic separation of these lepidocrocite clusters would be effective for their removal, enabling the further reutilization of the pressed slurry as a recirculated material in the main technological process. Some of the data in the literature indicate that the liquid magnetic separation of the iron hydroxide fine microscopic particles is possible [19,36]. Thus, water treatment could be improved with the intercalation of a magnetic selector at the pretreatment station. This is a challenge for future developments in the ceramic industry and the way in which used water is treated.

Our findings prove that both FTUW and WTUW samples contain significant amounts of kaolinite nanoplatelets; these strongly bind with the surrounding submicron and micro particles. The binding effect of those nanoplatelets is enhanced by the pressing process; this proves that the pressed slurry might be useful material for applications in which an amount of 8% wt. lepidocrocite does not affect the utilization. Some of our preliminary studies proved that the compressive strength of the pressed slurry can be increased by a proper compaction of the microstructure [37]. The study must be continued, and an investigation of the effect of the conditioning of the kaolinite matrix on the flexural strength of the pressed samples would be valuable. There are two prospective development pathways regarding the production of ecological bricks by low-temperature firing or geopolymer development; this is because of the high amount of alkaline fine particles [38,39] and the blending of pressed slurry, with an ecological stabilizing agent that can be used to promote spatial reticulation within the mineral structure [40] to achieve plaster for ecological buildings.

5. Conclusions

The obtained results were discussed in relation to the data in the literature; these discussions allow us to draw several important conclusions. These may be valuable for practitioners in ceramic research and industry.

- Both FTS and WTS sediments contain quartz, mullite, and kaolinite; FTS is slightly rich in mullite and WTS is rich in kaolinite. Traces of Fe were found in the FTS and WTS sediments, but there was no proof of crystallization.
- The efficacy of the filtering system was proven on the basis of the targeted particulate matters; these ensure the recirculation of water into technological processes. The sludge that is obtained after the filtration process presents with a dense grainy structure of sediment particles. These contain quartz, mullite, and kaolinite, along with traces of iron hydroxide crystallized as lepidocrocite.
- The pressed slurry cannot be reused in the technological flux because of the presence of iron hydroxide, which causes glaze staining; however, the observed microstructure, along with the mineralogical composition, indicates that it could be used for other applications, like ecological bricks or plasters. These potential applications will be investigated further.
- The liquid magnetic selective removal of the observed lepidocrocite clusters could offer a significant improvement in the process of used water pretreatment; this will allow the direct recirculation of pressed slurry into the main technological process.

Author Contributions: Conceptualization, S.E.A. and I.P.; methodology, S.E.A. and I.P.; software, G.B.; validation, S.E.A. and I.P.; formal analysis, B.V.B.; investigation, S.E.A., L.B.T., G.B., B.V.B. and I.P.; resources, S.E.A. and B.V.B.; data curation, I.P.; writing—original draft preparation, S.E.A. and I.P.; writing—review and editing, I.P.; visualization, L.B.T. and I.P.; supervision, S.E.A. and I.P.; project administration, S.E.A. and I.P. All authors have read and agreed to the published version of the manuscript.

Funding: The XRD diffractometer maintenance which was supported by the Ministry of Research, Innovation, and Digitization through Program 1-Development of the National Research and Development System, Subprogram 1.2-Institutional Performance-Funding Projects for Excellence in RDI, Contract No. 37PFE/30.12.2021.

Data Availability Statement: Data are available on request from the corresponding author.

Conflicts of Interest: The authors declare no conflicts of interest.

References

1. Vieira, A.V.; Rosso, L.S.; Demarch, A.; Pasini, P.; Ruzza, S.P.; Arcaro, S.; Ribeiro, M.J.; Angioletto, E. Life cycle assessment in the ceramic tile industry: A review. *J. Mater. Res. Technol.* **2023**, *23*, 3904–3915. [CrossRef]
2. Manikandan, K.P.; Nanthakumar, P.; Balachandar, M.; Gowri Shankar, D.; Vijayakumari, G. Partial replacement of aggregate with ceramic tile in concrete. *Mater. Today Proc.* **2023**, *in press*. [CrossRef]
3. Zanelli, C.; Conte, S.; Molinari, C.; Soldati, R.; Dondi, M. Waste recycling in ceramic tiles: A technological outlook. *Resour. Conserv. Recycl.* **2012**, *168*, 105289. [CrossRef]
4. Ziman, N. Contributions on Wall Tiles Technology with Implications in the Quality and Prices. Ph.D. Thesis, Romanian, University "Politehnica" of Timisoara, Timisoara, Romania, 2006.
5. Baricza, A.; Bajnóczi, B.; Máté, S.; Tóth, M.; Bendő, Z.; Szabo, C. Deterioration of glazed architectural ceramics due to environmental factors: A comparative study of two buildings in Budapest. *Carpathian J. Earth Environ. Sci.* **2016**, *11*, 449–462.
6. Balint, R.; Calotescu, L.G.; Boero, V.; Ajmone-Marsan, F. Clay minerals-clinoptilolite assemblages in two romanian tuffs. *Carpathian J. Earth Environ. Sci.* **2015**, *10*, 27–39.
7. Ilyina, V.; Klimovskaya, E.; Bubnova, T. Ceramic Materials Based on Clay and Soapstone Waste: Thermo-Mechanical Properties and Application. *Minerals* **2023**, *13*, 1376. [CrossRef]
8. Hosu Prack, G.A.; Petean, I.; Arghir, G.; Bobos, L.D.; Tomoaia-Cotisel, M. Nano-scale particulate matters found in urban street dust in Cluj–Napoca, Romania. *Carpathian J. Earth Environ. Sci.* **2016**, *11*, 539–546.
9. Avram, S.E.; Filip, M.R.; Barbu Tudoran, L.; Borodi, G.; Petean, I. Investigation of ferrous conglomerate particles found in carwashslurry and their environmental implications. *Stud. UBB Chem.* **2023**, *68*, 57–70. [CrossRef]
10. Rus, L.; Avram, S.E.; Micle, V. Determination and assessments of physicochemical parameters of the water from anthropo-saline lakes located in the protected area "Salina Turda", Romania. *Stud. UBB Chem.* **2020**, *65*, 257–268. [CrossRef]

11. Avram, S.E.; Rus, L.; Micle, V.; Hola, S.S. Evaluation and Evolution of the Physico-Chemical Parameters of Ocnei and Rotund Lakes Located near the "Salina Turda" Mine, Romania. *Water* **2022**, *14*, 2366. [CrossRef]
12. Mohd, H.I.; Mohd, I.M.; Anis, N.I.; Syarifah, A.I.; Nabihah, O. Development of porous glass-ceramic using silica sand for wall tiles application. *Mater. Today Proc.* **2023**, *in press*. [CrossRef]
13. Mourou, C.; Zamorano, M.; Ruiz, D.P.; Martín-Morales, M. Characterization of ceramic tiles coated with recycled waste glass particles to be used for cool roof applications. *Constr. Build. Mater.* **2023**, *398*, 132489. [CrossRef]
14. Alves, C.L.; Skorych, V.; De Noni, A.; Hotza, D.; Gómez González, S.Y.; Heinrich, S.; Dosta, M. Improving the sustainability of porcelain tile manufacture by flowsheet simulation. *Ceram. Int.* **2023**, *49*, 24581–24597. [CrossRef]
15. Contini, G.; Peruzzini, M.; Bulgarelli, S.; Bosi, G. Developing key performance indicators for monitoring sustainability in the ceramic industry: The role of digitalization and industry 4.0 technologies. *J. Clean. Prod.* **2023**, *414*, 137664. [CrossRef]
16. Wang, Y.; Yu, S.; Chu, J.; Chen, D.; Chen, J. Study on the copper and iron coexisted coloring glaze and the mechanism of the fambe. *J. Eur. Ceram. Soc.* **2018**, *38*, 3681–3688. [CrossRef]
17. Gol, F.; Saritas, Z.G.; Cıbuk, S.; Ture, C.; Kacar, E.; Yilmaz, Y.; Arslan, M.; Sen, F. Coloring effect of iron oxide content on ceramic glazes and their comparison with the similar waste containing materials. *Ceram. Int.* **2022**, *48*, 2241–2249. [CrossRef]
18. Muntean, D.F.; Ristoiu, D.; Arghir, G.; Campean, R.F.; Petean, I. Iron hydroxides occurrence in winter air particulate matters suspensions in Cluj–Napoca, Romania. *Carpathian J. Earth Environ. Sci.* **2012**, *7*, 175–182.
19. Muntean, D.F.; Ivan, I.; Muresan, L. Environmental Implications Concerning the Chemical Composition and Particle Distribution of Anti–Skid Material. *Stud. UBB Chem.* **2015**, *60*, 207–218.
20. Caulfield, J.T.; Tomlinson, E.L.; Chew, D.M.; Marks, M.A.V.; McKenna, C.A.; Ubide, T.; Smith, V.C. Microanalysis of Cl, Br and I in apatite, scapolite and silicate glass by LA-ICP-MS. *Chem. Geol.* **2020**, *557*, 119854. [CrossRef]
21. Kim, J.M.; Kim, H.S. Glass-ceramic produced from a municipal waste incinerator fly ash with high Cl content. *J. Eur. Ceram. Soc.* **2004**, *24*, 2373–2382. [CrossRef]
22. Ramaswamy, S.; Raghavan, P. Significance of Impurity Mineral Identification in the Value Addition of Kaolin—A Case Study with Reference to an Acidic Kaolin from India. *J. Miner. Mater. Charact. Eng.* **2011**, *10*, 1007–1025. [CrossRef]
23. Segura, J.C.F.; Reyes Cruz, V.E.; Jesús Pérez Bueno, J.; Lozada Ascencio, E.M.; García, F.L. Characterization and electrochemical treatment of a kaolin. *Appl. Clay Sci.* **2017**, *146*, 264–269. [CrossRef]
24. Wang, Y.; Shi, Y.; Xu, X.; Zhu, Y. A Study on the Efficiency of Green Technology Innovation in Listed Chinese Water Environment Treatment Companies. *Water* **2024**, *16*, 510. [CrossRef]
25. Santos, A.B.d.; Giacobbo, A.; Rodrigues, M.A.S.; Bernardes, A.M. Electrodeionization for Wastewater Reuse in Petrochemical Plants. *Water* **2024**, *16*, 401. [CrossRef]
26. Boschi, G.; Bonvicini, G.; Masi, G.; Bignozzi, M.C. Recycling insight into the ceramic tile manufacturing industry. *Open Ceram.* **2023**, *16*, 100471. [CrossRef]
27. Ding, K.; Li, A.; Lv, J.; Gu, F. Decarbonizing ceramic industry: Technological routes and cost assessment. *J. Clean. Prod.* **2023**, *419*, 138278. [CrossRef]
28. Rao, F.; Ramirez-Acosta, F.J.; Sanchez-Leija, R.J.; Song, S.; Lopez-Valdivieso, A. Stability of kaolinite dispersions in the presence of sodium and aluminum ions. *Appl. Clay Sci.* **2011**, *51*, 38–42. [CrossRef]
29. Dwari, R.K.; Mishra, B.K. Evaluation of flocculation characteristics of kaolinite dispersion system using guar gum: A green flocculant. *Int. J. Min. Sci. Technol.* **2019**, *29*, 745–755. [CrossRef]
30. Markowski, Ł.; Kotliński, K.; Ostrowska, A. Sustainable Consumption and Production in the European Union—An Attempt to Assess Changes and Convergence from the Perspective of Central and Eastern European Countries. *Sustainability* **2023**, *15*, 16485. [CrossRef]
31. Alresheedi, M.T.; Haider, H.; Albuaymi, A.M.; AlSaleem, S.S.; Shafiquzzaman, M.; Alharbi, A.; Ahsan, A. Sustainability of a Low-Cost Decentralized Treatment System for Wastewater Reuse: Resident Perception-Based Evaluation for Arid Regions. *Water* **2023**, *15*, 3458. [CrossRef]
32. Chen, D.; Yen, M.; Lin, P.; Groff, S.; Lampo, R.; McInerney, M.; Ryan, J. A Corrosion Sensor for Monitoring the Early-Stage Environmental Corrosion of A36 Carbon Steel. *Materials* **2014**, *7*, 5746–5760. [CrossRef] [PubMed]
33. Xu, X.; Song, J.; Li, Y.; Wu, J.; Liu, X.; Zhang, C. The microstructure and properties of ceramic tiles from solid wastes of Bayer red muds. *Constr. Build. Mater.* **2019**, *212*, 266–274. [CrossRef]
34. Pérez-Villarejo, L.; Corpas-Iglesias, F.A.; Martínez-Martínez, S.; Artiaga, R.; Pascual-Cosp, J. Manufacturing new ceramic materials from clay and red mud derived from the aluminium industry. *Constr. Build. Mater.* **2012**, *35*, 656–665. [CrossRef]
35. Kuma, K.; Suzuki, Y.; Matsunaga, K. Solubility and dissolution rate of colloidal γ-FeOOH in seawater. *Water Res.* **1993**, *27*, 651–657. [CrossRef]
36. Arghir, G.; Petean, I.; Munteanu, D.F.; Muresan, L.; Suciu, C. Raw material for powder metallurgy obtained from environmental iron hydroxides. *Powder Metall. Prog.* **2011**, *3–4*, 340–346.
37. Avram, S.E.; Barbu Tudoran, L.; Cuc, S.; Borodi, G.; Birle, B.V.; Petean, I. Microstructural Investigations Regarding Sustainable Recycling of Ceramic Slurry Collected from Industrial Waste Waters. *Sustainability* **2024**, *16*, 1123. [CrossRef]
38. Romero Quidel, G.; Soto Acuña, M.J.; Rojas Herrera, C.J.; Rodríguez Neira, K.; Cárdenas-Ramírez, J.P. Assessment of Modular Construction System Made with Low Environmental Impact Construction Materials for Achieving Sustainable Housing Projects. *Sustainability* **2023**, *15*, 8386. [CrossRef]

39. Horabik, J.; Jozefaciuk, G. Structure and strength of kaolinite–soil silt aggregates: Measurements and modeling. *Geoderma* **2021**, *382*, 114687. [CrossRef]
40. Kaminskas, R.; Savickaite, B. Expanded Clay Production Waste as Supplementary Cementitious Material. *Sustainability* **2023**, *15*, 11850. [CrossRef]

Disclaimer/Publisher's Note: The statements, opinions and data contained in all publications are solely those of the individual author(s) and contributor(s) and not of MDPI and/or the editor(s). MDPI and/or the editor(s) disclaim responsibility for any injury to people or property resulting from any ideas, methods, instructions or products referred to in the content.

water

MDPI

Article

A Spatial–Seasonal Study on the Danube River in the Adjacent Danube Delta Area: Case Study—Monitored Heavy Metals

Catalina Topa [1,2], Gabriel Murariu [1,2,*], Valentina Calmuc [1,2], Madalina Calmuc [1,2], Maxim Arseni [1,2], Cecila Serban [2,3], Carmen Chitescu [2,4] and Lucian Georgescu [1,2,*]

[1] Faculty of Sciences and Environment, "Dunarea de Jos" University of Galati, 800008 Galati, Romania; catalina.topa@ugal.ro (C.T.); valentina.calmuc@ugal.ro (V.C.); madalina.calmuc@ugal.ro (M.C.); maxim.arseni@ugal.ro (M.A.)

[2] Rexdan Research Infrastructure, "Dunarea de Jos" University of Galati, 800008 Galati, Romania; cecilia.serban@ugal.ro (C.S.); carmen.chitescu@ugal.ro (C.C.)

[3] Department of Applied Sciences, Cross-Border Faculty, "Dunarea de Jos" University of Galati, 800008 Galati, Romania

[4] Faculty of Medicine and Pharmacy, "Dunarea de Jos" University of Galati, A.I. Cuza 35, 800010 Galati, Romania

* Correspondence: gabriel.murariu@ugal.ro (G.M.); lucian.georgescu@ugal.ro (L.G.)

Abstract: Monitoring and protecting flowing watercourses is a complex and challenging task that requires the collaboration and coordination of various stakeholders such as governments, industries, farmers, consumers and environmental groups. The study of the dynamics of the concentration of polluting factors and especially the concentrations of heavy metals and highlighting a seasonal variation is a necessary element from this point of view. In this article, we present the results of our analyses carried out in two measurement campaigns executed in 10 monitoring points along the Danube River, between Braila city and Isaccea city in the pre-deltaic area, during the summer season and autumn season 2022. The importance of this area is given by the fact that the Danube Delta is part of the UNESCO heritage, and the monitoring of polluting factors is a necessity in the desire to protect this area. The data measured during the July and August 2022 campaign cover a wide range of chemical species: Phosphate, CCO, CBO_5, NH_4^+, $N\text{-}NO_2$, $N\text{-}NO_3^-$, N-Total, $P\text{-}PO_4{}^{3-}$, $SO_4{}^{2-}$, Cl^-, phenols, as well as metals with a harmful effect: Al, As, Cd, Cr, Fe. The study includes an evaluation based on the statistical approach of the results to highlight the significant correlations and differences identified between the two data sets. Next, to highlight the obtained results, a numerical model was considered using HEC-RAS and ESRI ArcGIS applications in a two-dimensional unsteady flow model in order to obtain the non-homogenous concentrations' distributions in the studied area. These two-dimensional models have been less studied in the specialized literature. In this way, interesting results could be obtained, and prediction methods regarding the dynamics of metal concentrations could be structured. The data obtained were used for the terrain model from the USGS service, and the flows of the Danube and its two tributaries were simulated using the data provided by the national services. In this work, we present the results obtained for the dynamics of the concentrations of the metals Al, As, Cd, Cr and Fe and the evaluation of the specific absorption coefficients for the explanation and correlation with the results of the measurements. Except for the numerical model presented, we would like to highlight the existence of some contributions of the main tributaries of the Danube in the study area. Such a systematic study has not been carried out due to conditions imposed by the border authorities. From this point of view, this study has an element of originality. The study is part of a more complex project in which the spatio-temporal distribution of the polluting factors in the water was evaluated, and the habitats in the study area were inventoried—especially those of community interest. In this way, we were able to expose the self-purification capacity of the Danube and highlight the existence of a concentration reduction gradient along the course of the river. The aspects related to the influence of the distribution of polluting factors on the state of health will be the subject of another article.

Keywords: aluminum; arsenic; cadmium; chromium; iron; absorption coefficients; numerical approach; flow simulation; HEC-RAS; model unsteady

1. Introduction

The problem of pollution of running water courses is a serious and widespread problem [1], affecting the health and welfare of people [2], animals and plants [3]. Flowing watercourses such as streams, rivers, lakes and estuaries are vital sources of drinking water, irrigation, recreation and biodiversity [4]. However, they are also vulnerable to contamination from various sources [5], such as industrial waste [2], agricultural runoff [6], sewage, oil spills, and radioactive materials [5]. These pollutants can alter the physical, chemical and biological properties of water, making it unsuitable for its intended uses and damaging natural ecosystems [4]. Some of the negative effects of water pollution include: reduced oxygen levels [7], which can cause fish kills and affect aquatic life; increased nutrient levels that can lead to eutrophication and algal blooms [6], resulting in decreased water quality and increased toxicity [8]; increased sedimentation, which can reduce water clarity and clog the gills of fish and other organisms [9]; increased acidity [10], which can corrode metal pipes and damage aquatic life [11]; increasing temperatures, which can affect the metabolism and reproduction of aquatic organisms and increase the risk of invasive species and disease; increased presence of pathogens such as bacteria, viruses and parasites that can cause waterborne diseases and infections in humans and animals; the increased presence of toxic substances such as heavy metals, pesticides and pharmaceuticals, which can accumulate in the food chain and pose health risks to humans and wildlife [11,12].

Some examples of water pollutants are: pathogens, such as bacteria, viruses and parasites, which can cause waterborne diseases and infections in humans and animals [13]. For example, mercury can accumulate in the food chain and pose health risks to humans and wildlife [5]; organic materials such as pesticides, pharmaceuticals [6] and petroleum products can degrade water quality and increase water toxicity [1,14]. For example, methyl tert-butyl ether (MTBE) is a volatile organic chemical that was previously used as an air-cleaning gas additive and can cause leukemia, lymphoma and tumors in humans and animals; macroscopic pollutants such as plastic waste, wood, metal and shipwrecks [6,13], which can reduce water clarity and clog waterways. For example, plastic waste can form large patches in the oceans and harm marine life.

There are many examples of successful river restoration projects around the world [15,16] that have improved the ecological, social and economic benefits of rivers [17]: the Herring River Restoration Project in Cape Cod, Massachusetts, which aims to restore the natural tidal flow and the salinity of the Herring River estuary, which was altered by a levee built in 1909; the Reedy River Master Plan in Greenville, South Carolina, is a comprehensive vision for the revitalization of the urban river corridor, which has been degraded by industrial pollution and urban development, etc.

For these reasons, the study of stream water monitoring is a vital and challenging field that requires the use of various methods and technologies to collect, analyze and interpret water quality and quantity data [18,19]. Some of the most innovative study methods in this field are: IoT-based water monitoring systems [20,21]—these systems use sensors, wireless networks, cloud computing and data analysis to measure and transmit water parameters in real-time, such as temperature, pH, turbidity, dissolved oxygen and conductivity [22]. These systems can provide accurate, timely and cost-effective information on water quality and quantity, as well as enable remote control and management of water resources.

Studying the Danube in the Danube Delta area is important for several reasons [2,23]. The Danube Delta is one of the continent's most valuable habitats for delta-specific wildlife and biodiversity [6,24,25]. It is the second largest delta in Europe and the best preserved of the European deltas [18]. It is also a UNESCO World Heritage Site and a Ramsar site, meaning it is a wetland of international importance [1]. The Danube Delta is home to over

300 species of birds and 45 species of freshwater fish in all of its lakes and marshes [6]. They also provide important ecosystem services such as drinking water, food, flood protection, nutrient recycling and recreation [5,6].

The Danube Delta is a unique and valuable ecosystem, but it faces many challenges that threaten its sustainability [8,24]. Some of the challenges are:

- Nutrient overload: the Danube River carries large amounts of nutrients from agricultural runoff [11], sewage and industrial waste [2], which cause eutrophication and algal blooms in the Delta [8]. This reduces water quality, oxygen levels and the biodiversity of aquatic habitats [26].
- Biodiversity loss: the Delta is home to many rare and endangered species of plants and animals [8], but these are threatened by habitat loss [5], fragmentation, degradation, invasive species, overfishing [13], hunting and climate change. Some of the species at risk include the sturgeon [27], pelican, pygmy cormorant and red-breasted goose [28].
- Land use transformations: the Delta has been altered by human activities such as drainage, irrigation, rehabilitation, deforestation and urbanization [5], which have reduced natural wetlands [8] and floodplains and increased the risk of erosion [29], salinization and desertification. These changes have also affected the livelihoods and culture of local communities that depend on the Delta's resources [19].
- Organic pollution and microplastics: the Danube River and its Delta are polluted with various organic contaminants [24,28], such as pesticides [6], pharmaceuticals, hormones and personal care products, which can have negative effects on the health of organisms and the ecosystem [27]. In addition, the river and Delta are also contaminated with microplastics [28], which are small plastic particles that can accumulate in the environment and food chain and pose a threat to wildlife and human health [6].

These are some of the challenges facing the Danube Delta, but there are also many opportunities and initiatives to address them and promote the conservation and restoration of this precious natural heritage. The distribution and dispersion of heavy metals in the Danube Course is a complex and important environmental issue [2], as the Danube River is the second-longest river in Europe and flows through 10 countries, affecting millions of people and ecosystems. Heavy metals can originate from various sources, such as natural weathering, industrial activities, agricultural runoff, urban wastewater, and mining operations. Some of the heavy metals, such as cadmium (Cd), and arsenic (As), etc. [24], are toxic and can accumulate in the aquatic organisms and sediments, posing health risks to humans and wildlife [8]. In this respect, the aim is to assess the seasonal variation in the concentrations of heavy metals (such as cadmium, chromium, arsenic, etc.) in the aquatic ecosystems of the Danube River Braila-Galati area. The aim of the study is to investigate the distributions and dispersions of heavy metals on the Danube Course.

2. Materials and Methods

2.1. Data Collection

In order to determine the polluted areas in the Lower Danube region in Romania, we carried out a monitoring study of the physico-chemical parameters of the aquatic ecosystems in the study area (the Lower Danube and the adjacent Siret and Prut rivers) in order to assess and establish the degree of their pollution, according to the Water Framework Directive (https://environment.ec.europa.eu/topics/water/water-framework-directive_en, accessed on 24 February 2020). As it is known, the 'water framework directive' is a law that aims to protect and improve the quality and quantity of water resources in the European Union [30]. It applies to rivers, lakes, groundwater, coastal and transitional waters. It requires member states to achieve good status for all water bodies by 2015, and to prevent deterioration. It also promotes an integrated approach to water management, based on river basin districts and cooperation among countries. In this respect some important parameters were considered: knowing the value of biochemical oxygen consumption per 5 days (CBO5) is extremely important in evaluating the degree of pollution of waste water expressing the amount of biodegradable organic substances contained in the respective water. The chemical oxygen consumption CCO-Cr (potassium bichromate method) represents the mass concentration of oxygen equivalent to the amount of potassium bichromate consumed for the oxidation in an acidic environment of dissolved and suspended organic matter present in wastewater.

Water samples were analyzed from 10 sampling stations (P1–P10)—see Table 1, whose distribution includes the warm spots of the Danube, Siret, Prut (Figure 1). There are very important industrial objectives: the Galati steel plant (P3), the mineral port of the steel plant (P3), the Damen shipyard (P5), the international port at the confluence of the Prut River with the Danube River (P8), the Ukrainian port of Reni (P9), etc.

Table 1. Monitoring stations list.

Code/Name	Station	GPS Coordinates [d.d.]
P1	City water Danube Pump Station—on the Danube course—upstream of the mineral port of the steel complex	45.37261173; 28.02879381
P2	Siret River monitoring point—on the Siret River upstream course	45.40016436; 27.9973168
P3	Siret River confluence with Danube River monitoring point	45.40848419; 28.027358
P4	Libertatea restaurant monitoring point—on the Danube course—next to the municipal tourist area	45.42957954; 28.05900221
P5	Damen Shipyard downstream monitoring point—on the Danube course	45.43628888; 28.13124235
P6	Cotul Pisicii recreation area monitoring point—on the Danube course	45.41873878; 28.19135779
P7	Prut River—Giurgiulesti monitoring point—on the Prut River upstream course	45.48016; 28.185536
P8	Prut River confluence with Danube River monitoring point	45.46528806; 28.23220795
P9	Ukrainian shipyard Reni monitoring point—on the Danube course	45.38315546; 28.29554717
P10	Ukrainian passing border Isaccea monitoring point—on the Danube course	45.28405785; 28.45693996

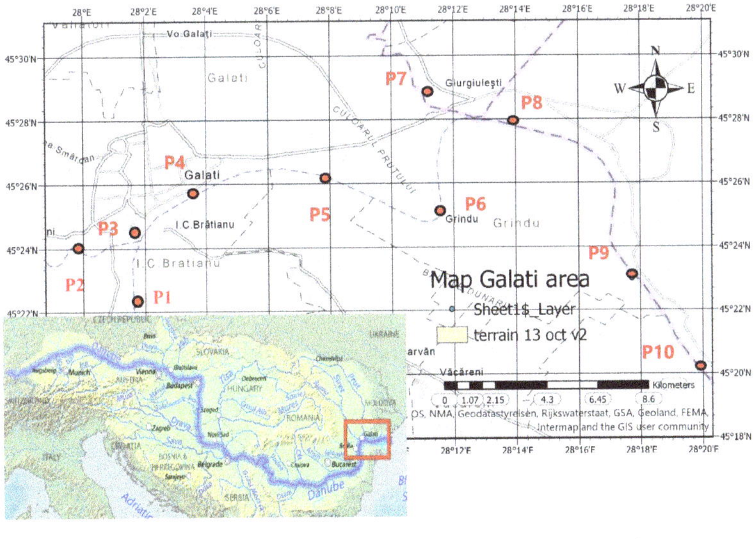

Figure 1. The map of monitoring points configuration and a broader regional map.

Below, we present the main data analysis methods and procedures.

A total of 17 physicochemical parameters were determined: pH, conductivity, CCO, CBO_5, NH_4^+, N-NO_2^-, N-NO_3^-, N-Total, P-PO_4^{3-}, SO_4^{2-}, Cl^-, phenols, Al, As, Be, Cd, Cr. Some of the parameters were analyzed in situ, namely pH, conductivity, and TsOD/BOD, using a portable multiparameter equipped with a multisensory electrode and the turbidity using Turb 430 T—Turbidimeter, Xylem Analytics Germany GmbH (www.rexdan.ugal.ro).

Sampling and sample analyses for physico-chemical parameters were made according to the following Standard methods (https://www.asro.ro/lista-standarde-calitatea-apei/, accessed on 2 August 2024): SR EN 1899-2:2002 for BOD determination, SR EN ISO 11732:2005 for ammonium determination, SR EN ISO 6878:2005 for total phosphorus determination, SR EN 26777:2006 for nitrogen from nitrite determination, SR EN ISO 11905-1:2003 for nitrogen from nitrate determination, SR ISO 6332:1996/C91:2006—for iron total content, SR ISO 9297/2001—for chloride content, SR ISO 10566:2001—for lead content, SR ISO 9174-98—for Cr total, SR EN 16503:2015—Guide standard for the evaluation of the hydro morphological characteristics of transient and coastal waters, SR EN ISO 10523:2012—Water quality—Determination of pH; SR EN 16479:2014 Water quality. Performance requirements and compliance test procedures for water monitoring equipment. Automatic sampling devices (samplers) for water and waste water; SR EN ISO 5667-1:2007—Sampling—Part 1: General guide for establishing sampling programs and techniques SR EN ISO 5667-1:2007/AC:2007—Water quality—Sampling. Part 1: General guide for establishing sampling programs and techniques; SR EN ISO 5667-14:2017—Water quality—Sampling—Part 14: Guide for quality assurance and quality control in the sampling and treatment of water from the environment, etc. The research center is ISO-accredited and is obliged to comply with European standards.

The remaining parameters were analyzed in the laboratory in accordance with the standards in force, as follows: metals were analyzed using the inductively coupled plasma–mass spectrometry (ICP-MS) technique; nutrients, chlorides, sulfates, CCO and phenols were analyzed spectrophotometrically using Spectroquant® test kits; biochemical oxygen demand (CBO_5) was determined using VELP a respirometric BOD Sensor System—VELP Scientifica, and FOC 200E Cooled Incubator (www.rexdan.ugal.ro).

2.2. Methodology

2.2.1. Water and Sediment Sampling Methodology

Water samples were taken from 10 sampling stations placed on the Danube, Siret and Prut rivers using the sampling in monitoring points illustrated in Figure 1. Water samples were collected in polyethylene containers (PE) and were transported to the CREDENTIAL laboratory that is part of the European Center of Excellence for the Environment (ECEE) of "Dunarea de Jos" University of Galati [14]. The water sample collection methods were carried out as in most campaigns described in the literature: the samples were collected mainly in the minor basin, and the samples were taken from a minimum depth of 1.5 m. The samples were collected for each point every 48 h. For this reason, the data showed a tight clustering. No storm or rain samples were collected—being a dry year. The instrument used was—ORION VERSA STAR 90 ThermoFisher Scientific—more details presented at the address—https://www.rexdan.ugal.ro/index.php/en/rexdan-ri/rexdan-research-center (accessed on 24 February 2020)—see The REXDAN Research Center (includes 9 laboratories) and, respectively, The REXDAN Research Vessel (includes 9 laboratories). Collected samples were preserved at a temperature of 4 °C and analyzed within 24 h. Surface sediment samples were collected using a Van Veen Grab Sampler (KC Denmark A/S, Silkeborg, Denmark) from the first 10 cm of the sediment surface and stored in polyethylene recipients. The sediment samples were preserved at 4 °C [24].

Metals from the IARC list (Al, As, Be, Cd, Cr) were analyzed using ICP-MS Nexion 2000C (PerkinElmer, Waltham, MA, USA) [24] (https://www.rexdan.ugal.ro/index.php/en/rexdan-ri/rexdan-research-center/the-chromatography-laboratory, 24 February 2020), in the Spectrometry Laboratory of the REXDAN Research Infrastructure, "Dunarea de

Jos" University of Galati (https://rexdan.ugal.ro/index.php/en/, 24 February 2020). The quantification limit for each metal was 0.001 µg/L. The detection limits for the 5 analyzed metals were the following: Al (0.01 ng/L), As (0.3 ng/L), Be (0.5 ng/L), Cd (0.06 ng/L), Cr (0.05 ng/L). For the analysis of water samples, the blank was water acidified 5% with nitric acid 65% (Suprapur, Darmstadt, Germany) [5]. The International Agency for Research on Cancer (IARC) selects metals for analysis based on several criteria. Primarily, they consider the evidence of human exposure and the potential or known carcinogenicity of the substances. Metals like aluminum, arsenic, beryllium, cadmium, and chromium have been included in their analyses because of their significant presence in the environment and their potential health impacts. The calibration curve for each element was made in the concentration range of 0–15 µg/L. Also, calibration curves are linear with a correlation coefficient (R^2) greater than 0.999 for the five trace metals analyzed [13].

The operating conditions were as follows: nebulizer gas (argon) flow rate: 1.1 L/min, auxiliary gas (argon) flow 1.2 L/min, plasma (Argon) gas flow: 15 L/min, Pulse Stge Voltage 800 V, ICP RF power: 1600 W, CeO/Ce $\leq$ 0.022 and Ce++/Ce+ $\leq$ 0.012, Dual Detector Mode: Pulse, Aqc. Dead Time (ns): 35, Current Dead Time (ns): 35, Torch Z position (mm): 0.00. Method parameters were the following: Sweeps/Reading 20, Readings/Replicate = 1, Replicates = 3 [24]. Three water samples were taken from each station, and a mixed sample was taken and analyzed. The water samples were taken in two seasons: summer and autumn.

2.2.2. Methodology for Statistical Analysis

The statistical analysis methods used were: the Pearson correlation method, the univariate parametric and non-parametric ANOVA analysis method, the Principal Components Analysis (PCA) and Factor Analysis (FA) methods [2]. The software used was TIBCO Statistica V13 and MATLAB (R2013b).

2.2.3. Methodology for Numerical Analysis

The terrain map was constructed using the OpenTopography data service (Figure 2). The spatial resolution using the USGS database (https://earthexplorer.usgs.gov/) was enough to build an appropriate hydrological model [19].

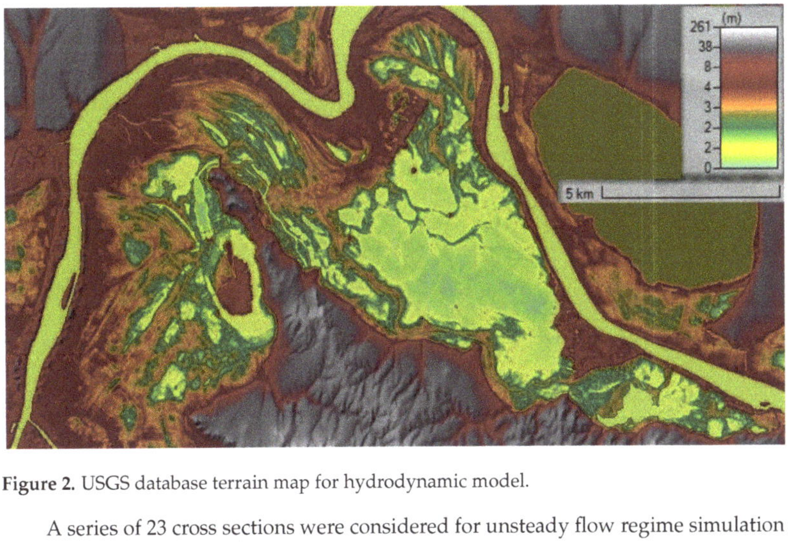

Figure 2. USGS database terrain map for hydrodynamic model.

A series of 23 cross sections were considered for unsteady flow regime simulation and diffusion modeling [19,29].

The used hydrodynamic model was constructed, calibrated and validated with the 1D HEC-RAS (version 6.4.1) software under an unsteady flow regime [29] (Figure 3), with hydrologic at the two upstream and downstream gauging stations [31]. The procedure followed was the one used in the literature in our previous studies with the same team [29]. The step of applying the model was 12 h, in non-stationary mode, with graphic presentation every 7 days and the flow data of the Danube that were measured by the authorities were used [19,29,32].

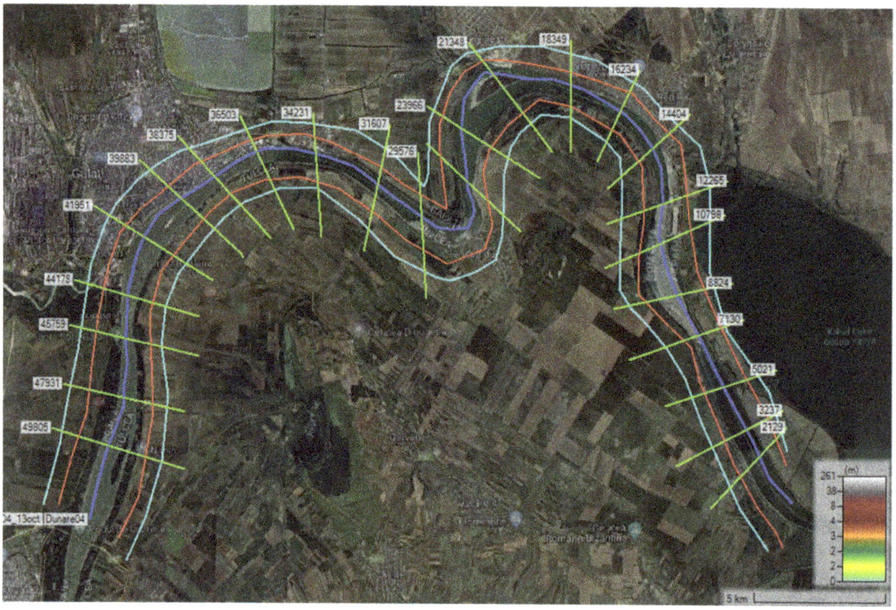

Figure 3. The HEC-RAS hydrological terrain map with transversal cross sections for numerical model.

In order to optimize the results (using, for example, R^2 criterion), a series of successive simulations were performed in order to refine the results. The numerical interpolation algorithm for the sources identified in the study area was described by the following steps:

Step 1: A definition of the parameters that describe the distribution of pollution sources—the initial distribution—is made.

Step 2: A monitoring point is considered—starting with the first upstream point and continuing successively downstream.

Step 3: Enter the parameters related to the local sources and perform the numerical simulation—HEC—RAS two-dimensional.

Step 4: The downstream concentrations are evaluated, and the comparison is made with the measured ones, determining the quantities to be applied for correction.

Step 5: Change the concentrations from the upstream points to adjust the difference and repeat step 3.

If the difference between the concentrations in the considered point has reached a minimum, move to the next downstream point and repeat step 3.

3. Results

3.1. Data Analysis

In the following, the statistical and numerical approaches' results will be provided. In Table 2 are presented the distribution values for July measurement campaign.

Table 2. Results of the parameters from the IARC list analyzed from the water samples taken in July 2022.

Station	Aluminum (μg/L)	Arsenic (μg/L)	Beryllium (μg/L)	Cadmium (μg/L)	Chromium (μg/L)	Phosphate—P-PO4 (mg/L)
P1	14.468	0.002	<LOD (<0.5 ng/L)	0.008	1.027	0.080
P2	19.670	0.001	<LOD (<0.5 ng/L)	0.004	0.764	0.040
P3	46.890	0.002	<LOD (<0.5 ng/L)	0.021	1.280	0.070
P4	15.704	0.002	<LOD (<0.5 ng/L)	0.016	1.605	0.140
P5	38.868	0.002	<LOD (<0.5 ng/L)	0.016	1.239	0.090
P6	17.214	0.001	<LOD (<0.5 ng/L)	0.008	1.123	0.070
P7	24.168	0.001	<LOD (<0.5 ng/L)	0.038	1.291	0.050
P8	8.665	0.001	<LOD (<0.5 ng/L)	0.004	0.771	0.070
P9	28.889	0.001	<LOD (<0.5 ng/L)	0.020	2.318	0.080
P10	32.650	0.002	<LOD (<0.5 ng/L)	0.019	1.694	0.080

3.2. *Statistical Approach*

Descriptive Statistics

In this section, the elementary statistical results are presented. In this respect, Tables 3 and 4 show the elements of the preliminary description of chemical species.

Table 3. Descriptive Statistics (July–October databases). The units of measure are those indicated above (see the Table 2).

		July Database						October Database					
Variable	Valid N	Mean	Median	Min.	Max.	Std. Dev.	Std. Error	Mean	Median	Min.	Max.	Std. Dev.	Std. Error
Phosphate	10	0.07700	0.07500	0.0400	0.14000	0.02669	0.008439	0.05400	0.0600	0.0100	0.070	0.01713	0.005416
CCO	10	13.59000	10.9000	7.5000	33.40000	7.99617	2.528612	12.76000	10.200	7.1000	30.300	7.23559	2.288095
CBO_5	10	25.46000	24.7500	22.400	29.70000	2.59410	0.820325	23.11000	23.2500	20.300	27.100	2.16920	0.685962
NH_4^+	10	0.83300	0.88500	0.340	1.00000	0.19166	0.060609	0.79700	0.7700	0.6400	1.060	0.13392	0.042349
N-NO_2	10	0.01630	0.01500	0.0130	0.02000	0.00267	0.000844	0.00860	0.0085	0.0070	0.0110	0.00126	0.000400
N-NO_3^-	10	0.64000	0.65000	0.500	0.80000	0.13499	0.042687	0.30000	0.3000	0.1000	0.500	0.13333	0.042164
N-Total	10	1.81000	1.80000	1.500	2.00000	0.16633	0.052599	2.40000	2.5500	0.5000	3.700	1.08115	0.341890
P-PO_4^{3-}	10	0.07700	0.07500	0.040	0.14000	0.02669	0.008439	0.05400	0.0600	0.0100	0.070	0.01713	0.005416
SO_4^{2-}	10	27.40000	21.0000	19.000	61.00000	14.04121	4.440220	44.70000	41.000	38.000	71.00	9.91127	3.134220
Cl^-	10	28.70000	25.0000	22.000	57.00000	10.45679	3.306727	39.90000	41.500	22.000	57.000	10.26807	3.247050
phenols	10	0.03600	0.03000	0.030	0.08000	0.01578	0.004989	0.05800	0.060	0.010	0.100	0.02741	0.008667

Table 4. T-test for Independent Samples (July–October databases). Note: Variables were treated as independent samples—P 2-sided refers to the *p*-value for a two-tailed test.

Group 1—July vs. Group 2—Oct	Mean Group 1	Mean Group 2	t-Value	df	*p*	t Separ. Var. Est.	df	P 2-Sided	Std. Dev. Group 1	Std. Dev. Group 2	F-Ratio Variances	*p* Variances
Phosphate July–Phosphate Oct	0.07700	0.05400	2.2936	18	0.034064	2.2936	15.33828	0.036325	0.02669	0.01713	2	0.202485
CCO July vs. CCO Oct	13.59000	12.76000	0.2434	18	0.810456	0.2434	17.82312	0.810483	7.99617	7.23559	1	0.770730
CBO_5 July vs. CBO_5 Oct	25.46000	23.11000	2.1976	18	0.041303	2.1976	17.45324	0.041738	2.59410	2.16920	1	0.602631
NH4+ July vs. NH4+ Oct	0.83300	0.79700	0.4869	18	0.632211	0.4869	16.09645	0.632899	0.19166	0.13392	2	0.300431
N-NO_2 July vs. N-NO_2 Oct	0.01630	0.00860	8.2447	18	0.000000	8.2447	12.84941	0.000002	0.00267	0.00126	4	0.036529
N-NO_3^- July vs. N-NO_3^- Oct	0.64000	0.30000	5.6667	18	0.000022	5.6667	17.99726	0.000022	0.13499	0.13333	1	0.971262
N-Total July vs. N-Total Oct	1.81000	2.40000	−1.7056	18	0.105272	−1.7056	9.42581	0.120742	0.16633	1.08115	42	0.000005
P-PO_4^{3-} July vs. P-PO_4^{3-} Oct	0.07700	0.05400	2.2936	18	0.034064	2.2936	15.33828	0.036325	0.02669	0.01713	2	0.202485
SO_4^{2-} July vs. SO_4^{2-} Oct	27.40000	44.70000	−3.1831	18	0.005150	−3.1831	16.18487	0.005713	14.04121	9.91127	2	0.314085
Cl^- July vs. Cl^- Oct	28.70000	39.90000	−2.4167	18	0.026502	−2.4167	17.99403	0.026505	10.45679	10.26807	1	0.957620
phenols July vs. phenols Oct	0.03600	0.05800	−2.2000	18	0.041109	−2.2000	14.37439	0.044629	0.01578	0.02741	3	0.115426

Note: Red values correspond to significant differences.

Table 3 shows the descriptive statistics for the chemical species that were measured in the two monitoring campaigns. Mean values and standard deviations and errors are determined. There are also a number of significant differences. These significant differences are presented in Table 4.

The table above shows the results of the Fisher *t*-test [33,34] that evaluates the different sets of measurements from the two measurement campaigns. The normal distribution was tested as in all our previous papers [2,25]. It is observed that for phosphate, CBO_5, $N\text{-}NO_2$, $N\text{-}NO_3$, $P\text{-}PO_4{}^{3-}$, $SO_4{}^{2-}$, Cl^-, and phenols, the differences are significant. This fact is probably due to the mechanisms and processes that influence these factors and that have different effects in the two studied seasons. For the studied set of metallic species, the descriptive statistics of the measurements from the two measurement campaigns are presented in Table 5.

Table 5. Descriptive Statistics for metals (July–October database final). The units of measure are those in Table 2.

		July Database						October Database					
Variable	**Valid N**	**Mean**	**Median**	**Minim.**	**Maxim.**	**Std. Dev.**	**Std. Error**	**Mean**	**Median**	**Minim.**	**Maxim.**	**Std. Dev.**	**Std. Error**
Al	10	24.71860	21.91900	8.665000	46.89000	12.01038	3.798014	16.01210	14.56050	9.000000	26.00000	6.656686	2.105029
As	10	0.00150	0.00150	0.001000	0.00200	0.00053	0.000167	0.01010	0.00900	0.005000	0.01900	0.003755	0.001187
Cd	10	0.01540	0.01600	0.004000	0.03800	0.01025	0.003243	0.01330	0.01400	0.005000	0.01800	0.003713	0.001174
Cr	10	1.31120	1.25950	0.764000	2.31800	0.46673	0.147593	0.58920	0.61350	0.137000	0.96200	0.343044	0.108480
Fe	10	0.030000	0.021500	0.013000	0.063000	0.018203	0.020600	0.013000	0.006000	0.082000	0.022751	0.007194	0.020600

Note: The values describing the preliminary statistics of the values measured in the monitoring campaigns.

Necessarily, a comparison between the two data sets could be presented to highlight the differences and the seasonal variability.

Thus, for aluminum, arsenic, cadmium, chromium and phosphate, the results of the Fisher type T-tests were calculated to highlight differences and to identify significant differences (Table 6).

Table 6. T-test for Independent Samples (01 July–October IARC database). Note: Variables were treated as independent samples—The units of measure are those in Table 2.

Group 1 vs. Group 2	Mean Group 1	Mean Group 2	t-Value	df	p	t Separ. Var. Est.	df	*p* 2-Sided	Std. Dev. Group 1	Std. Dev. Group 2	F-Ratio Vari-ances	*p* Variances
Al. Jul vs. Al. Oct	24.71860	16.01210	2.005017	18	0.060237	2.005017	14.05258	0.064618	12.01038	6.656686	3.255346	0.093547
As Jul vs. As Oct	0.001500	0.010100	−7.17220	18	0.000001	−7.17220	9.354472	0.000043	0.000527	0.003755	50.76000	0.000002
Cd. Jul vs. Cd Oct	0.015400	0.013300	0.608902	18	0.550198	0.608902	11.32041	0.554612	0.010255	0.003713	7.626108	0.005756
Cr Jul. vs. Cr oct	1.311200	0.589200	3.941673	18	0.000956	3.941673	16.52723	0.001104	0.466729	0.343044	1.851099	0.372571
Fe Jul vs. Fe Oct	0.030000	0.020600	1.020213	18	0.321144	1.020213	17.17325	0.321785	10	10	0.018203	0.022751
Phosphate Jul vs. Phosphate Oct	0.07700	0.05400	2.2936	18	0.034064	2.2936	15.33828	0.036325	0.02669	0.01713	2	0.202485

Note: Red values correspond to significant differences.

It is observed that in the case of arsenic ($p < 0.001$) and chromium ($p < 0.001$), there are significant differences between the two data sets. In the case of cadmium and aluminum, the differences are insignificant, with approximately constant values recorded during the two seasons. Cadmium (Cd) could suggest natural sources because cadmium occurs naturally in the Earth's crust and can be released through geological processes. At the same time, Cd could suggest agricultural runoff or anthropogenic sources. Considering the boxplot representation (Figure 4c), the last source case could be more credible. The same discussion could be carried out for aluminum.

Table 7 shows the values of the correlation coefficients between the elements that make up the entire database:

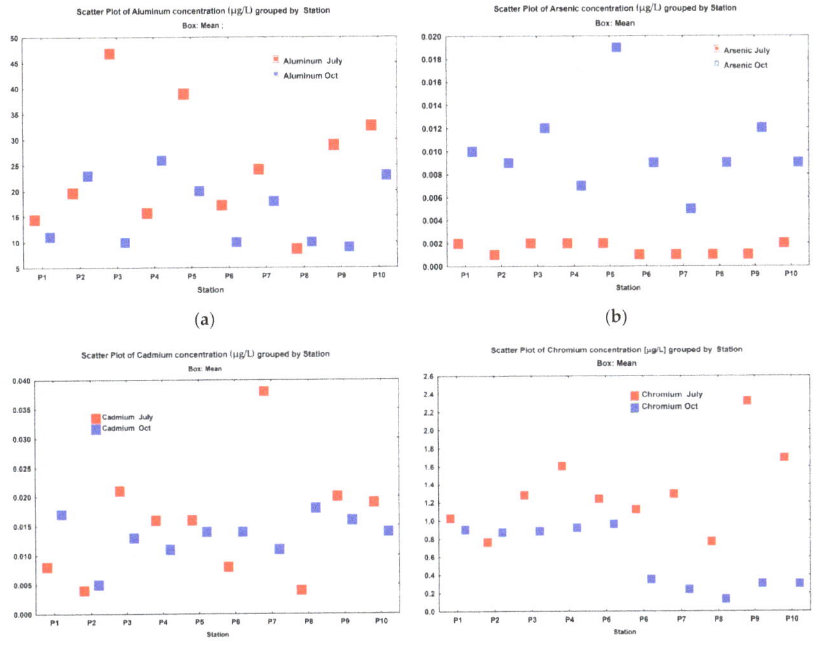

(a) (b) (c) (d)

Figure 4. Scatter plot representation for heavy metals and phenols in the two measurement campaigns in the Danube River study area: (**a**) scatterplot representation for Al concentration seasonally measured in the 10 monitoring points; (**b**) scatter plot representation for As concentration seasonally measured in the 10 monitoring points; (**c**) scatter plot representation for the Cd concentration seasonally measured in the 10 monitoring points; (**d**) scatter plot representation for the Cr concentration seasonally measured in the 10 monitoring points.

Table 7 shows the values of the correlation coefficients for the values measured during the summer season. It is observed that the number of significant correlations is relatively few—except the link between the concentration of Cr and CCO—the link is easy to understand from the literature: as it is known [2], the link between the concentration of chromium (Cr) and CCO-Cr, in the context of water quality, is primarily concerned with the potential contamination of water by chromium compounds and their impact on human health and the environment [35]. As it is known, chromium can exist in several oxidation states, but the most common and stable forms are Cr(III) and Cr(VI). Cr(VI) is particularly harmful due to its high solubility and mobility, which makes it more bioavailable and toxic [15]. Water quality assessments often measure the concentration of total chromium, and specifically Cr(VI), to ensure that they are below the permissible limits set by regulatory bodies [17]. In summary, the link between Cr concentration and CCO-Cr in water quality is a measure of the potential toxicity and pollution levels in water, which is critical for ensuring safe drinking water and maintaining ecological balance [1]. There are also links between the concentration of chlorine ions and the concentrations of total Nitrogen—N_{total} and the SO_4 group. These aspects have been identified in the literature and have been explained [2,36]. The fact that during the summer season the influences between the parameters are less and this is largely due to the high flow of the Danube River (Figure 3) ensures a significant dilution of the concentrations of the studied chemical species.

Table 7. Pearson correlation (2-tailed) between studied parameters—Correlations table for the IARC July database.

Variable	Al	As	Cd	Cr	Fe	CCO	CBO$_5$	NH$_4^+$	N-NO$_2$	N-NO$_3^-$	N-Total	P-PO$_4^{3-}$	SO$_4^{2-}$	Cl$^-$	Phenols
Al	1.0000	0.4386	0.4822	0.3696	0.2077	0.3736	−0.0869	−0.3660	0.2185	−0.1587	0.1489	−0.0579	−0.1131	−0.1075	0.3626
	p = ---	p = 0.0205	p = 0.158	p = 0.293	p = 0.565	p = 0.288	p = 0.811	p = 0.298	p = 0.544	p = 0.661	p = 0.681	p = 0.874	p = 0.0756	p = 0.768	p = 0.303
As	0.4386	1.0000	0.0617	0.1305	0.1274	−0.1911	−0.5364	−0.2805	−0.0395	−0.6247	0.4436	0.5925	−0.4805	−0.4335	0.4009
	p = 0.205	p = ---	p = 0.866	p = 0.719	p = 0.726	p = 0.597	p = 0.110	p = 0.432	p = 0.914	p = 0.053	p = 0.199	p = 0.071	p = 0.160	p = 0.211	p = 0.251
Cd	0.4822	0.0617	10.0000	0.4969	0.3720	0.1694	0.3799	−0.0781	0.4864	−0.4944	0.0886	−0.0195	0.4834	−0.1407	0.1621
	p = 0.158	p = 0.866	p = ---	p = 0.144	p = 0.290	p = 0.640	p = 0.279	p = 0.830	p = 0.154	p = 0.146	p = 0.808	p = 0.957	p = 0.157	p = 0.698	p = 0.655
Cr	0.3696	0.1305	0.4969	1.0000	−0.2396	0.7130	0.1156	0.2550	−0.0617	−0.1081	0.2698	0.4192	−0.2223	−0.3824	0.2841
	p = 0.293	p = 0.719	p = 0.144	p = ---	p = 0.505	p = 0.021	p = 0.750	p = 0.477	p = 0.866	p = 0.766	p = 0.451	p = 0.228	p = 0.537	p = 0.276	p = 0.426
Fe	0.2077	0.1274	0.3720	−0.2396	1.0000	−0.0137	0.1610	−0.7392	−0.0526	−0.4477	−0.4184	−0.2059	0.4286	0.1185	−0.3521
	p = 0.565	p = 0.726	p = 0.290	p = 0.505	p = ---	p = 0.970	p = 0.657	p = 0.015	p = 0.885	p = 0.195	p = 0.229	p = 0.568	p = 0.216	p = 0.744	p = 0.318
CCO	0.3736	−0.1911	0.1694	0.7130	−0.0137	1.0000	0.1856	−0.0485	−0.2571	0.2691	−0.0784	0.0623	−0.2814	−0.2178	−0.0426
	p = 0.288	p = 0.597	p = 0.640	p = 0.021	p = 0.970	p = ---	p = 0.608	p = 0.894	p = 0.473	p = 0.452	p = 0.829	p = 0.864	p = 0.431	p = 0.545	p = 0.907
CBO5	−0.0869	−0.5364	0.3799	0.1156	0.1610	0.1856	1.0000	0.1306	0.2122	0.0336	0.0860	−0.4144	0.4239	−0.0132	0.1043
	p = 0.811	p = 0.110	p = 0.279	p = 0.750	p = 0.657	p = 0.608	p = ---	p = 0.719	p = 0.556	p = 0.927	p = 0.813	p = 0.234	p = 0.222	p = 0.971	p = 0.774
NH4$^+$	−0.3660	−0.2805	−0.0781	0.2550	−0.7392	−0.0485	0.1306	1.0000	0.1153	0.3942	0.2290	−0.0263	−0.1665	−0.2850	0.2286
	p = 0.298	p = 0.432	p = 0.830	p = 0.477	p = 0.015	p = 0.894	p = 0.719	p = ---	p = 0.751	p = 0.260	p = 0.525	p = 0.943	p = 0.646	p = 0.425	p = 0.525
N-NO$_2$	0.2185	−0.0395	0.4864	−0.0617	−0.0526	−0.2571	0.2122	0.1153	1.0000	−0.3763	0.1927	−0.2668	0.3582	−0.0362	0.5331
	p = 0.544	p = 0.914	p = 0.154	p = 0.866	p = 0.885	p = 0.473	p = 0.556	p = 0.751	p = ---	p = 0.284	p = 0.594	p = 0.456	p = 0.309	p = 0.921	p = 0.113
N-NO$_3^-$	−0.1587	−0.6247	−0.4944	−0.1081	−0.4477	0.2691	0.0336	0.3942	−0.3763	1.0000	−0.3167	−0.4565	−0.1266	0.2928	−0.3339
	p = 0.661	p = 0.053	p = 0.146	p = 0.766	p = 0.195	p = 0.452	p = 0.927	p = 0.260	p = 0.284	p = ---	p = 0.373	p = 0.185	p = 0.727	p = 0.412	p = 0.346
N-Total	0.1489	0.4436	0.0886	0.2698	−0.4184	−0.0784	0.0860	0.2290	0.1927	−0.3167	1.0000	0.6082	−0.5538	−0.6944	0.4404
	p = 0.681	p = 0.199	p = 0.808	p = 0.451	p = 0.229	p = 0.829	p = 0.813	p = 0.525	p = 0.594	p = 0.373	p = ---	p = 0.062	p = 0.097	p = 0.026	p = 0.203
P-PO$_4^{3-}$	−0.0579	0.5925	−0.0195	0.4192	−0.2059	0.0623	−0.4144	−0.0263	−0.2668	−0.4565	0.6082	1.0000	−0.5776	−0.5570	0.0211
	p = 0.874	p = 0.071	p = 0.957	p = 0.228	p = 0.568	p = 0.864	p = 0.234	p = 0.943	p = 0.456	p = 0.185	p = 0.062	p = ---	p = 0.080	p = 0.094	p = 0.954
SO$_4^{2-}$	−0.1131	−0.4805	0.4834	−0.2223	0.4286	−0.2814	0.4239	−0.1665	0.3582	−0.1266	−0.5538	−0.5776	1.0000	0.6744	−0.1525
	p = 0.756	p = 0.160	p = 0.157	p = 0.537	p = 0.216	p = 0.431	p = 0.222	p = 0.646	p = 0.309	p = 0.727	p = 0.097	p = 0.080	p = ---	p = 0.032	p = 0.674
Cl$^-$	−0.1075	−0.4335	−0.1407	−0.3824	0.1185	−0.2178	−0.0132	−0.2850	−0.0362	0.2928	−0.6944	−0.5570	0.6744	1.0000	−0.1899
	p = 0.768	p = 0.211	p = 0.698	p = 0.276	p = 0.744	p = 0.545	p = 0.971	p = 0.425	p = 0.921	p = 0.412	p = 0.026	p = 0.094	p = 0.032	p = ---	p = 0.599
phenols	0.3626	0.4009	0.1621	0.2841	−0.3521	−0.0426	0.1043	0.2286	0.5331	−0.3339	0.4404	0.0211	−0.1525	−0.1899	1.0000
	p = 0.303	p = 0.251	p = 0.655	p = 0.426	p = 0.318	p = 0.907	p = 0.774	p = 0.525	p = 0.113	p = 0.346	p = 0.203	p = 0.954	p = 0.674	p = 0.599	p = ---

Note: Red values correspond to significant differences.

Table 8 shows the values of the correlation coefficients for the values measured during the autumn campaign. In this case, the number of significant correlations is significantly higher than in the previous case [2]. A situation of this kind was encountered in previous studies and was scientifically explained [2,14]. Thus, in the case of the autumn season, due to the low flow of the Danube, the influences between the different chemical elements become easier to identify. The above table shows the values of the correlation coefficients for the measurements made in the autumn season. It is observed that the number of significant correlations is significantly higher mainly due to the low recorded flow of the Danube. In these conditions of low flow in the autumn season, the influences between the measured quantities are easier to highlight.

Thus, the links between the concentration of aluminum and cadmium ions can be highlighted (r = −0.6543, p = 0.04). We can also emphasize the correlation between the concentration of Fe and As ions (r = 0.6954, p = 0.025), a fact that could be determined by the existence of a common source. Also, there is a correlation between cadmium and phosphate concentrations (r = 0.7652, p = 0.010) and the link between Cd and $N\text{-}NO_2$ (r = 0.7522, p = 0.027). The fact that there is obviously a seasonal variability for the concentration of $N\text{-}NO_2$ was also highlighted in the literature [36]. The existence of a link between the two polluting factors is an original aspect that has not been highlighted until now. There is also an extremely interesting correlation factor between the concentration of Cr and that of $N\text{-}NO_3$ (r = 0.6923, p = 0.027), respectively, that between the concentration of Cr and the concentration of Ntotal (r = 0.8094, p = 0.005). These connections between the concentration of chromium ions and the other two chemical species represent a specific aspect. Our previous study [25] shed light on the complex interactions between chromium ions, nitrate nitrogen ($N\text{-}NO_3$), and total nitrogen (Ntotal) in the Danube River, emphasizing the need for vigilant monitoring and sustainable water management practices [10,24].

Last but not least, the links between CBO_5 and NH_4 (r = 0.6358, p = 0.048), CBO_5 and Ntotal (r = 0.6813, p = 0.030) and between NH_4 and SO_4 (r = 0.6907, p = 0.027) should be listed, which links were also noticed in the literature [2]. Further on, we present in the form of a boxplot the identification of the seasonal dynamics of the parameters that are included in the IARC database that studies heavy metals [24].

Figure 4 shows the scatter plot representations for heavy metal concentrations measured in the 10 monitoring points. Thus, Figure 4a shows the concentration of Al ions measured in the 10 monitoring points. The existence of insignificant differences can be observed (see Table 4). It is observed that during the summer season, the concentrations are significantly higher (mainly due to the increased reactivity of aluminum ions and the obvious higher contribution recorded near monitoring point P3. Necessarily, this aspect was reported in the specialized literature and probably is due to the activity of the steel industry in the area [2,25]. By analyzing the spatial distribution of Al concentrations, higher levels of aluminum downstream of industrial or urban areas as compared to upstream can indicate human influence. Figure 4b shows the concentration of As. The differences recorded in the two monitoring seasons are significant (Table 4), and the values measured in the warm season are systematically lower than those measured in the cold season, as is a ubiquitous toxic metal belonging to the metalloid group of the periodic system of the main group II, which occurs naturally in the lithosphere, hydrosphere and atmosphere, as well as generally in the biosphere [18]. The higher value in the cold season could be due to local industries' seasonal cycles of production and reduced water flow values.

Table 8. Pearson correlation (2-tailed) between studied parameters—Correlations table for October database.

Variable	Al	As	Cd	Cr	Fe	CCO	CBO_5	NH_4^+	N-NO_2	N-NO_3^-	N-Total	P-PO_4^{3-}	SO_4^{2-}	Cl^-	Phenols
Al	1.0000	−0.1423	−0.6543	0.3350	0.3737	−0.2228	−0.2869	−0.0909	0.1558	−0.0155	−0.0174	−0.2819	0.1452	0.4777	0.2096
	p = ---	p = 0.695	p = 0.040	p = 0.344	p = 0.287	p = 0.536	p = 0.422	p = 0.803	p = 0.667	p = 0.966	p = 0.962	p = 0.430	p = 0.689	p = 0.163	p = 0.561
As	−0.1423	1.0000	0.2526	0.4151	0.6964	0.5281	−0.0315	−0.1253	−0.1076	−0.2441	0.3421	0.2177	−0.5693	−0.2475	0.1641
	p = 0.695	p = ---	p = 0.481	p = 0.233	p = 0.025	p = 0.117	p = 0.931	p = 0.730	p = 0.767	p = 0.497	p = 0.333	p = 0.546	p = 0.086	p = 0.490	p = 0.651
Cd	−0.6543	0.2526	1.0000	−0.3699	−0.0839	0.3284	0.2796	−0.2214	−0.7522	0.1346	−0.1661	0.7652	−0.4501	−0.5557	−0.5284
	p = 0.040	p = 0.481	p = ---	p = 0.293	p = 0.818	p = 0.354	p = 0.434	p = 0.539	p = 0.012	p = 0.711	p = 0.647	p = 0.010	p = 0.192	p = 0.095	p = 0.116
Cr	0.3350	0.4151	−0.3699	1.0000	0.4802	−0.2653	−0.5368	−0.3392	0.2875	−0.6923	0.8094	−0.0756	−0.3728	−0.3436	0.1409
	p = 0.344	p = 0.233	p = 0.293	p = ---	p = 0.160	p = 0.459	p = 0.110	p = 0.338	p = 0.421	p = 0.027	p = 0.005	p = 0.836	p = 0.289	p = 0.331	p = 0.698
Fe	0.3737	0.6964	−0.0839	0.4802	1.0000	0.1544	0.0260	0.1082	−0.0448	0.0037	0.1974	0.1528	−0.1647	−0.0506	0.0752
	p = 0.287	p = 0.025	p = 0.818	p = 0.160	p = ---	p = 0.670	p = 0.943	p = 0.766	p = 0.902	p = 0.992	p = 0.585	p = 0.673	p = 0.649	p = 0.890	p = 0.836
CCO	−0.2228	0.5281	0.3284	−0.2653	0.1544	1.0000	0.1611	−0.0473	−0.0906	−0.0656	−0.1426	0.2274	−0.2014	0.1353	0.1732
	p = 0.536	p = 0.117	p = 0.354	p = 0.459	p = 0.670	p = ---	p = 0.657	p = 0.897	p = 0.804	p = 0.857	p = 0.694	p = 0.528	p = 0.577	p = 0.709	p = 0.632
CBO_5	−0.2869	−0.0315	0.2796	−0.5368	0.0260	0.1611	1.0000	0.6358	−0.4884	0.3304	−0.6813	0.0825	0.3676	0.1288	−0.1230
	p = 0.422	p = 0.931	p = 0.434	p = 0.110	p = 0.943	p = 0.657	p = ---	p = 0.048	p = 0.152	p = 0.351	p = 0.030	p = 0.821	p = 0.296	p = 0.723	p = 0.735
NH_4^+	−0.0909	−0.1253	−0.2214	−0.3392	0.1082	−0.0473	0.6358	1.0000	0.1430	0.5600	−0.5372	−0.2122	0.6907	0.1290	0.4220
	p = 0.803	p = 0.730	p = 0.539	p = 0.338	p = 0.766	p = 0.897	p = 0.048	p = ---	p = 0.694	p = 0.092	p = 0.109	p = 0.556	p = 0.027	p = 0.722	p = 0.224
N-NO_2	0.1558	−0.1076	−0.7522	0.2875	−0.0448	−0.0906	−0.4884	0.1430	1.0000	−0.1318	0.4062	−0.6360	0.3439	0.3217	0.5192
	p = 0.667	p = 0.767	p = 0.012	p = 0.421	p = 0.902	p = 0.804	p = 0.152	p = 0.694	p = ---	p = 0.717	p = 0.244	p = 0.048	p = 0.331	p = 0.365	p = 0.124
N-NO_3^-	−0.0155	−0.2441	0.1346	−0.6923	0.0037	−0.0656	0.3304	0.5600	−0.1318	1.0000	−0.7322	−0.0000	0.4540	0.2191	0.0304
	p = 0.966	p = 0.497	p = 0.711	p = 0.027	p = 0.992	p = 0.857	p = 0.351	p = 0.092	p = 0.717	p = ---	p = 0.016	p = 10.00	p = 0.187	p = 0.543	p = 0.934
N-Total	−0.0174	0.3421	−0.1661	0.8094	0.1974	−0.1426	−0.6813	−0.5372	0.4062	−0.7322	1.0000	0.0960	−0.5641	−0.3383	0.1087
	p = 0.962	p = 0.333	p = 0.647	p = 0.005	p = 0.585	p = 0.694	p = 0.030	p = 0.109	p = 0.244	p = 0.016	p = ---	p = 0.792	p = 0.089	p = 0.339	p = 0.765
P-PO_4^{3-}	−0.2819	0.2177	0.7652	−0.0756	0.1528	0.2274	0.0825	−0.2122	−0.6360	−0.0000	0.0960	1.0000	−0.5420	−0.5598	−0.2414
	p = 0.430	p = 0.546	p = 0.010	p = 0.836	p = 0.673	p = 0.528	p = 0.821	p = 0.556	p = 0.048	p = 10.00	p = 0.792	p = ---	p = 0.106	p = 0.092	p = 0.502
SO_4^{2-}	0.1452	−0.5693	−0.4501	−0.3728	−0.1647	−0.2014	0.3676	0.6907	0.3439	0.4540	−0.5641	−0.5420	1.0000	0.3414	0.0344
	p = 0.689	p = 0.086	p = 0.192	p = 0.289	p = 0.649	p = 0.577	p = 0.296	p = 0.027	p = 0.331	p = 0.187	p = 0.089	p = 0.106	p = ---	p = 0.334	p = 0.925
Cl^-	0.4777	−0.2475	−0.5557	−0.3436	−0.0506	0.1353	0.1288	0.1290	0.3217	0.2191	−0.3383	−0.5598	0.3414	1.0000	0.2756
	p = 0.163	p = 0.490	p = 0.095	p = 0.331	p = 0.890	p = 0.709	p = 0.723	p = 0.722	p = 0.365	p = 0.543	p = 0.339	p = 0.092	p = 0.334	p = ---	p = 0.441
phenols	0.2096	0.1641	−0.5284	0.1409	0.0752	0.1732	−0.1230	0.4220	0.5192	0.0304	0.1087	−0.2414	0.0344	0.2756	1.0000
	p = 0.561	p = 0.651	p = 0.116	p = 0.698	p = 0.836	p = 0.632	p = 0.735	p = 0.224	p = 0.124	p = 0.934	p = 0.765	p = 0.502	p = 0.925	p = 0.441	p = ---

Note: Red values correspond to significant differences.

Both organic and inorganic forms of arsenic exist in nature (mostly in the form of complexes); various environmental transport routes have been identified, and fairly high concentrations (mainly in water sources) have been reported in several regions around the world. Mainly, the increased concentration during the cold season could be explained by the reduced reactivity in the case of lower temperatures. In Figure 4c, they are represented in the form of a scatter plot for Cd and in Figure 4d are represented in the form of a scatter plot for the measured values of Cr. In the case of Cd, the differences measured in the two seasons do not differ significantly (Table 4). The high contribution in the case of monitoring point P3 is also clearly observed, and an exponential decrease in the concentration of Cr along the water course is also observed. Regarding the Cr distribution, the differences recorded in the two seasons are obviously significant (Table 4), and significantly high values are also observed near monitoring points P4 and P7, respectively. However, it has to be emphasized the comprehensiveness and adequacy of existing data from previous studies and monitoring efforts that have consistently shown metal concentrations within acceptable limits [5,14].

Figure 5 shows the scatter plot representations for phosphate, CCO, CBO_5 and NH_4 concentrations measured in the 10 monitoring points. Thus, Figure 5a shows the concentration of the phosphate group measured in the 10 monitoring points. The existence of insignificant differences between the two sets of data is observed. The distribution is of a bimodal type, with a more intense contribution near point P3 with an exponential decreasing tendency—characteristic of the diffusion phenomenon. It is observed that during the summer season, the concentrations are significantly higher (mainly due to the increased reactivity of the phosphate group and the obvious higher contribution recorded near monitoring point P3). Necessarily, this aspect was reported in the literature of specialty and is mainly due to human activity [2,25].

Figure 5b shows the concentration of the CCO group. The differences recorded in the two monitoring seasons are insignificant, and the values measured in the warm season are systematically higher than those measured in the cold season. Mainly, the concentration of CCO—the mass concentration of oxygen equivalent to the amount of potassium bichromate consumed for the oxidation in an acid environment of the dissolved and suspended organic matter present in the wastewater—is lower during the cold season [1,2]. This fact could probably be explained by the reduced chemical activity at lower temperatures. In Figure 5c, they are represented in the form of a scatter boxplot for CBO_5, and in Figure 5d, they are represented in the form of a boxplot for the measured values of NH4. In the case of CBO5—an index that represents a measure of the organic impurity of wastewater and the amount of oxygen (mg/mass) that is required for the oxidative degradation by microorganisms of the organic substances contained in the water volume humidity—the measured differences in the two seasons are not significantly different. Also, the high contribution in the case of monitoring points P6 and P7 is clearly observed, and an exponential decrease in concentration along the water course is also observed. Regarding the distribution of NH_4, the differences recorded in the two seasons are obviously insignificant, and significantly high values are also observed near monitoring points P6 and P7, respectively. These aspects obviously represent original aspects, because such measurements with such a resolution could not be carried out until now.

Figure 6 shows the scatter plot representations for the concentrations of N_{total}, N-NO_2, N-NO_3 and P-PO_4 measured in the 10 monitoring points. Thus, Figure 6a shows the concentration of the N-NO_2 group measured in the 10 monitoring points. There are significant differences between the two data sets. The distribution is of a bimodal type, with a more intense contribution near point P3 with an exponential decreasing tendency—characteristic of the diffusion phenomenon. It is observed that during the summer season, the concentrations are significantly higher (mainly due to the increased reactivity of the N-NO_2 group and the obvious higher contribution recorded during the hot season). Necessarily, this aspect was reported in the literature of specialty and is mainly due to anthropogenic activity. Figure 6b shows the concentration of the N-NO_3 group.

The differences recorded in the two monitoring seasons are insignificant, and the values measured in the warm season are systematically higher than those measured in the cold season. In Figure 6c, they are represented in the form of a scatter plot for Ntotal, and in Figure 5d, they are represented in the form of a scatter plot for the measured values of P-PO_4, and the differences are insignificant.

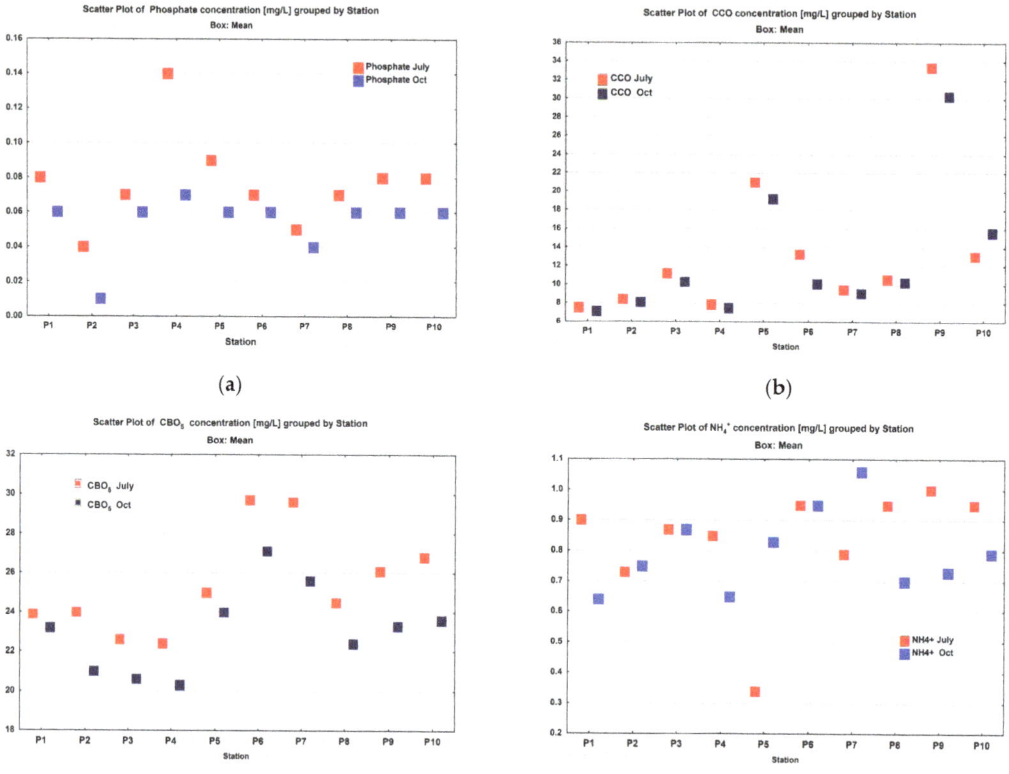

(a) (b) (c) (d)

Figure 5. Scatter plot representation for the two measurement campaigns in the Danube River study area: (**a**) scatter plot representation for phosphonate concentration seasonally measured in the 10 monitoring points; (**b**) scatter plot representation for CCO concentration seasonally measured in the 10 monitoring points; (**c**) scatter plot representation for CBO_5 concentration seasonally measured in the 10 monitoring points; (**d**) scatter plot representation for NH_4 concentration seasonally measured in the 10 monitoring points.

Figure 7 shows the scatter plot representations for SO_4, Cl, phenols and Fe concentrations measured in the 10 monitoring points. Thus, Figure 7a shows the concentration of the SO_4 group measured in the 10 monitoring points. There are significant differences between the two data sets. The distribution is of a bimodal type, with a more intense contribution near point P7 with an exponential decreasing tendency—characteristic of the diffusion phenomenon. It is observed that during the summer season, the concentrations are significantly lower (mainly due to the increased reactivity of the SO_4 group and the obviously higher contribution recorded during the hot season). Necessarily, this aspect has been reported in the specialized literature and is mainly due to anthropogenic activity. Figure 7b shows the Cl^- concentration. The differences recorded in the two monitoring seasons are significant, and the values measured in the warm season are systematically lower than those measured in the cold season. In Figure 7c, they are represented in the form of a scatter plot for phenols, and in Figure 7d, the measured values for Fe are represented in the form of a scatter plot. The differences are insignificant.

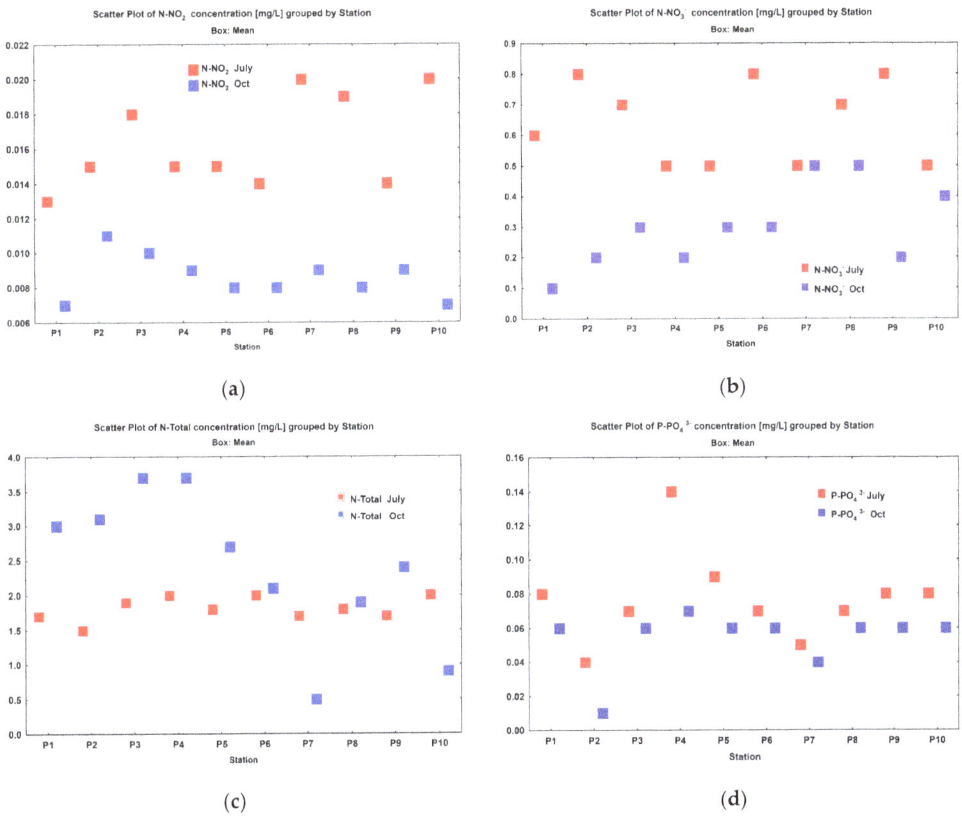

Figure 6. Scatter plot representation for the two measurement campaigns in the Danube River study area: (**a**) scatter plot representation for the N-NO_2 concentration seasonally measured in the 10 monitoring points; (**b**) scatter plot representation for the N-NO_2 concentration seasonally measured in the 10 monitoring points; (**c**) scatter plot representation for the N-total concentration seasonally measured in the 10 monitoring points; (**d**) scatter plot representation for the P-PO_4 concentration seasonally measured in the 10 monitoring points.

3.3. PCA Method Analysis

The Principal Factors method is a classic method, often found in specialized literature, and is a versatile statistical method for reducing a cases-by-variables data table to its essential features, called principal components [2,25]. Thus, several variables or studied parameters can be grouped in order to categorize them into factors/groups of influence on the variability of the data set. This method was also used in our previous studies [2,25], carried out over a period of 10 years. In that article, the classification of the factors and the identification of related sources were presented. Interestingly, unlike previous studies that revealed a low number of main influence groups, in the case of this investigation, the number of factors that were identified as main factors (i.e., with determined eigenvalues greater than unity) was 6 (see Table 9). The theorem known as the Kaiser criterion in (PCA) stated that the stable solution corresponding to eigenvalues greater than 1 should be considered. Thus, for the summer season campaign, 6 main factors are identified that can explain 91.82% of the variability of the entire data set. In the case of the autumn measurement session, the main significant factors identified are also 6, thus covering more than 91% of the variability of the collected data set. This aspect is extremely interesting because, in the studies carried out between 1990 and 2000 in the study area, the number of significant factors was a maximum of 4 [2].

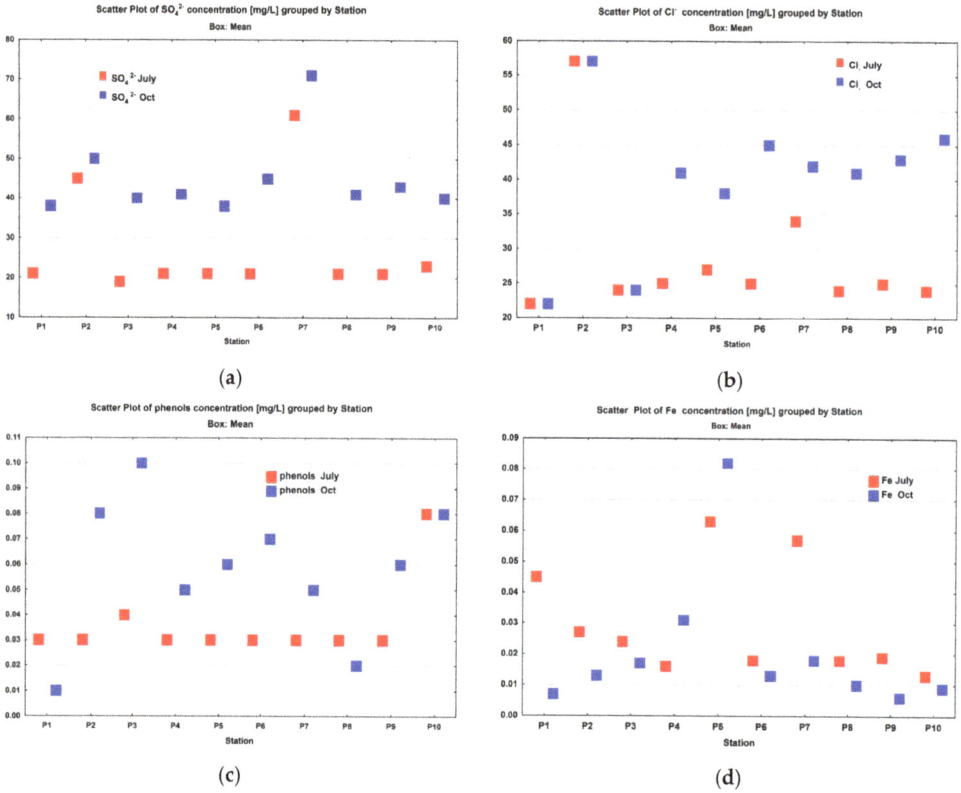

Figure 7. Scatter plot representation for the two measurement campaigns in the Danube River study area: (**a**) scatter plot representation for SO_4 concentration seasonally measured in the 10 monitoring points; (**b**) scatter plot representation for Cl concentration seasonally measured in the 10 monitoring points; (**c**) scatter plot representation for phenols concentration seasonally measured in the 10 monitoring points; (**d**) scatter plot representation for Fe concentration seasonally measured in the 10 monitoring points.

Next, we will present the results of the PCA analysis applied to the database containing the concentration of heavy metals (Figure 8). Thus, in the case of these two measurement campaigns, it is observed that the data sets are grouped differently. If in the summer campaign the data set contains a number of 2 main factors (Table 9), the same can be said for the autumn campaign (Table 9). This derives from the fact that the eigenvalues that are identified and that have the higher value as a unit determine the consideration of the first 2 factors (Figure 8a). The main factor 1 corresponds to the groups formed by Al and Cd and Cr between which there is a fairly strong correlation, a fact that leads us to the idea of a common source. Figure 8b shows the fact that there is a fairly strong connection between the measured aluminum concentration and the minimum surface obtained based on the Cd and Cr concentration values. In the case of the autumn measurement campaign, the configuration of the PCA diagram changes, with Al being negatively correlated in relation to the concentrations of the two metals Cd and Cr. This fact is also observable from the surface plot diagram in Figure 8d.

Table 9. Eigenvalues of correlation matrix and related statistics (01 July–October database). Active variables only.

Value Number	July Monitoring Campaign Eigenvalue	% Total Variance	Cumulative Eigenvalue	Cumulative %	October Monitoring Campaign Eigenvalue	% Total variance	Cumulative Eigenvalue	Cumulative %
1	6.204681 *	34.47045	6.20468	34.4705	5.628626 *	31.27015	5.62863	31.2701
2	3.07987 *	17.11041	9.28455	51.5809	4.444705 *	24.69281	10.07333	55.9630
3	2.625296 *	14.58498	11.90985	66.1658	2.093713 *	11.63174	12.16705	67.5947
4	2.063760 *	11.46534	13.97361	77.6312	1.828940 *	10.16078	13.99599	77.7555
5	1.401129 *	7.78405	15.37474	85.4152	1.312745 *	7.29303	15.30873	85.0485
6	1.153074 *	6.40597	16.52781	91.8212	1.198617 *	6.65898	16.50735	91.7075
7	0.603383	3.35213	17.13120	95.1733	0.687419	3.81899	17.19477	95.5265
8	0.460789	2.55994	17.59199	97.7333	0.469000	2.60556	17.66377	98.1320
9	0.408013	2.26674	18.00000	100.0000	0.336233	1.86796	18.00000	100.0000

Note: * Marked red values are those with significant eigenvalues, non * marked red values have no statistical significance.

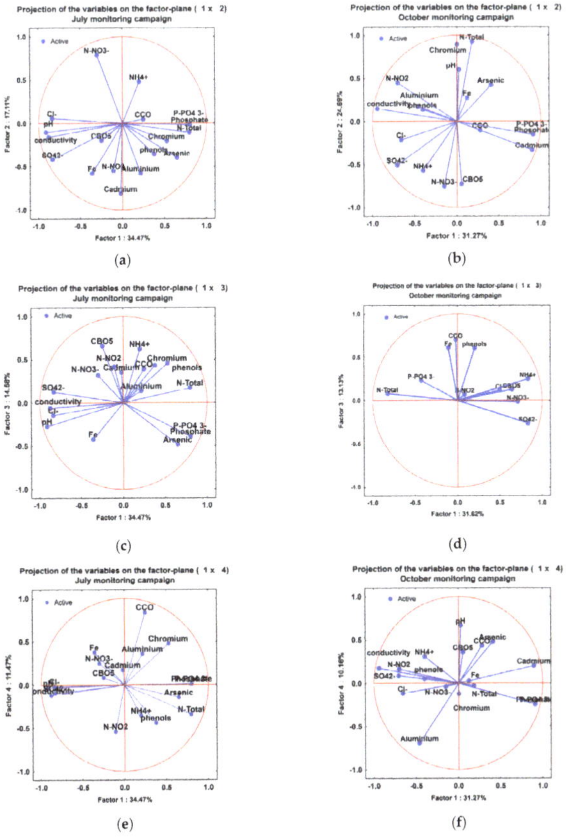

Figure 8. PCA representation of the two measurement campaigns in the Danube River study area: (**a**) PCA representation with main factors 1 and 2 for the July campaign; (**b**) PCA representation with main factors 1 and 2 for the October campaign; (**c**) PCA representation with main factors 1 and 3 for the July campaign; (**d**) PCA representation with main factors 1 and 3 for the October campaign; (**e**) PCA representation with main factors 1 and 4 for the July campaign; (**f**) PCA representation with main factors 1 and 4 for the October campaign.

Figure 8a shows the elements that form the first main factor in the set of data measured in the summer campaign. This factor, which is responsible for explaining 34.47% of the variance of the data set (Figure 8a), includes the elements Al, Cd, phosphate, pH, N-NO_2, N-NO_3, the SO_4 group, chlorine and conductivity. This factor which could be attributed more to the influence of anthropic activity brings together elements that are related to industrial activities in the area.

Factor 2, which is responsible for explaining 17.17% of the variation in the data set, includes elements such as Cr, CBO_5, NH_4, N-NO_3, N-total, which would represent a factor related to pollution due to biological factors in the study area. Factor 3, which explains approximately 14.58% of the variance of the data set from the July campaign, includes elements such as As and Fe. Factor 3 (Figure 8c), which explains approximately 14.58% of the variation in the data group, includes elements such as Al and pH—a factor that includes components with high chemical influence. Factor 4 (Figure 8e), which explains approximately 11.44%, respectively, factors 5 and 6 which explain 7.78% and 6.41%, include CCO—factor 5 and phenols (factor 6).

In the case of the autumn campaign, the situation has changed quite significantly (Figure 8b). Thus, the factor that is responsible for explaining 31.17% (Table 9) of the variance of the October data set (Figure 8b) includes the elements Cd, phosphate, pH, N-NO_2, the SO_4 group, chlorine and conductivity. Note the elimination of the Al element from this factor. This factor could also be attributed to the influence of human activity in the area. The main factor 2 that explains 24.70% of the data set variant includes Cr, CBO_5, NH_4, N-NO_3, and N-total, which would represent a factor related to pollution due to biological factors in the study area. It is observed that the elimination of the N-NO_2 group within this factor is found in the group of factor 3. Factor 3 explains approximately 14.58%. Factor 3 (Figure 8d), which explains approximately 11.63% of the variation in the data group, includes elements such as As and Fe—a factor that includes components with similar chemical reactivity. This aspect represents a difference from the previous case. Factor 4 (Figure 8d), which explains approximately 10.16% of the variation in the data group, includes elements such as Al and pH—a factor that includes components with high chemical influence. Factors 5 and 6, which explain 7.30% and 6.41%, respectively, include CCO—factor 5 and phenols, respectively (factor 6).

3.4. Cluster Method Analysis

In the following, the results of the cluster analysis are presented to identify, depending on the geometric distance between the elements, the way of grouping and ranking between the sizes determined in the two campaigns. This topological analysis method was chosen in order to highlight more easily the difference between the two data sets [37]. The main reason to use cluster analysis in statistical analysis is to group similar data points together based on their characteristics—topological distance. This unsupervised machine learning technique helps to identify patterns and structure in data, making it easier to understand and analyze [33]. By organizing data into clusters, we can uncover hidden insights and make predictions about future data points. In this sense, the Euclidean distance calculation method and the weighted-pair centroid method were chosen—a method often found in specialized literature [38]. Figure 9a,b show the analysis results of these classifications based on distances. It is observed the existence of some changes regarding the closest group—respectively, the group in which Cd was replaced with As in the group located at the smallest distance. There was also a permutation in the group formed by the radical SO_4, CBO_5 and, respectively, Cl.

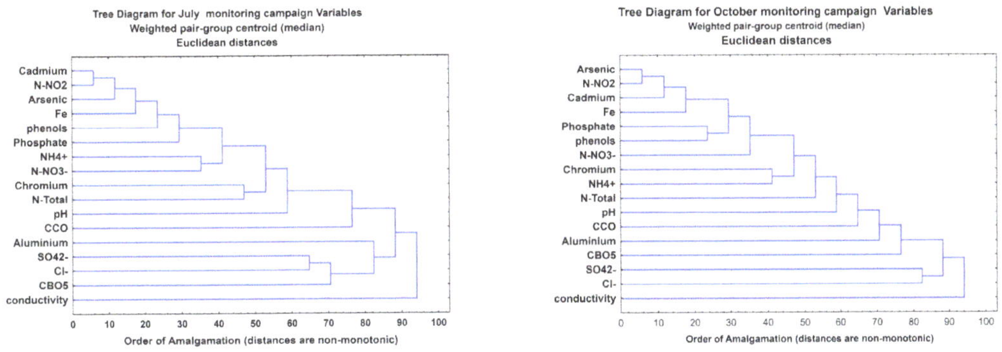

(a) (b)

Figure 9. (**a**) Tree Diagram for monitored Variables—Weighted-pair group centroid (median)—Euclidean distances—July monitoring campaign; (**b**) Tree Diagram for monitored Variables—Weighted-pair group centroid (median)—Euclidean distances—October monitoring campaign.

In this sense, we can state that the structure of the topological cluster model for the Danube is given in Figure 9 [37,38]. In this way, the fact that there is an absolutely striking similarity in the two sets of data was observed. Identified clusters have a similar component with two small exceptions. The method allows the classification of groups that seem to be closely related using the topological criterion. For example—the first cluster is formed by Cd, As and N-NO_2. A cluster with Cd, As and nitrite (N-NO_2) could indicate a common source or similar behavior of these substances in the water environment. Here are some possible explanations for such a cluster: (a) industrial pollution: these substances are often byproducts of industrial processes such as in the studied area—in the study area, there are a number of important industrial objectives; (b) agricultural runoff: Cd and As can be present in pesticides and fertilizers used in agriculture, and nitrites are a common component of nitrogen-based fertilizers. Runoff from agricultural fields can carry these substances into water bodies; (c) natural occurrence: While often associated with human activity, these substances can also occur naturally. As, for instance, can be released into groundwater through the erosion of natural deposits. Further research needs to be developed.

3.5. PCA Method Analysis for Metals Samples

In what follows, we will consider the PCA method applied only to the metal species identified and studied in the two measurement campaigns in the Danube River water. Table 10 shows the coordinates of the variables according to the main factors in the July database. It is observed that factor 1 (Figure 10a), which describes the variant of over 41.84% of the total set, includes Al and Cd. Factor 2, which describes the variation in over 24.67% of the total set, includes Cr and Fe. The last factor, which is responsible for 20.37% of the total variation, includes As (Figure 10c).

Table 10. Eigenvalues of correlation matrix and related statistics for IARC database of metals.

Value number	July Monitoring Campaign				October Monitoring Campaign			
	Eigenvalue	% Total Variance	Cumulative Eigenvalue	Cumulative %	Eigenvalue	% Total Variance	Cumulative Eigenvalue	Cumulative %
1	2.091981 *	41.83961	2.091981	41.8396	2.276505 *	45.53009	2.276505	45.5301
2	1.233531 *	24.67062	3.325512	66.5102	1.724148 *	34.48295	4.000652	80.0130
3	1.018576 *	20.37151	4.344087	86.8817	0.581033	11.62066	4.581685	91.6337
4	0.423145	8.46291	4.767233	95.3447	0.282902	5.65804	4.864587	97.2917
5	0.232767	4.65535	5.000000	100.0000	0.135413	2.70825	5.000000	100.0000

Note: * Marked red values are those with significant eigenvalues.

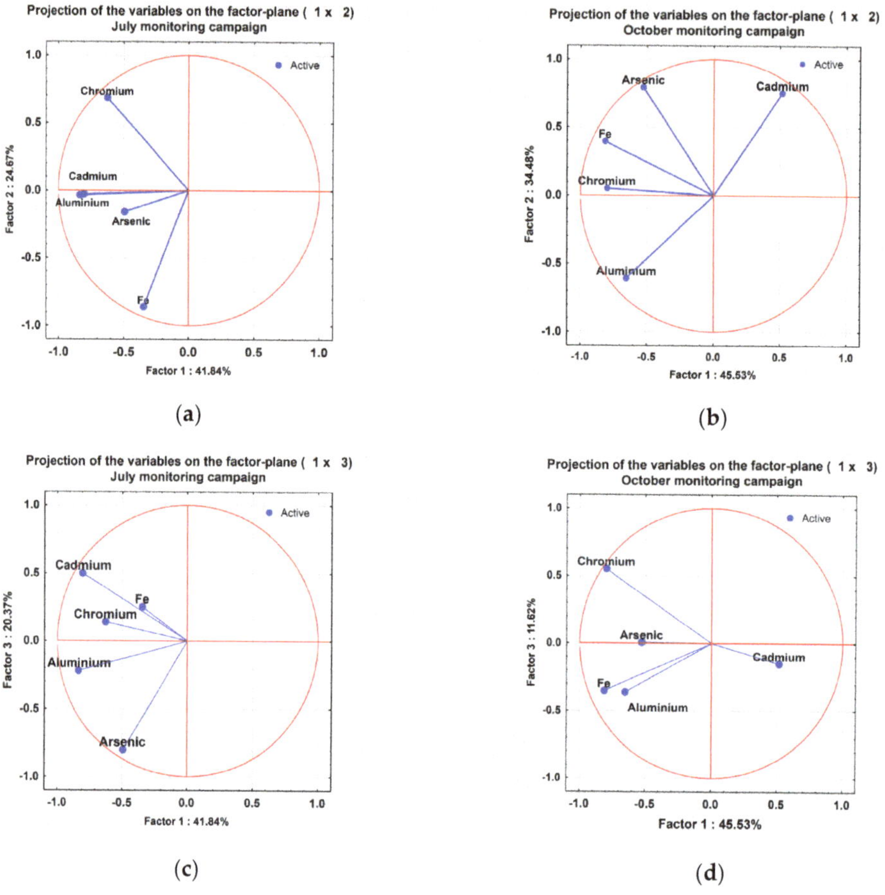

Figure 10. PCA representation of the two measurement campaigns in the Danube River study area: (**a**) PCA representation with main factors 1 and 2 for the July campaign; (**b**) PCA representation with main factors 1 and 3 for the July campaign; (**c**) PCA representation with main factors 1 and 2 for the October campaign; (**d**) PCA representation with main factors 1 and 3 for the October campaign.

In the case of the October database, things are a little different. Within this data set, there are only two significant main values. It can be observed that factor 1, which describes the variation of more than 45.53% of the total set, includes Al, Cr and Fe (Figure 10b). Factor 2 describes the variation of over 34.48% of the total set, including As and Cd (Figure 10d).

Figure 10a shows that factor 1, which describes the variation of more than 50% of the total set, includes Al, Cr and Cd.

At the same time, Figure 10b shows that factor 1, which describes the variant of over 50% of the total set, includes Al, Cr and Cd.

3.6. Numerical Analysis

For a numerical simulation and an evaluation of the distribution of the concentrations of the studied metals in the investigation area, we considered the possibility of using dedicated software programs, often used in the literature [1,6,23]. HEC-RAS is a program that allows, in the new version, to solve diffusion-type equations in the 2D case, such as the equations below [18]:

$$\frac{\partial H}{\partial t} + \frac{\partial H}{\partial x} + \frac{\partial H}{\partial y} = r \tag{1}$$

$$\frac{\partial p}{\partial t} + \frac{\partial}{\partial x}\left(\frac{p^2}{h}\right) + \frac{\partial}{\partial y}\left(\frac{pq}{h}\right) = -\frac{\eta^2 pg\sqrt{p^2+q^2}}{h^2} - gh\frac{\partial H}{\partial x} + pf + \frac{\partial}{\rho\partial x}(h\tau_{xx}) + \frac{\partial}{\rho\partial y}(h\tau_{xy}) + A(x,y)$$

$$\frac{\partial p}{\partial t} + \frac{\partial}{\partial x}\left(\frac{pq}{h}\right) + \frac{\partial}{\partial y}\left(\frac{q^2}{h}\right) = -\frac{\eta^2 pg\sqrt{p^2+q^2}}{h^2} - gh\frac{\partial H}{\partial y} + qf + \frac{\partial}{\rho\partial x}(h\tau_{xy}) + \frac{\partial}{\rho\partial y}(h\tau_{yy}) + A(x,y)$$

where $H(x,y,t) = z(x,y) + h(x,y,t)$ is the height of the surface (in m), $z = z(x,y)$ is the elevation of the cell in the Cartesian coordinate system,(x,y), h—is the water depth (in m), $p = hu$ and $q = hv$ are the specific flows along the axes x and, respectively, y; u and v are the specific speeds along the axes x and, respectively, y; g—is the gravitational acceleration (in m/s^2), η is the Manning roughness coefficient (in s/m^3),ρ—is the density (in kg/m^3), iar $\tau_{x,x}$, $\tau_{x,y}$, $\tau_{y,y}$—represents the components of the flow tensor and f is the Coriolis parameter (in s^{-1}). These equations can be solved within HEC-RAS using finite element calculation [32,39,40]. The term $A(x,y)$ represents the attenuation term—correlated with the transfer phenomenon. In the context of the HEC-RAS bidimensional model, an additive terms can be significant in order to obtain a correlation for concentration variability by two reasons: (a) model complexity: an additive term can account for additional physical processes or factors that affect concentration variability but are not captured by the primary terms of the model; (b) accuracy and precision: an additive term can improve the accuracy and precision of the model by fine-tuning the representation of the transport and diffusion of substances in water which is the main aim; (c) calibration and validation: These terms can be crucial for the calibration and validation of the model, ensuring that the simulated data match observed data as closely as possible.

The modeler in HEC-RAS allowed the development of geometric data either by primary drawing in the river system scheme using the Geometric Data Window (Figure 3) or by developing a scheme in HEC-RAS Mapper (HEC-RAS Mapper). Thus, the river system diagram is a diagram of how the water flow system is connected together. In the present case, 2D models were considered because the existence of weirs in the river bed make the evaluated concentrations differ and there are differences between areas related to the two banks. The interpolation and simulation procedure used by our team was described in the specialized literature and several examples of such simulations were presented.

In the present case, the 2D simulations brought an element of novelty [19,24]. The calibration of the hydraulic model was first performed upstream of the study area, more precisely in the Braila sector for Danube River, and later downstream, at the mouth of the Ceatal branch in the Reni—Tulcea area. The calibration of the model in the HEC-RAS software was made by alternate values of the Manning roughness coefficients, between 0.025 and 0.03, and by applying the rating curve data from the Tulcea hydrometric station [13,19]. In this respect, by considering that the disadvantage of the static method is that the simulations cannot be calibrated with field data, the dynamic methodology was used, which has the advantage of introducing a much more complex input data set, including calibration data, which gives a higher confidence in the results obtained.

In general, the time of concentration (TC) is a parameter in runoff estimation, used for the study and design of different types of hydrographic systems (Pierfranco Costabile, Masih Zolghadr). Any error in the TC calculation leads to an inaccurate flux estimate, which can lead to over- or under-dimensioning of the results. Therefore, choosing the correct time estimation method for an unsteady flow regime (Figure 11) is of great importance.

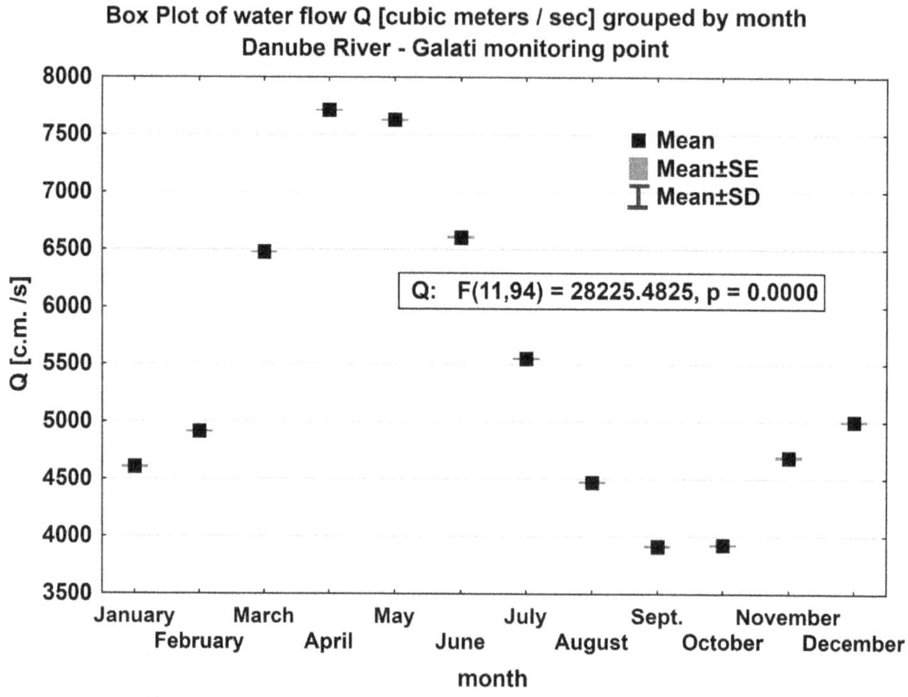

Figure 11. The graph representation of the monthly evolution of the Danube flow in the monitored area—for the non-stationary study.

In the present case, it was considered unsteady modeling using the data on the monthly flow of the Danube throughout the year 2022, with a sampling rate of 1 h, an average rate of 1 week and a graphic evaluation with a rate of 1 month. In this way, they were considered the model data that are the closest to reality. Also, the monthly flow data were entered for the Siret and Prut rivers according to https://www.hidro.ro/prognoze/ (accessed on 26 October 2021).

Considering the ratio between the flows of the tributaries and that of the Danube, the contribution related to the transport attributed to the tributaries is significant only in the areas close to the confluence points, manifesting a dilution along the main flow.

For each chemical species, a number of simulations were carried out, the scenarios being selected based on the comparison between the measured values and those evaluated within the 2D models used.

Figure 12 shows the results of the 2D diffusion models for the two measurement campaigns—from July (Figure 12a) and from October (Figure 12b), respectively, for iron. Thus, in Figure 12a, because the flow of the Danube is higher, the flood zones that are recorded near the right bank are also considered [1,32,40]. The measured values and those obtained after the evaluation are in agreement. Thus, during the summer, the values obtained are systematically higher—an observation made in other previous studies [2,36]. The most interesting aspect is given by the possibility of predicting the downstream values, although the distances between the monitoring points cover a distance of 57.8 km. The quality of fit between the measured and evaluated values is in good agreement (R^2 = 0.8715).

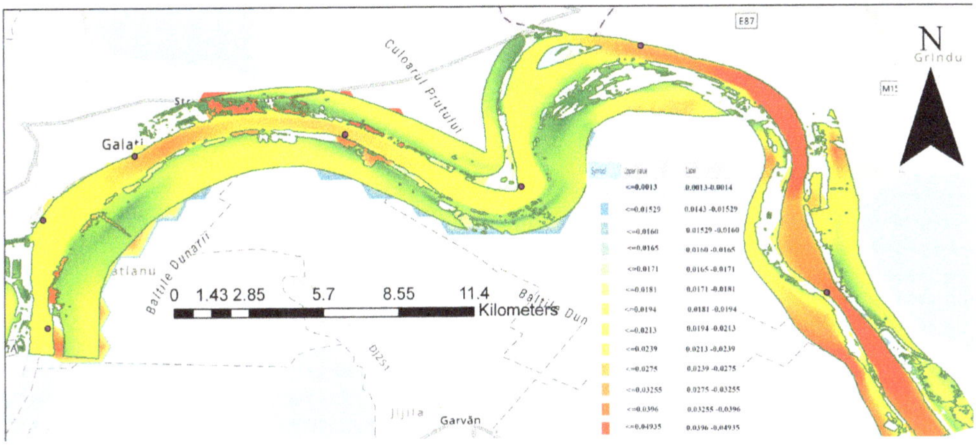

(**a**)

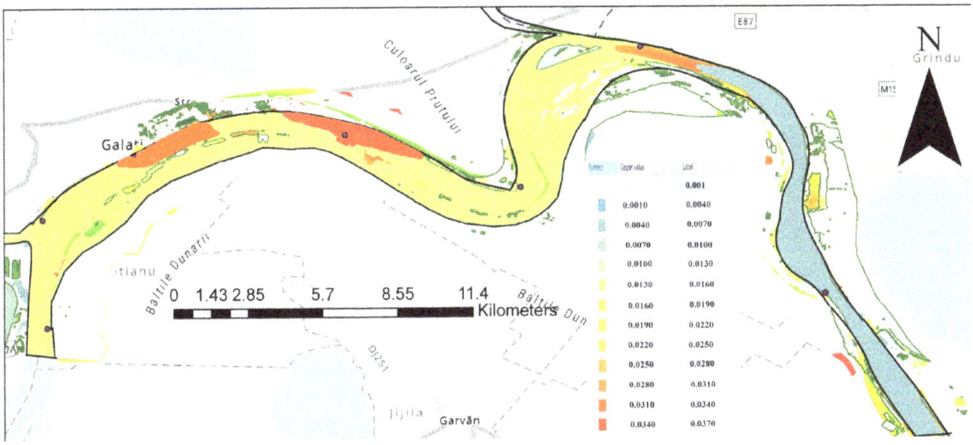

(**b**)

Figure 12. The representation of distribution maps: (**a**) representation of distribution maps for Fe concentration—July campaign; (**b**) representation of distribution maps for Fe concentration—October campaign.

Figure 13 shows the results of the 2D diffusion models for the two measurement campaigns—from July (Figure 13a) and from October (Figure 13b), respectively, for chromium. Thus, in Figure 13a, the evaluated and measured values have a constant trend with minimal variations, a maximum being recorded next to monitoring points P4 and P9, respectively (Figure 13b). This aspect may be related to the industrial activities near the two points. Similarly, during the autumn campaign, the tendency to keep a constant value is respected with a tendency to reduce values downstream. The measured values and those obtained after the evaluation are in accordance with the measured values ($R^2 > 0.80$). This is relatively easy to understand, considering the fact that there are values that tend to maintain a constant trend along the studied 57.8 km.

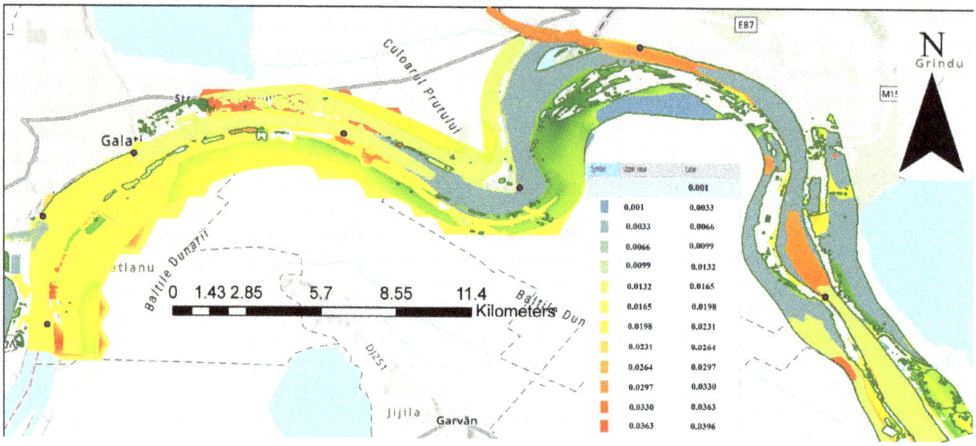

(**a**)

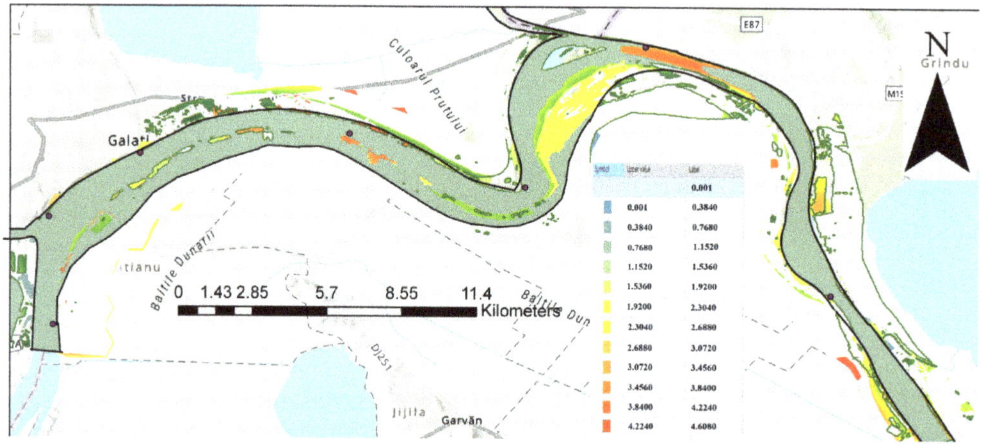

(**b**)

Figure 13. The representation of distribution maps: (**a**) representation of distribution maps for Cr concentration—July campaign; (**b**) representation of distribution maps for Cr concentration—October campaign.

Figure 14 shows the results of the 2D diffusion models for the two measurement campaigns—from July (Figure 14a) and from October (Figure 14b), respectively, for cadmium. In Figure 14a, the evaluated and measured values have a constant trend with minimal variations, a maximum being recorded near monitoring point P7 (Figure 14a). This aspect may be related to the industrial activities near this point. Similarly, during the autumn campaign, the tendency to maintain a constant value is respected. The measured values and those obtained after the evaluation are in accordance with the measured values ($R^2 > 0.67$).

This is relatively easy to understand, considering the fact that there are values that tend to maintain a constant trend along the studied 57.8 km.

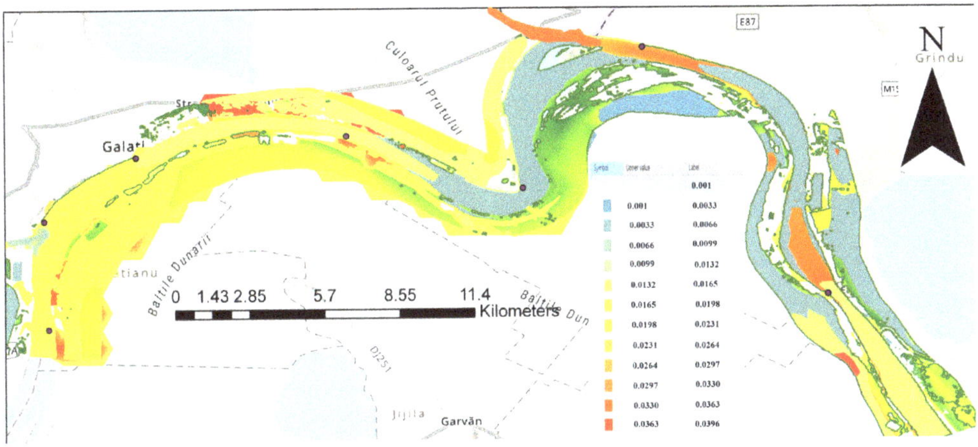

(a)

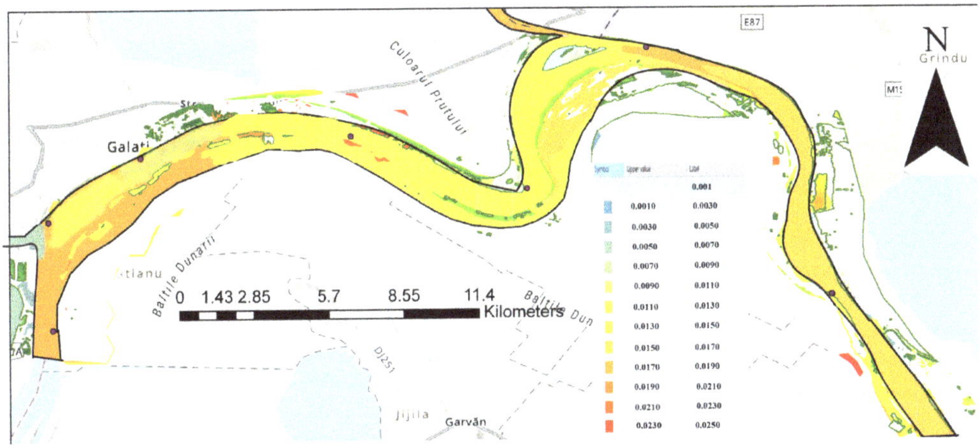

(b)

Figure 14. The representation of distribution maps: (**a**) representation of distribution maps for Cd concentration—July campaign; (**b**) representation of distribution maps for Cd concentration—October campaign.

Figure 15 shows the results of the 2D diffusion models for the two measurement campaigns—from July (Figure 15a) and from October (Figure 15b), respectively, for arsenic. Thus, in Figure 15a, the evaluated and measured values have a constant tendency; however, being maximum near monitoring points P5 and P9, respectively (Figure 15a). Similarly, during the autumn campaign, the tendency to keep a constant value is respected with a tendency to reduce values downstream. The measured and evaluated values are in agreement, the values from the autumn campaign being significantly higher than those during the summer. This was also observed by other authors [1,32]. Figure 16 shows the results of the 2D diffusion models for the two measurement campaigns—from July (Figure 16a) and from October (Figure 16b), respectively, for Al. One can see, in the July campaign, a series of areas with a higher concentration.

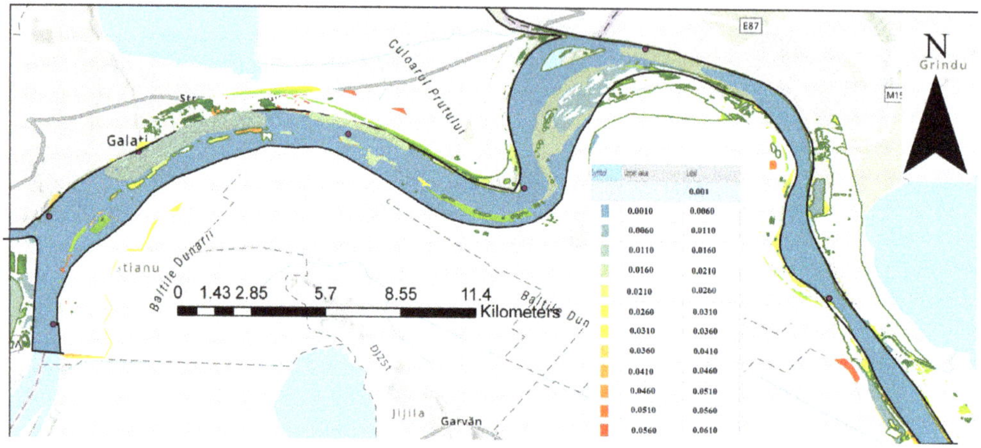

(**a**)

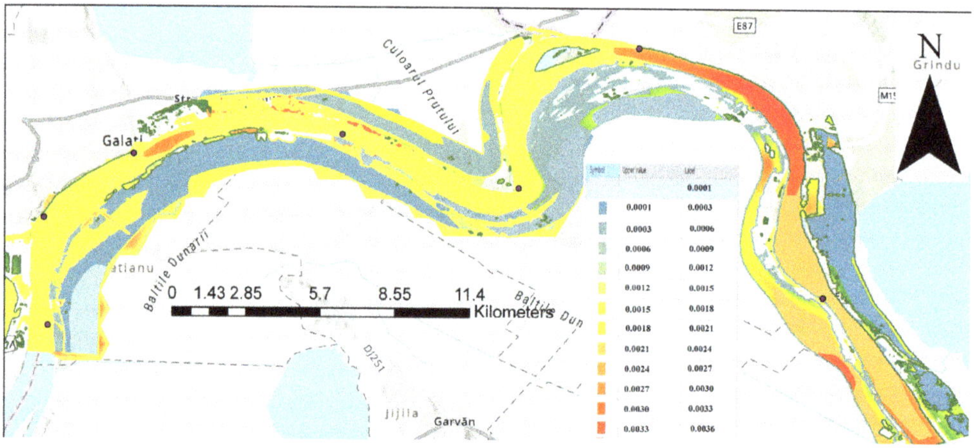

(**b**)

Figure 15. The representation of distribution maps: (**a**) representation of distribution maps for As concentration—July campaign; (**b**) representation of distribution maps for As concentration—October campaign.

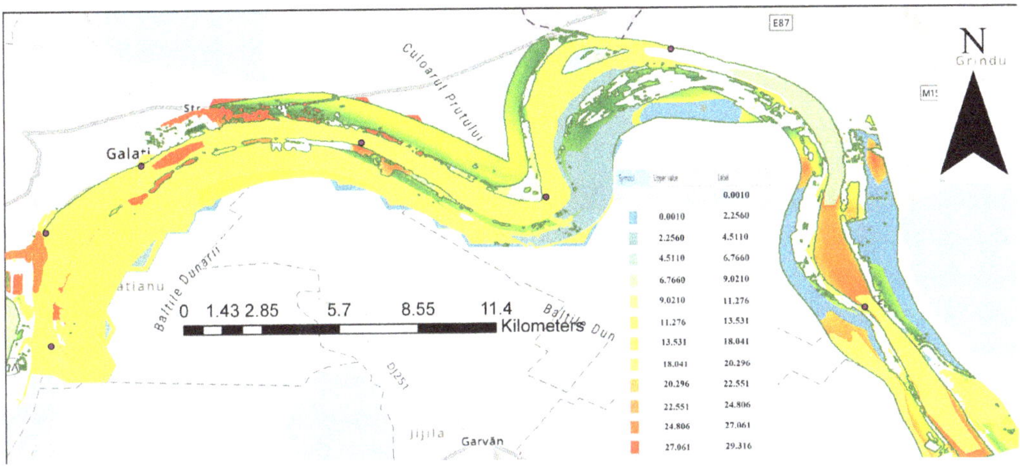

(**a**)

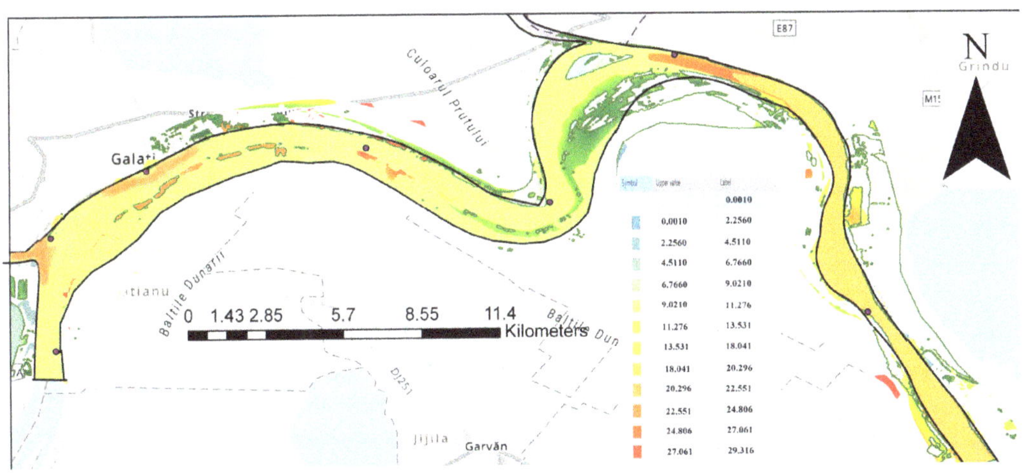

(**b**)

Figure 16. The representation of distribution maps: (**a**) representation of distribution maps for Al concentration—July campaign; (**b**) representation of distribution maps for Al concentration—October campaign.

4. Discussion

The aims of this paper were the distributions and dispersions structures of heavy metals in the Danube Course—in this complex an important environmental issue [26,41].

In the present study, the influences of river–tributary connectivity on metal migration were analyzed, and their contributions were negligible. The average concentrations of Cr, Al, Cd, Fe and As were considered as statistical and dynamic approaches.

In our previous studies, we analyzed the seasonal variation in water quality parameters in the area of the Municipality of Galati City [2,14,25]. Thus, through statistical approaches [2], significant seasonal differences were highlighted using Factor Analysis, Principal Component Analysis and ANOVA methods. Following the studies carried out, the existence of a self-purification capacity was found.

In this respect, in the current campaign, each measured parameter presents a seasonal variation (Tables 3 and 4). Comparing with previous studies, we can highlight some aspects. As an example, for $N-NO_3$, the values are much lower than those recorded in our previous campaign carried out between 1990 and 1998 [2]. The trend is still respected, i.e., the values in summer are lower than those in autumn. The smaller amounts of NO_3^- are attributed to the rivers' intense photosynthesis that occurs in summer. The large amounts of NO_3^- are attributed to the water discharges of municipal waste [2,18].

Generally, the measured parameters' variations in the current campaign are similarly the variations found in our previous study during 1990–1998 [24,27].

4.1. PCA Method Analysis

In the related section, the PCA results obtained by this specific statistical method are based on the correlations. In this respect, we identified six important factors that directly influence the state parameters of the Danube River [15,16,42]. These main key identified factors [2,43,44] show different seasonal weights, based on a clear seasonal variability [2,16,25]:—in the summer, we highlighted the anthropogenic footprint on water quality ecosystem [42,45]. In the autumn season, reducing or even missing the natural biological activity leads to decrease in pollution caused by natural sources of the region [39,46].

The most important factor group includes the elements Al, Cd, phosphate, pH, $N-NO_2$, $N-NO_3$, the SO_4 group, chlorine and conductivity (Figure 8a,b). This factor, which could be attributed more to the influence of anthropic activity, brings together elements that are related to industrial activities in the area [42,47,48]. The second factor, which includes elements such as chromium, CBO_5, NH_4, $N-NO_3$ and N-total, would represent a factor related to pollution due to biological factors in the study area [13,35,47].

From the point of view of the PCA method, the summer campaign allows the identification of some elements that form the six main factors [6,42,49]: thus, the main factor F1—the factor: N-Total, $P-PO_4{}^{3-}$, $N-NO_2$, $SO_4{}^{2-}$, Cl^-, As, respectively; the main factor F2: cadmium, N-NO2, N-NO3-, Fe, chromium, aluminum (Figure 8a,b). These factors can be attributed to a combination of natural geological processes [42–44] and human activities [50], which contributes to the simultaneous presence of these chemical substances in the Danube River area [5,18,36,49]. For the autumn campaign, the PCA method presents a set of six main factors: thus, for the main factor F1, which includes Al, Cr, CCO, CBO_5, NH_4^+, $N-NO_3^-$ and N-Total, respectively, and factor F2, which includes Cd, $N-NO_2$, phenols and $P-PO_4{}^{3-}$. The seasonal variation can be explained by: Agricultural Runoff and Wastewater Discharges (for N-Total); Wastewater Discharges, Biological Processes and Urban Runoff (for $P-PO_4{}^{3-}$); Biological Processes and Wastewater Discharges (for $N-NO_2$); Natural Sources and Industrial Processes (for SO_4); Industrial Discharges (for Cl^-), respectively, Natural Sources and Agricultural Runoff (for As).

In this way, they can be identified and explained through these groupings and through the variation in the water flow, the seasonal variation.

What is interesting is that the cluster type method presents a relatively simpler way of grouping sources with a relatively point-like character (Figure 9a,b): thus, the easiest cluster to identify—Cd, $N-NO_2$ and As—presents practically a Similarity Metrics: because there were calculated similarity metrics between Cd, $N-NO_2$, and As concentrations which lead us to consider that there are common sources [51] (e.g., industrial discharges) for these substances with influence from biological processes [52]. This fact is easy to highlight considering the position of the points where the maximum values are recorded. What is important is that Fe is adherent, correlated with a greater distance to this cluster (Figure 9a,b) [53]. The possible source is near monitoring point P5 (Figure 1), and the adherence from the topological point of view to this kernel can be explained mainly by the decreasing trend for the downstream areas (Figure 9a,b). Mainly, the downward trend has the same configuration, which explains this cluster membership [54].

Another important aspect is that presented by the cluster formed by phosphates and phenols [55]. It can be seen that there is an accession to the previous group in the case of autumn, but the configuration remains similar [2]. The distributions from the summer campaign have a bimodal configuration and a decreasing trend for the downstream points, with local maxima near points P3 and P9. In the case of phenols, the higher values in the case of the cooler season determine the fact that phenols adhere differently.

The group formed by N-NH_4 and N-NO_3 (Figure 9a,b) during the summer campaign is easy to understand through the trimodal configuration shown (Figures 5d and 6c). As is known, if NH_4 and NO_3 tend to co-occur or exhibit similar patterns, they may form a cluster, which is the case.

The last group consisting of Cr, N-total and N-NH_4 (Figure 9a,b) can be easily understood through the prism of environmental context by considering the sources of these substances and their transport mechanisms, respectively, seasonal variations and spatial distribution also play a role.

4.2. Numerical Approach

In the new conditions, the studies carried out in different areas on the course of the Danube illustrate the existence of certain spatial distributions, especially for the concentrations of As, Cr, Fe, etc., in the surface water of the lower basin of the Danube [19,26,29].

In this sense, a numerical approach to highlight the dynamics, using numerical methods of hydraulic modeling, proved to be necessary. We chose a two-dimensional approach [56] because previous evaluations were one-dimensional and cannot explain specific aspects [32].

In the articles from the literature, sets of in situ measurements were taken, and one-dimensional modeling [55] was carried out, not highlighting point sources or areas where the concentrations present obvious differences along the course [31,56]. In the case of the present study, we took water samples from the main tributaries of the Danube in the study area [35]. Considering the ratio between the flows of the tributaries and those of the Danube, the influences proved to be without obvious significance.

However, it was necessary to create a non-stationary 2D model because certain values of the measured concentrations can only be explained by certain plume effects inside the water flow [22,39].

The sampling points were carefully selected, being the hypothesis of the existence of point sources that increase the concentration of metals on the Danube course in the studied area.

A trend of increasing pollution with aluminum ions can be observed along the course of the Danube, especially in the downstream area of the studied area [23]. This aspect can only be explained by the influence of human activity in the area of the last three study points. Also, a similar trend is observed regarding the concentration of Cr. The existence of some ports/shipyards could be the cause of such consequences [57].

For example, the Cd concentration has a maximum recorded around point P7, and an attenuation of the concentration is observed in the downstream area, an aspect that could be deepened through sediment analyses for the evaluation of a mitigation process.

The PCA and Pearson correlation analysis showed that As and Cr originate mainly from the natural environment with anthropic additional contribution; Cd, Al and Fe were affected by industry and agriculture throughout the studied route.

These aspects are obviously original and underline the main characteristic of the river and specify its self-purification capacity

Metals being difficult to diffuse, there is a need for monitoring during precipitation events, and the relationship between soil properties and the degree of use will be investigated in future experiments.

An important aspect that emerged from the study is the fact that, despite the fact that there is an uneven distribution of the concentrations of heavy metals studied in the river bed, in the areas where there are no point sources, the concentrations are relatively low, and the habitats of community interest are not under significant pressure.

A further study on heavy metals on the Danube Course should include the following elements:

- A complex analysis of the bioavailability of heavy metals in sediments using high-performance analytical methods such as inductively coupled plasma–mass spectrometry (ICP-MS) or atomic absorption spectroscopy (AAS) from our laboratory.
- An assessment of factors influencing the sorption and mobility of heavy metals in soils and their mechanisms of accumulation.
- A comparison of different technologies for the remediation of soils polluted with heavy metals, such as in situ or ex situ physico-chemical processes, or bioremediation using microorganisms or plants.

5. Conclusions

1. In conclusion, the present paper presents the possibility of modeling the distribution of heavy metal concentrations over a much greater distance than in previous cases.
2. It was possible to identify some locations of point pollution sources, and the existence of a downstream mitigation phenomenon could be highlighted.
3. It was possible to highlight the self-purification capacity of the Danube, which was also demonstrated in the studies carried out 15 years ago.
4. An evaluation of the coefficients for reducing the concentrations of polluting factors could be carried out.

Author Contributions: Conceptualization, C.T., L.G. and G.M.; methodology, C.T., L.G. and G.M.; software, M.A. and G.M.; validation, M.A., G.M. and M.C.; formal analysis, V.C. and G.M.; investigation, V.C. and M.C.; resources, C.S., C.C. and G.M.; data curation, V.C. and M.C.; writing—original draft preparation, C.S. and C.C.; writing—review and editing, M.A., L.G. and G.M.; visualization, C.S. and C.C.; supervision, V.C. and M.C.; project administration, C.T., L.G. and G.M.; funding acquisition, C.T. All authors have read and agreed to the published version of the manuscript.

Funding: This study was supported by the project "Advanced nanotechnology based approaches to waste water purification form organic pollutants and their monitoring in water bodies", 2SOFT/1.2/139 (https://keep.eu/projects/24656/Advanced-nanotechnology-bas-EN/, accessed on 2 August 2024).

Data Availability Statement: Data is contained within the article.

Conflicts of Interest: The authors declare no conflicts of interest.

References

1. Popescu, F.; Trumić, M.; Cioabla, A.E.; Vujić, B.; Stoica, V.; Trumić, M.; Opris, C.; Bogdanović, G.; Trif-Tordai, G. Analysis of Surface Water Quality and Sediments Content on Danube Basin in Djerdap—Iron Gate Protected Areas. *Water* **2022**, *14*, 2991. [CrossRef]
2. Popa, P.; Murariu, G.; Timofti, M.; Georgescu, L.P. Multivariate Statistical Analyses of Danube River Water Quality at Galati, Romania. *Environ. Eng. Manag. J.* **2018**, *17*, 1249–1266. [CrossRef]
3. Fang, T.; Lu, W.; Cui, K.; Li, J.; Yang, K.; Zhao, X.; Liang, Y.; Li, H. Distribution, Bioaccumulation and Trophic Transfer of Trace Metals in the Food Web of Chaohu Lake, Anhui, China. *Chemosphere* **2019**, *218*, 1122–1130. [CrossRef] [PubMed]
4. Hakanson, L. An Ecological Risk Index for Aquatic Pollution Control.a Sedimentological Approach. *Water Res.* **1980**, *14*, 975–1001. [CrossRef]
5. Simionov, I.-A.; Călmuc, M.; Iticescu, C.; Călmuc, V.; Georgescu, P.-L.; Faggio, C.; Petrea, Ş.-M. Human Health Risk Assessment of Potentially Toxic Elements and Microplastics Accumulation in Products from the Danube River Basin Fish Market. *Environ. Toxicol. Pharmacol.* **2023**, *104*, 104307. [CrossRef]
6. Teodorof, L.; Ene, A.; Burada, A.; Despina, C.; Seceleanu-Odor, D.; Trifanov, C.; Ibram, O.; Bratfanof, E.; Tudor, M.-I.; Tudor, M.; et al. Integrated Assessment of Surface Water Quality in Danube River Chilia Branch. *Appl. Sci.* **2021**, *11*, 9172. [CrossRef]
7. Cordeli, A.N.; Oprea, L.; Crețu, M.; Dediu, L.; Coadă, M.T.; Mînzală, D.-N. Bioaccumulation of Metals in Some Fish Species from the Romanian Danube River: A Review. *Fishes* **2023**, *8*, 387. [CrossRef]
8. Matache, M.L.; Marin, C.; Rozylowicz, L.; Tudorache, A. Plants Accumulating Heavy Metals in the Danube River Wetlands. *J. Environ. Health Sci. Eng.* **2013**, *11*, 39. [CrossRef]
9. Georgescu, P.-L.; Moldovanu, S.; Iticescu, C.; Calmuc, M.; Calmuc, V.; Topa, C.; Moraru, L. Assessing and Forecasting Water Quality in the Danube River by Using Neural Network Approaches. *Sci. Total Environ.* **2023**, *879*, 162998. [CrossRef]
10. Bąk, Ł.; Szeląg, B.; Sałata, A.; Studziński, J. Modeling of Heavy Metal (Ni, Mn, Co, Zn, Cu, Pb, and Fe) and PAH Content in Stormwater Sediments Based on Weather and Physico-Geographical Characteristics of the Catchment-Data-Mining Approach. *Water* **2019**, *11*, 626. [CrossRef]

11. Iticescu, C.; Georgescu, P.-L.; Arseni, M.; Rosu, A.; Timofti, M.; Carp, G.; Cioca, L.-I. Optimal Solutions for the Use of Sewage Sludge on Agricultural Lands. *Water* **2021**, *13*, 585. [CrossRef]
12. Sun, C.; Zhang, Z.; Cao, H.; Xu, M.; Xu, L. Concentrations, Speciation, and Ecological Risk of Heavy Metals in the Sediment of the Songhua River in an Urban Area with Petrochemical Industries. *Chemosphere* **2019**, *219*, 538–545. [CrossRef] [PubMed]
13. Simionov, I.-A.; Cristea, D.S.; Petrea, Ş.-M.; Mogodan, A.; Jijie, R.; Ciornea, E.; Nicoară, M.; Turek Rahoveanu, M.M.; Cristea, V. Predictive Innovative Methods for Aquatic Heavy Metals Pollution Based on Bioindicators in Support of Blue Economy in the Danube River Basin. *Sustainability* **2021**, *13*, 8936. [CrossRef]
14. Calmuc, M.; Calmuc, V.; Arseni, M.; Topa, C.; Timofti, M.; Georgescu, L.P.; Iticescu, C. A Comparative Approach to a Series of Physico-Chemical Quality Indices Used in Assessing Water Quality in the Lower Danube. *Water* **2020**, *12*, 3239. [CrossRef]
15. Dong, D.; Zhao, X.; Hua, X.; Liu, J.; Gao, M. Investigation of the Potential Mobility of Pb, Cd and Cr(VI) from Moderately Contaminated Farmland Soil to Groundwater in Northeast, China. *J. Hazard. Mater.* **2009**, *162*, 1261–1268. [CrossRef]
16. Liu, Y.; Xu, Z.; Hu, X.; Zhang, N.; Chen, T.; Ding, Z. Sorption of Pb(II) and Cu(II) on the Colloid of Black Soil, Red Soil and Fine Powder Kaolinite: Effects of pH, Ionic Strength and Organic Matter. *Environ. Pollut. Bioavailab.* **2019**, *31*, 85–93. [CrossRef]
17. Resz, M.-A.; Roman, C.; Senila, M.; Török, A.I.; Kovacs, E. A Comprehensive Approach to the Chemistry, Pollution Impact and Risk Assessment of Drinking Water Sources in a Former Industrialized Area of Romania. *Water* **2023**, *15*, 1180. [CrossRef]
18. Takić, L.; Mladenović-Ranisavljević, I.; Vasović, D.; Đorđević, L. The Assessment of the Danube River Water Pollution in Serbia. *Water Air Soil Pollut* **2017**, *228*, 380. [CrossRef]
19. Banescu, A.; Arseni, M.; Georgescu, L.P.; Rusu, E.; Iticescu, C. Evaluation of Different Simulation Methods for Analyzing Flood Scenarios in the Danube Delta. *Appl. Sci.* **2020**, *10*, 8327. [CrossRef]
20. Zhang, W.; Long, J.; Zhang, X.; Shen, W.; Wei, Z. Pollution and Ecological Risk Evaluation of Heavy Metals in the Soil and Sediment around the HTM Tailings Pond, Northeastern China. *Int. J. Environ. Res. Public Health* **2020**, *17*, 7072. [CrossRef]
21. Da Silva Júnior, R.O.; Almeida, H.P.; Da Silva, M.S.; França, A.C.; Balleroni, E.; Dos Santos, N.; Vilela, P.H.; De Melo, A.M.Q.; Guimarães, J.T.F. Methodological Approach for an Online Water Quality Monitoring System in an Iron Ore Tailing Dam. *Water* **2023**, *15*, 3663. [CrossRef]
22. Pham Van, C.; Chua, V. Numerical Simulation of Hydrodynamic Characteristics and Bedload Transport in Cross Sections of Two Gravel-Bed Rivers Based on One-Dimensional Lateral Distribution Method. *Int. J. Sediment Res.* **2020**, *35*, 203–216. [CrossRef]
23. Wang, H.; Yuan, W.; Zeng, Y.; Liang, D.; Deng, Y.; Zhang, X.; Li, Y. How Does Three Gorges Dam Regulate Heavy Metal Footprints in the Largest Freshwater Lake of China. *Environ. Pollut.* **2022**, *292*, 118313. [CrossRef] [PubMed]
24. Calmuc, V.A.; Calmuc, M.; Arseni, M.; Topa, C.M.; Timofti, M.; Burada, A.; Iticescu, C.; Georgescu, L.P. Assessment of Heavy Metal Pollution Levels in Sediments and of Ecological Risk by Quality Indices, Applying a Case Study: The Lower Danube River, Romania. *Water* **2021**, *13*, 1801. [CrossRef]
25. Iticescu, C.; Georgescu, L.P.; Murariu, G.; Topa, C.; Timofti, M.; Pintilie, V.; Arseni, M. Lower Danube Water Quality Quantified through WQI and Multivariate Analysis. *Water* **2019**, *11*, 1305. [CrossRef]
26. Saeed, O.; Székács, A.; Jordán, G.; Mörtl, M.; Abukhadra, M.R.; Eid, M.H. Investigating the Impacts of Heavy Metal (Loid)s on Ecology and Human Health in the Lower Basin of Hungary's Danube River: A Python and Monte Carlo Simulation-Based Study. *Environ. Geochem. Health* **2023**, *45*, 9757–9784. [CrossRef]
27. Murariu, G.; Timofti, M.; Savinuta, I.; Murariu, A.; Iticescu, C.; Gerogescu, L.; Tudor, M.; Mocanu, G.D. Seasonal Statistical Analysis on the Quality of Underwater Supplies in South-Eastern Romania. *Rev. Chim.* **2019**, *70*, 820–827. [CrossRef]
28. Drasovean, R.; Murariu, G.; Constantinescu, G.; Circiumaru, A. Assessment of Surface Water Quality of Danube River in Terms of Usual Parameters and Correlation Analyses. *Rev. Chim.* **2019**, *70*, 398–406. [CrossRef]
29. Arseni, M.; Roșu, A.; Murariu, G.; Georgescu, L.P.; Iticescu, C.; Calmuc, M.; Calmuc, V.-A. The Role of River Channel Roughness for Water Level Modeling during the 2005 Year Flood on Siret River Using HEC-RAS Model. *Ann. Dunarea Univ. Galati. Fascicle II Math. Phys. Theor. Mech.* **2019**, *42*, 68–76. [CrossRef]
30. Water Directive. Available online: https://environment.ec.europa.eu/topics/water/water-framework-directive_en (accessed on 27 October 2023).
31. Barrero-Moreno, M.C.; Diaz-Vargas, C.A.; Restrepo-Parra, E. Computational Simulation of Filters Used in the Removal of Heavy Metals Using Rice Husks. *Agriculture* **2021**, *11*, 146. [CrossRef]
32. Zolghadr, M.; Rafiee, M.R.; Esmaeilmanesh, F.; Fathi, A.; Tripathi, R.P.; Rathnayake, U.; Gunakala, S.R.; Azamathulla, H.M. Computation of Time of Concentration Based on Two-Dimensional Hydraulic Simulation. *Water* **2022**, *14*, 3155. [CrossRef]
33. Mocanu, G.D.; Murariu, G.; Georgescu, L.; Sandu, I. Investigating the Attitudes of First-Year Students of the Faculty of Physical Education and Sports of Galati towards Online Teaching Activities during the COVID-19 Pandemic. *Appl. Sci.* **2021**, *11*, 6328. [CrossRef]
34. Mocanu, G.D.; Murariu, G.; Iordan, D.A.; Sandu, I. Analysis of the Influence of Age Stages on Static Plantar Pressure Indicators for Karate Do Practitioners (Preliminary Report). *Appl. Sci.* **2021**, *11*, 7320. [CrossRef]
35. Maxim, C.; Filote, C.; Cojocaru, D.C. Comparative Study of Physical Indicators and Those of Their Regime of Oxygen, on Water Quality of the Great Ȕomuzu River, in the Year 2009. *Arch. Environ. Prot.* **2011**, *42*, 41–48.
36. Mocanu, G.D.; Murariu, G.; Iordan, D.A.; Sandu, I.; Munteanu, M.O.A. The Perception of the Online Teaching Process during the COVID-19 Pandemic for the Students of the Physical Education and Sports Domain. *Appl. Sci.* **2021**, *11*, 5558. [CrossRef]

37. Crişan, V.-E.; Dincă, L.; Bragă, C.; Murariu, G.; Tupu, E.; Mocanu, G.D.; Drasovean, R. The Configuration of Romanian Carpathians Landscape Controls the Volume Diversity of *Picea Abies* (L.) Stands. *Land* **2023**, *12*, 406. [CrossRef]
38. Murariu, G.; Dinca, L.; Tudose, N.; Crisan, V.; Georgescu, L.; Munteanu, D.; Dragu, M.D.; Rosu, B.; Mocanu, G.D. Structural Characteristics of the Main Resinous Stands from Southern Carpathians, Romania. *Forests* **2021**, *12*, 1029. [CrossRef]
39. Jin, Y.; Zhou, Q.; Wang, X.; Zhang, H.; Yang, G.; Lei, T.; Mei, S.; Yang, H.; Liu, L.; Yang, H.; et al. Heavy Metals in the Mainstream Water of the Yangtze River Downstream: Distribution, Sources and Health Risk Assessment. *Int. J. Environ. Res. Public Health* **2022**, *19*, 6204. [CrossRef]
40. Costabile, P.; Costanzo, C.; Ferraro, D.; Macchione, F.; Petaccia, G. Performances of the New HEC-RAS Version 5 for 2-D Hydrodynamic-Based Rainfall-Runoff Simulations at Basin Scale: Comparison with a State-of-the Art Model. *Water* **2020**, *12*, 2326. [CrossRef]
41. Luo, K.; Liu, H.; Yu, E.; Tu, Y.; Gu, X.; Xu, M. Distribution and Release Mechanism of Heavy Metals in Sediments of Yelang Lake by DGT. *Stoch. Env. Res. Risk Assess.* **2020**, *34*, 793–805. [CrossRef]
42. Ha, M.; Schleiger, R. Environmental Science. Available online: https://bio.libretexts.org/Bookshelves/Ecology/Environmental_Science_(Ha_and_Schleiger) (accessed on 23 March 2022).
43. Githaiga, K.B.; Njuguna, S.M.; Gituru, R.W.; Yan, X. Water Quality Assessment, Multivariate Analysis and Human Health Risks of Heavy Metals in Eight Major Lakes in Kenya. *J. Environ. Manag.* **2021**, *297*, 113410. [CrossRef]
44. Feng, W.; Wang, T.; Zhu, Y.; Sun, F.; Giesy, J.P.; Wu, F. Chemical Composition, Sources, and Ecological Effect of Organic Phosphorus in Water Ecosystems: A Review. *Carbon. Res.* **2023**, *2*, 12. [CrossRef]
45. Xu, M.; Wang, R.; Yang, X.; Yang, H. Spatial Distribution and Ecological Risk Assessment of Heavy Metal Pollution in Surface Sediments from Shallow Lakes in East China. *J. Geochem. Explor.* **2020**, *213*, 106490. [CrossRef]
46. Murariu, G.; Iticescu, C.; Murariu, A.; Rosu, B.; Munteanu, D.; Buruiana, D.L. Assessment of Water Quality State Dynamics Using Adaptive Filtering Methods and Neural Networks Approaching. Case study—Danube River in Galati area. *Rev. Chim.* **2019**, *70*, 1914–1919. [CrossRef]
47. Pavel, A.; Durisch-Kaiser, E.; Balan, S.; Radan, S.; Sobek, S.; Wehrli, B. Sources and Emission of Greenhouse Gases in Danube Delta Lakes. *Environ. Sci. Pollut. Res.* **2009**, *16*, 86–91. [CrossRef] [PubMed]
48. Mocanu, G.D.; Murariu, G.; Munteanu, D. The Influence of Socio-Demographic Factors on the Forms of Leisure for the Students at the Faculty of Physical Education and Sports. *Int. J. Environ. Res. Public Health* **2021**, *18*, 12577. [CrossRef]
49. Bărbulescu, A.; Barbeş, L.; Dani, A. Statistical Analysis of the Quality Indicators of the Danube River Water (in Romania). In *Frontiers in Water-Energy-Nexus—Nature-Based Solutions, Advanced Technologies and Best Practices for Environmental Sustainability*; Naddeo, V., Balakrishnan, M., Choo, K.-H., Eds.; Advances in Science, Technology & Innovation; Springer International Publishing: Cham, Italy, 2020; pp. 277–279, ISBN 978-3-030-13067-1.
50. Mocanu, G.D.; Murariu, G.; Munteanu, D. Common Issues and Differences in Motivational Support and the Effects of Rugby for the U-15 and U-16 National Teams. *Sustainability* **2022**, *14*, 4535. [CrossRef]
51. Ultsch, A.; Lötsch, J. Euclidean Distance-Optimized Data Transformation for Cluster Analysis in Biomedical Data (EDOtrans). *BMC Bioinform.* **2022**, *23*, 233. [CrossRef]
52. Hemming, K.; Eldridge, S.; Forbes, G.; Weijer, C.; Taljaard, M. How to Design Efficient Cluster Randomised Trials. *BMJ* **2017**, *358*, j3064. [CrossRef]
53. Staples, L.; Ring, J.; Fontana, S.; Stradwick, C.; DeMaio, J.; Ray, H.; Zhang, Y.; Zhang, X. Reproducible Clustering with Non-Euclidean Distances: A Simulation and Case Study. *Int. J. Data Sci. Anal.* **2023**. [CrossRef]
54. Feio, M.J.; Hughes, R.M.; Callisto, M.; Nichols, S.J.; Odume, O.N.; Quintella, B.R.; Kuemmerlen, M.; Aguiar, F.C.; Almeida, S.F.P.; Alonso-EguíaLis, P.; et al. The Biological Assessment and Rehabilitation of the World's Rivers: An Overview. *Water* **2021**, *13*, 371. [CrossRef]
55. Mocanu, G.D.; Murariu, G. The Association of Gender and Body Mass Index on the Values of Static and Dynamic Balance of University Students (A Cross-Sectional Design Study). *Appl. Sci.* **2022**, *12*, 3770. [CrossRef]
56. Brunner, G. *Hec-Ras—Verification and Validation Tests*; US Army Corps of Engineers, Hydrologic Engineering Center: Davis, CA, USA, 1994.
57. Antohi, V.M.; Ionescu, R.V.; Zlati, M.L.; Iticescu, C.; Georgescu, P.L.; Calmuc, M. Regional Regression Correlation Model of Microplastic Water Pollution Control Using Circular Economy Tools. *Int. J. Environ. Res. Public Health* **2023**, *20*, 4014. [CrossRef] [PubMed]

Disclaimer/Publisher's Note: The statements, opinions and data contained in all publications are solely those of the individual author(s) and contributor(s) and not of MDPI and/or the editor(s). MDPI and/or the editor(s) disclaim responsibility for any injury to people or property resulting from any ideas, methods, instructions or products referred to in the content.

MDPI AG
Grosspeteranlage 5
4052 Basel
Switzerland
Tel.: +41 61 683 77 34

Water Editorial Office
E-mail: water@mdpi.com
www.mdpi.com/journal/water

Disclaimer/Publisher's Note: The title and front matter of this reprint are at the discretion of the Guest Editor. The publisher is not responsible for their content or any associated concerns. The statements, opinions and data contained in all individual articles are solely those of the individual Editor and contributors and not of MDPI. MDPI disclaims responsibility for any injury to people or property resulting from any ideas, methods, instructions or products referred to in the content.

www.ingramcontent.com/pod-product-compliance
Lightning Source LLC
LaVergne TN
LVHW070449061126
851858LV00011B/195